P9-AFA-352

# Basic Communications Electronics

## Analog Electronic Devices and Circuits

## How They Work and How They Are Used to Create Communication Systems

By
Jack Hudson
Jerry Luecke

*This book was developed and published by:*
Master Publishing, Inc.
Lincolnwood, Illinois

*Edited by:*
Pete Trotter, KB9SMG

*Printing by:*
Arby Graphic Service, Inc.
Lincolnwood, Illinois

Copyright © 1999
Master Publishing, Inc.
7101 N. Ridgeway Avenue
Lincolnwood, IL 60645
voice: 847/763-0916
fax: 847/763-0918
e-mail: MasterPubl@aol.com
All Rights Reserved
Printed in the United States

Visit Master Publishing on the Internet at:
www.masterpublishing.com

*Acknowledgements:*
Master Publishing, Inc., and the authors wish to thank the following individuals and organizations for their assistance with the development of this book:

- Motorola Incorporated; Texas Instruments Incorporated; and National Semiconductor Corporation for their permission to reproduce integrated circuit data sheet illustrations and information.
- The schematic appearing on page 81 has been reproduced with the express authorization of EXAR Corporation.
- Texas Instruments Incorporated for the photos illustrating the IC manufacturing process in Chapter 10.
- Joe Devore, Automotive Design Section Manager, MSDS Dept., Texas Instruments Incorporated, for his assistance with Chapter 10.

All photographs or illustrations that do not have a source identification are either courtesy of Radio Shack, the authors, or Master Publishing, Inc.

**REGARDING THESE BOOK MATERIALS**

Reproduction, publication, or duplication of this book, or any part thereof, in any manner, mechanically, electronically, photographically, is strictly prohibited without the express written permission of the publisher. For permission and other rights under this copyright, write Master Publishing, Inc.

**The Authors, Publisher, and Seller assume no liability with respect to the use of the information contained herein.**

9 8 7 6 5 4 3 2 1

# Table of Contents


| | | Page |
|---|---|---|
| | *Preface* | iv |
| Chapter 1 | Basic Communications Systems | 1 |
| Chapter 2 | Analog System Functions | 23 |
| Chapter 3 | A Refresher | 45 |
| Chapter 4 | Amplifiers & Oscillators | 59 |
| Chapter 5 | Modulation | 87 |
| Chapter 6 | Mixing & Heterodyning | 108 |
| Chapter 7 | Transmitters | 121 |
| Chapter 8 | Receiving – Including Detection | 137 |
| Chapter 9 | Transmission Links | 157 |
| Chapter 10 | Analog Integrated Circuits | 181 |
| Chapter 11 | Digital Signal Processing | 199 |
| | *Appendix* | 209 |
| | *Glossary* | 215 |
| | *Index* | 219 |

# Preface

*Basic Communications Electronics* explains how analog electronic devices and circuits are used to create communications systems. This books complements Master Publishing's other books in its *Basic* series, *Basic Electronics* and *Basic Digital Electronics.* It is intended to round-out the reader's knowledge about the basics of electronics, both analog and digital.

Today's electronic world has been revolutionized by digital devices, circuits, and systems. While digital now dominates the world of electronics, analog devices and circuits continue to play an important role in electronics, both to satisfy unique functions and to interface to digital circuits. *Basic Communications Electronics* will teach you the fundamentals of analog electronics.

Circuit and system designers first used electronic devices in analog systems — from simple gauges to the first computers — to create the foundation for today's electronic systems. Even though vacuum tubes have been replaced by transistors and discrete-component circuits by integrated circuits, the basic theory of how analog circuits work and are combined into communications systems using various electronic devices remains essentially the same. *Basic Communications Electronics* concentrates on semiconductor devices — bipolar and field-effect transistors — and integrated circuits using these devices.

*Basic Communications Electronics* teaches the basic analog functions; then shows how devices are used in circuits; and explains how circuits are combined into subsystems and complete systems. Our book begins by defining what makes up a communications system and explaining the difference between analog and digital systems. It describes, in general terms, the functions that are combined to make up different analog systems. Even though some readers will not require it, a refresher chapter is included to assure that everyone is up to par on the necessary definitions of electronic devices and some of the mathematics used in the circuit design examples.

The book continues by examining circuits that perform the functions using various devices, beginning with amplifiers and oscillators; and continuing with modulation, mixing and heterodyning. A discussion of transmitters and receivers follows to solidify how the functional circuits are combined to form subsystems or complete analog systems. More detail on wireless communication systems, in particular, is completed with a discussion of antennas and the wireless transmission link and ends with other transmission lines.

The book concludes its goal to teach the basics of analog systems with a chapter on analog integrated circuits, highlighting the major differences from digital, and provides a glimpse of the future with a chapter on digital signal processing.

We hope that you enjoy your learning experience. When you are finished with our book, you should have a solid foundation to further your electronics education!

JH, JL, MPI

# CHAPTER 1

# Basic Communications Systems

In *Basic Electronics*[1], the book discusses the nature of dc and ac electricity, how semiconductor devices — diodes and transistors — work, and explains how these devices are used in both analog and digital circuits to provide basic electronic functions.

*Basic Digital Electronics*[2] defines the difference between digital and analog systems, identifies the electronic functions needed by digital systems, and shows how circuits are used to build these functions as they are used in such digital systems as computers and compact disc players.

In this book, *Basic Communications Electronics*, we will discuss the functions needed by analog systems and show how electronic circuits are used to build these functions. Then we will explore how these analog circuits are put together to build communications systems.

## Analog Quantities

Analog quantities vary continuously. Digital quantities vary in discrete values — usually just two values, ON and OFF. Analog systems carry information using electrical signals that vary smoothly and continuously over a range. Digital systems carry information using combinations of ON-OFF electrical signals that are usually in the form of codes that represent the information.

Before the invention of the transistor and the development of integrated circuits, most electronic systems were analog. Since the transistor and, more importantly, the integrated circuit, digital circuits with their ON-OFF qualities, lower power usage, wider manufacturing tolerances, and smaller size, have taken over many electronic system functions. However, there are still many systems whose operation can be implemented much better and more efficiently using analog circuits and functions.

## Common Analog Systems

To begin, let's be sure we know what an analog system is. Here is an example from *Basic Digital Electronics* — it is a volume control circuit like those used in a stereo, clock radio, or TV set.

### A Volume Control

The circuit shown in *Figure 1-1a* is a partial block diagram of a radio receiver showing the volume control in greater detail. The volume control varies the amount of the output voltage ($V_O$ from the radio tuner) passed on to the audio amplifier as the control is changed. If the control is at the bottom (ground), there will be zero input signal voltage ($V_{IN}$) to the amplifier and no sound from the speaker. If the control is at the top

[1] *Basic Electronics*, G. McWhorter and A. J. Evans, ©1994, Master Publishing, Inc., Lincolnwood, IL
[2] *Basic Digital Electronics*, A. J. Evans, ©1996, Master Publishing, Inc., Lincolnwood, IL

(maximum rotation), the input signal voltage ($V_{IN}$) to the amplifier is the full signal voltage from the tuner ($V_O$) and maximum sound comes from the speaker. *Figure 1-1b* shows graphically the relationship between the input signal voltage to the amplifier ($V_{IN}$) and the amount the control is turned. Notice that the graph is a continuously varying line (in this case, a straight line) without jumps or breaks in it. The input signal to the amplifier ($V_{IN}$) is an analog of (analogous to) the output voltage from the radio tuner ($V_O$), even though the position of the control determines the amplitude. The input signal voltage ($V_{IN}$) is a continuously proportional amount of the output signal voltage ($V_O$) depending upon the position of the control. The larger $V_{IN}$ to the amplifier the greater the sound out from the speaker. Analog electronic functions are continuous, as in this analog signal example.

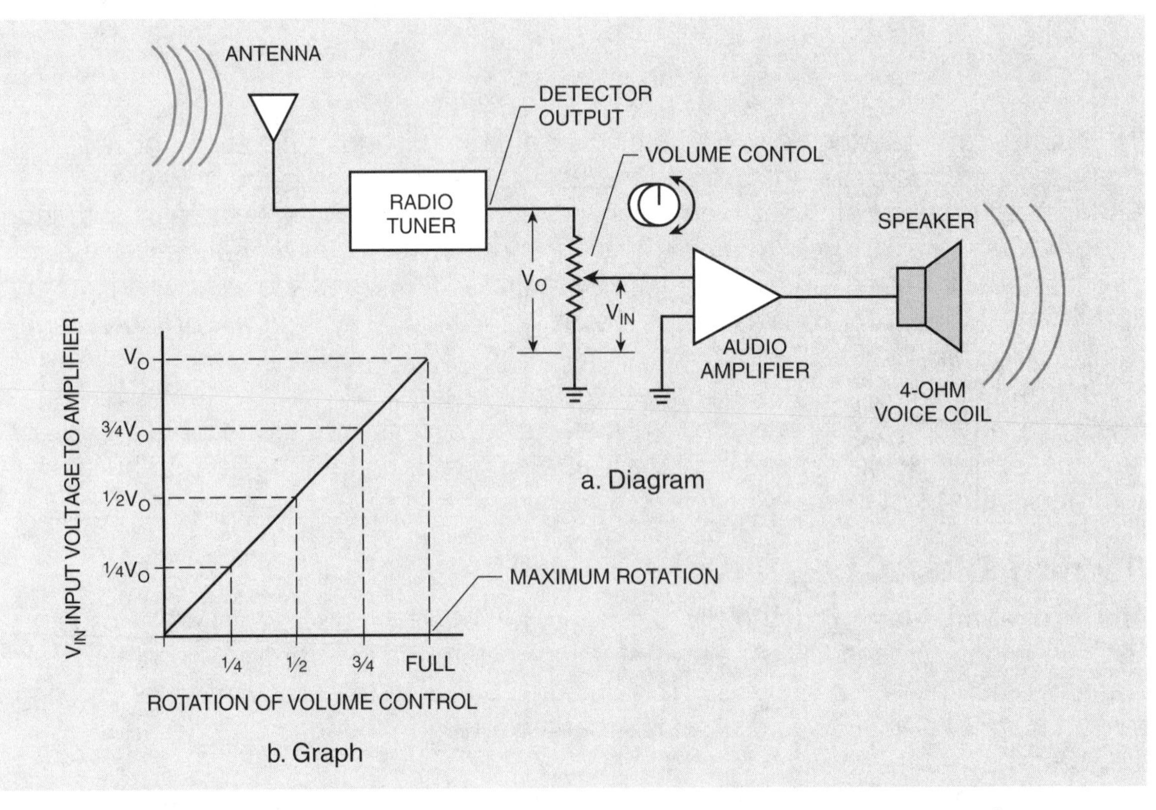


*Figure 1-1. Partial block diagram of radio receiver with schematic showing volume control in greater detail. Graph shows relationship between $V_{IN}$ to the audio amplifier and volume control rotation.*

*Source: Basic Digital Electronics, A.J. Evans, ©1996, Master Publishing, Inc., Lincolnwood, IL.*

## A Recording Thermometer

Let's look at another example — a recording thermometer to measure temperature in degrees Fahrenheit (°F). In this case, the outside summer temperature has been monitored for 72 hours. The temperature readings were recorded by a thermometer like that shown in *Figure 1-2a*. It has a tracing pen that continuously plots the readings on a paper tape against time, as shown in *Figure 1-2b*. Note that the thermometer changes gradually and continuously with time. There are no breaks in the data or sudden jumps — the information changes smoothly and continuously.

## A Fuel Gauge

An automobile fuel gauge is another good example of an analog system. As shown in *Figure 1-3*, a float attached to a variable resistor, a meter, some wire, and a battery are all the components that are required for the system.

a. Recording Thermometer

*Photo courtesy of Taylor Environmental Instruments.*

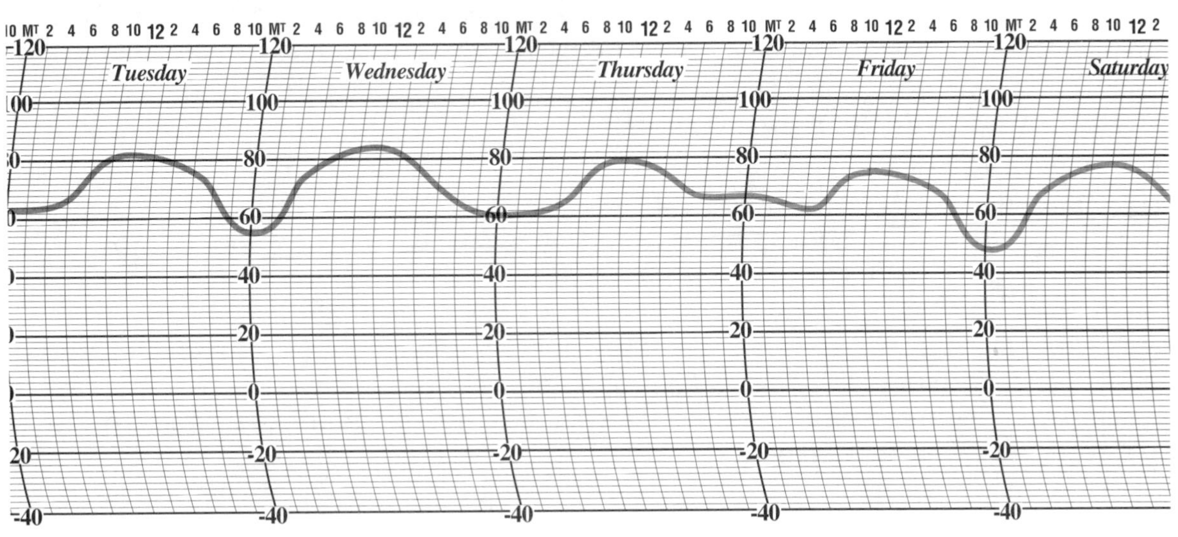


b. Plot of Daily Temperature Variations

*Figure 1-2. A recording thermometer — an analog system.*

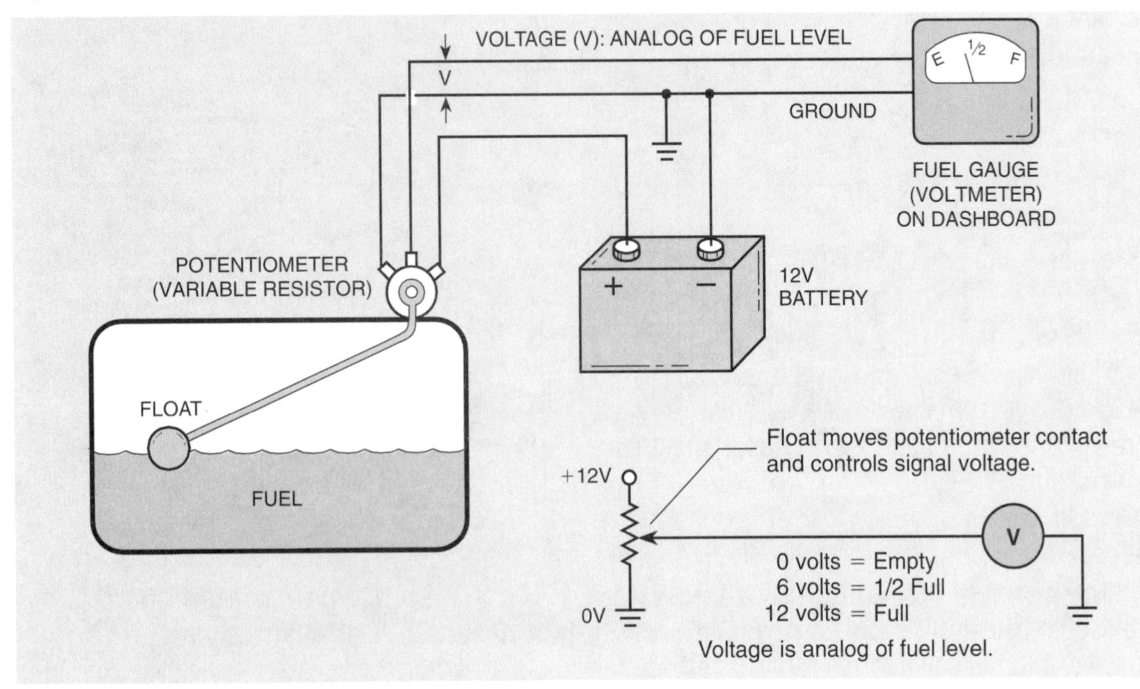


*Figure 1-3. An automobile fuel gauge — a simple system of a resistor, battery, meter, and wire — shows how a measurable electrical quantity carries information about something else. In this case, voltage is an analog of fuel level.*

*Source: Basic Electronics, G. McWhorter, A. Evans, ©1994, Master Publishing, Inc., Lincolnwood, IL.*

## The Choice of Communications Systems

Analog functions used in electronic systems cover a wide range of applications, from a simple fuel gauge in an automobile to a complex motor control system. In order to place some boundaries on the discussions in this book, we have chosen to focus on communications systems because the analog functions used to create these systems represent a large portion of all analog functions found in analog electronics. With an understanding of the operation of the analog functions used in communications systems and the way electronic components are used to build them, a great deal of the operation of other analog systems will be understood. Many such systems may only use one, two, or a combination of a select number of the analog functions found in a communications system. They may be employed in different combinations to meet specific system operation requirements. But their functions and how circuits are built to perform the function will be similar to those described in this book.

### What is a Communication System?

To answer this question, let's look at a very simple system that you may have used when you were a kid. It is shown in *Figure 1-4* — two tin cans with their bottoms directly coupled together with a tight string or wire. The person at the source shouts into the tin can. Sound waves cause changes in air pressure that move the bottom of the source tin can back and forth. This in-and-out movement is transmitted to the destination tin can by the tight string or wire. The motion in the string or wire created by the source tin can causes the bottom of the destination tin can to move in and out, thus creating air pressure sound waves that recreate the information from the source so it can be heard by the person at the destination.

*Figure 1-4. A very simple communication system.*

The purpose of a communication system is to transfer *information* from one location to another. Every communications system has three fundamental parts — a *source* (which is called the transmitter), a *transmission link* (in this case, tin cans directly-coupled by a tight string or wire), and a *destination* (which is called the receiver). Realize that the system in *Figure 1-4* is a very simple system, not very efficient in communicating information (it takes a lot of shouting). Obviously, the up-to-date techniques and equipment used in today's communications systems are much more advanced. But no matter how advanced, the communication system still has the same three basic parts — a transmitter, a transmission link, and a receiver.

## More Specific Examples

Normally, a stereo sound system would not be considered a communication system, but it is. Look at *Figure 1-5a*. The *information* to be transferred is the music stored on the CD, phonograph record, or cassette tape. The *source* is the combination of the CD, record player, or cassette tape player, the stereo amplifier, and the speakers. The *transmission link* consists of sound waves varying the pressure of the air. And the *destination*, or receiver, is a human ear.

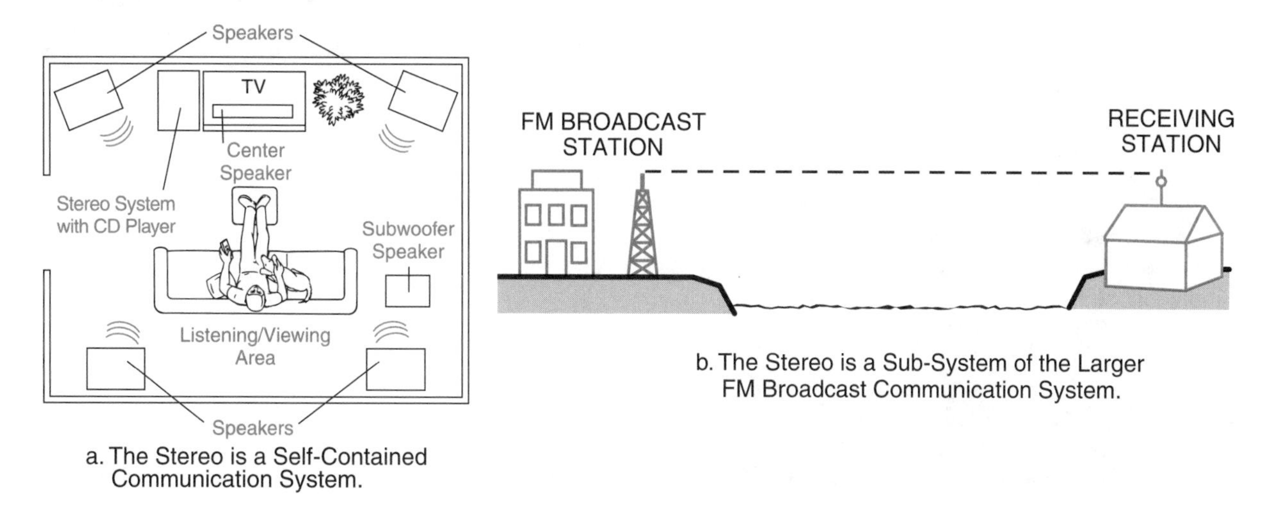


*Figure 1-5. A communication system within a communication system.*

Now look at the FM broadcasting system shown in *Figure 1-5b*. Here the *information* to be transferred is the broadcast station's programming material. The *source* is the FM station *transmitter,* the *transmission link* consists of radio waves from the station's antenna to the FM receiver's antenna, and the *destination* is the FM *receiver.* So the interpretation of a communication system varies as the size, complexity, and scope of the system changes. The individual stereo sound system that was our complete communication system is now a sub-system of the much larger FM broadcasting system.

Here are two important points about communication systems: First, one of the jobs of a transmitter is to change the information into a form that "fits" the transmission link so that the information can be properly transferred. Second, one of the receiver's jobs is to change the information back from the transmitted form in order to retrieve the original information.

## What Type Information Do We Have?

Today, information may be classified into many categories, mostly brought on by the "high-tech" age. We have chosen to use the following four categories as examples for comparison: text, data, audio, and video. Let's begin with text by looking at *Figure 1-6a*.

### Text

Text is information that is generated by typing on a typewriter, word processor, or a computer using a word processing program. It is a string of words, such as those written in a business letter or newspaper report, instructions in an owner's manual, a book, reference material, or general written information. It usually has a minimal amount of numerals. Special codes, like the ASCII Code, are used to store data in digital form just so computers can handle text easily.

a. Text

(Photo courtesy of *Chicago Tribune*, Chicago, Illinois)

b. Audio

(Photo courtesy of WGN, Radio 720 AM, Chicago, IL)

c. Video

(Photo courtesy of NBC 6, WTVJ, Miami, FL)

*Figure 1-6. Common types of information.*

## Data

Data, in general, is known information used to draw conclusions. For a more technical definition for computers and communication systems, in which text can be a subset, data usually is defined as information put into unique codes formatted according to agreed-upon specifications and standards so it can be operated on and manipulated by a digital system under the direction of a software program.

## Audio

Audio (*Figure 1-6b*) is information that is reproduced as sounds detectable by the human ear. Sounds — music, voice, noise — occur at different frequencies. We'll explain frequency in a minute, but recall that musical instruments like pianos, guitars, violins, and harps have strings that vibrate thus producing different sounds. Audio is defined as vibrations within the range of human hearing which we generally call sound — 10 cycles per second to 20,000 cycles per second.

## Video

Video (*Figure 1-6c*) represents pictorial information that includes still photographs, a painted picture, a magazine or book cover, sketches, drawings, schematics or, more generally, information classified as graphics. Video is information that is converted into an electron beam that varies in intensity to produce areas of light or no light on a phosphorous screen which, when a photograph is transmitted, reproduces the photograph. There is much more information to be transferred in a photograph than in lines of text. Thus, video is transferring much more information in a given amount of time than audio. We will look into why, but before we do we need to understand frequency and the relationship of signals of different frequencies.

## Frequency

A piano, harp, guitar or bass violin string, when plucked, vibrates back and forth as illustrated in *Figure 1-7a*. The amplitude (displacement) of the vibration, as shown in *Figure 1-7b*, decreases with each vibration until the string eventually returns to its initial position at rest. Note that the amplitude of the plucked string is first positive, then negative. When the vibration has gone from its beginning positive point, through zero, through negative, back through zero, and back to positive again, the vibration completes one cycle. Even though the amplitude changes, the time required to complete one cycle remains constant. The number of vibrations per second (cycles per second) determines the tone generated by the string. The *cycles per second* (vibrations per second) is the *frequency* of the tone. Different strings vibrating at different rates generate different frequency notes to produce the guitar music we enjoy. Electrical engineers have named *frequency* (cycles per second) after one of the leading researchers of electricity, Heinrich Hertz. Cycles per second are called hertz, which is abbreviated Hz.

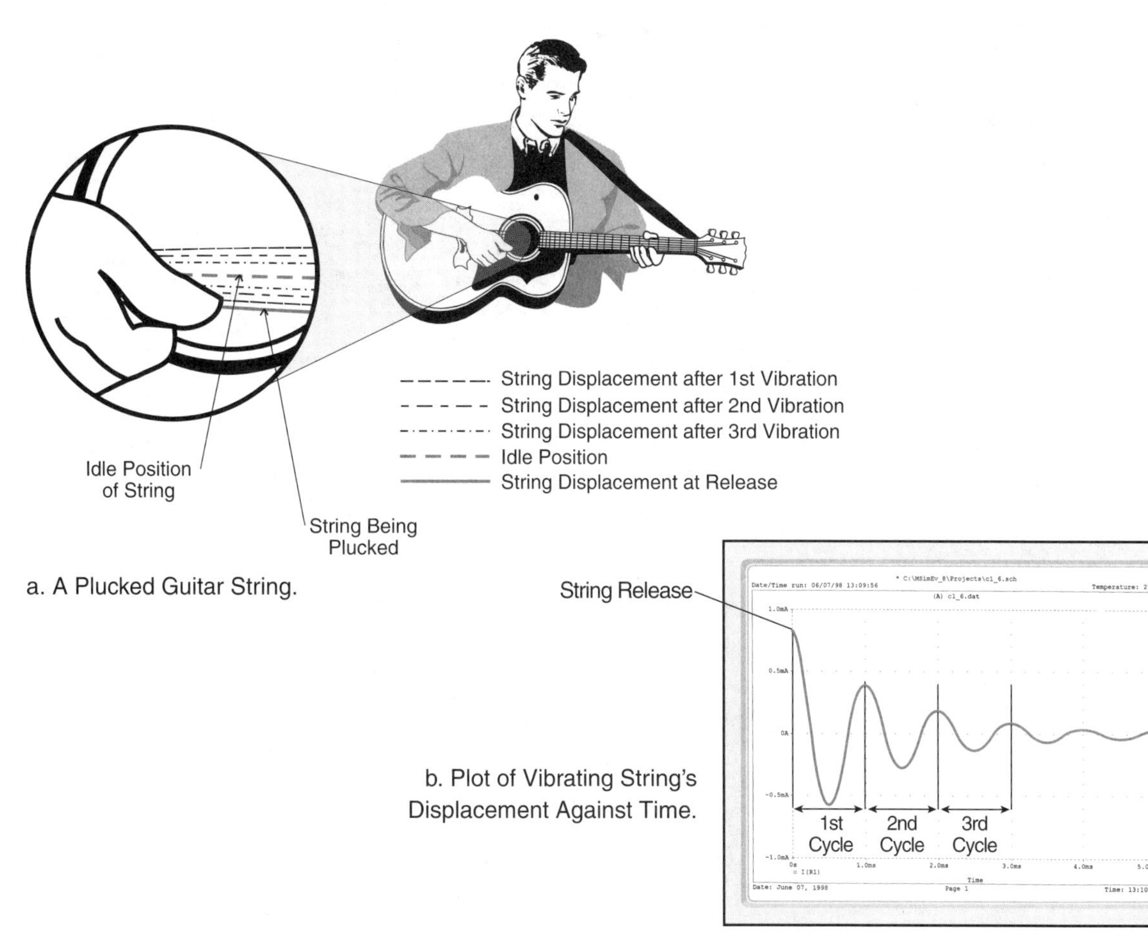


*Figure 1-7. The vibrations of a plucked guitar string.*

## The Sine Wave

Look at *Figure 1-8a*. It shows the definition of a vector. A rotating vector (sometimes called a phasor) is used by scientists and engineers to represent quantities that vary in time, in amplitude, and in phase. It is a line whose length, A, represents the signal amplitude and whose position from the x axis in x-y rectangular coordinates is represented by an angle of rotation, $\theta$. The angle $\theta$ (theta) represents the positions of the vector in time and, as we will find out later, represents the phase of the signal. The vector in an x-y coordinate system is the hypotenuse of the right triangle formed by the "Y-leg" and the "X-leg."

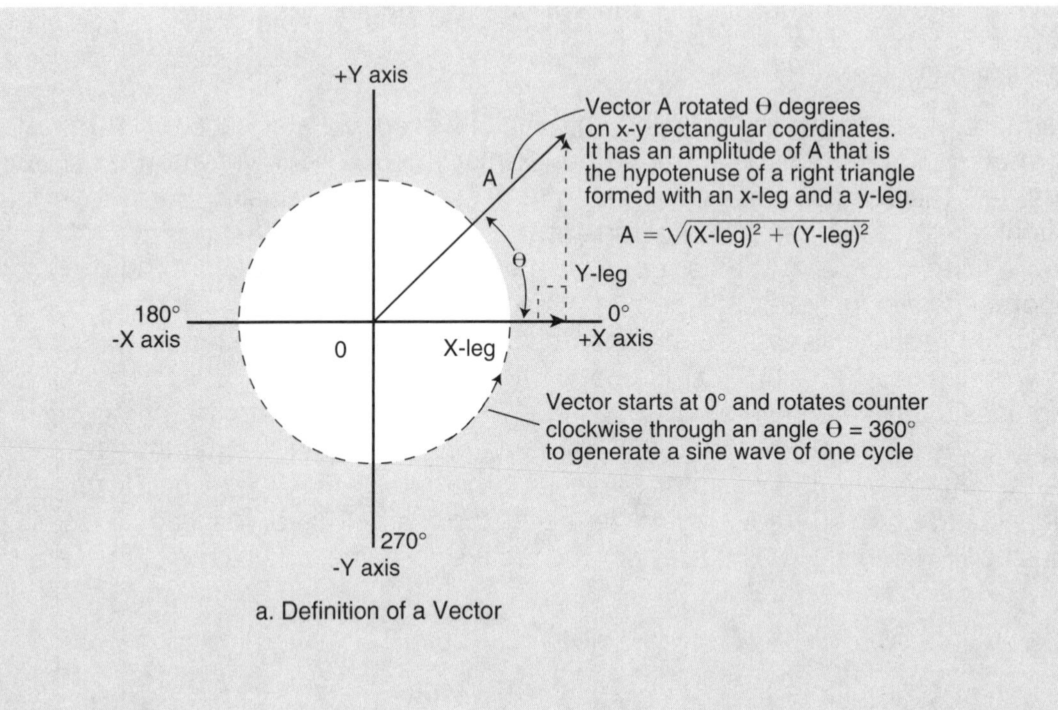


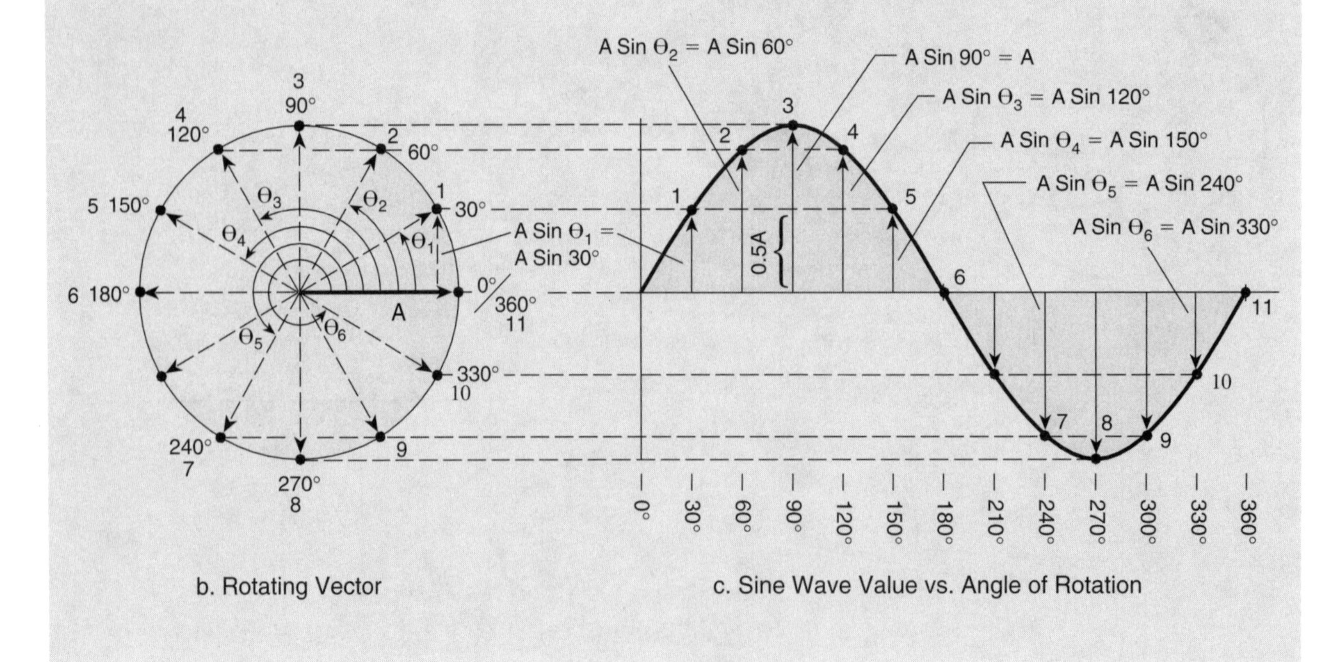


*Figure 1-8. A sine wave generated by the vector A sin θ.*

Now look at *Figure 1-8b* and c. Rotating a vector with amplitude A and plotting, over time, the displacement of the tip of the vector from the zero axis (0°) generates a waveform called a sine wave. Point 1 represents the displacement when the vector has rotated 30° counterclockwise. Point 4 is the amplitude when the vector has rotated 120°, point 7 when the rotation is 240°, and point 10 when the rotation is 330°. At point 6, the vector has rotated 180° and at point 11 the rotation is 360°. When the vector has rotated 360°, one cycle has been completed and the waveform begins to repeat itself. The number of cycles that the vector rotates each second (or the guitar string vibrates per second) is the *frequency*. Now note the waveform created by plotting the displacement of the vibrations of the plucked guitar string shown in *Figure 1-7b*. It can be approximated by the sine wave shown in *Figure 1-8c*, which reduces in amplitude as the cycles are generated. Sine waves are very important to engineers and scientists because they are used to describe signals and their frequency as they are transferred within a communications system. Most electronic analog signals — such as the tones generated by musical instruments, voice signals, and audio and video signals — can be represented by a single or a combination of a few or many sine waves.

### *Example 1. Sine Wave Amplitude at Different Rotational Angles*

Using *Figure 1-8*, what is the amplitude of the sine wave generated by the rotating vector when it has rotated through the following angles: 30°, 90°, 270°, and 330°?

***Solution:***

Looking at *Figure 1-8b*, the amplitude from the tip of the rotated vector down to the zero axis of the plot for the various angles is:

| **Angle** | **Amplitude** |
|---|---|
| 30° | + 0.5A |
| 90° | + A |
| 270° | - A |
| 330° | - 0.5A |

## Audio Signals

As explained earlier, an audio signal — when converted back to sound by a speaker — can be heard by humans. The signals may have a frequency from 10 cycles per second (10 Hz) to 20,000 cycles per second (20 kHz). Sounds or tones are characterized or classified by the many frequencies contained in the sound. For example, the tone we hear when a guitar string is plucked is not just the vibrating string itself but the additional sounds generated in the resonant chamber of the guitar body. The total sound is a combination of harmonics, or overtones, combined together to make the total signal. Waveforms generated by musical instruments often have harmonic frequencies, while voice waveforms almost never generate harmonics.

This combination of harmonics is illustrated in *Figure 1-9*. The lowest tone is called the fundamental, a tone twice the frequency is called the second harmonic, three times the frequency is the third harmonic, and four times the frequency is the fourth harmonic.

If a tone contained the fundamental, second, third, and fourth harmonics in the time and amplitude relationship shown in *Figure 1-9*, then the waveform of *Figure 1-9e* would represent the composite signal for this tone, which contains frequencies from 200 Hz to 800 Hz. The human voice, for example, contains frequencies from 80 Hz to 3000 Hz. A piano produces frequencies from about 40 Hz to 6000 Hz, and a pipe organ from about 30 Hz to 15,000 Hz. The combination of waveforms shown in *Figure 1-9* could vary in their amplitudes and in their time relationships. If they did, the composite waveform would change but the frequency content would remain the same.

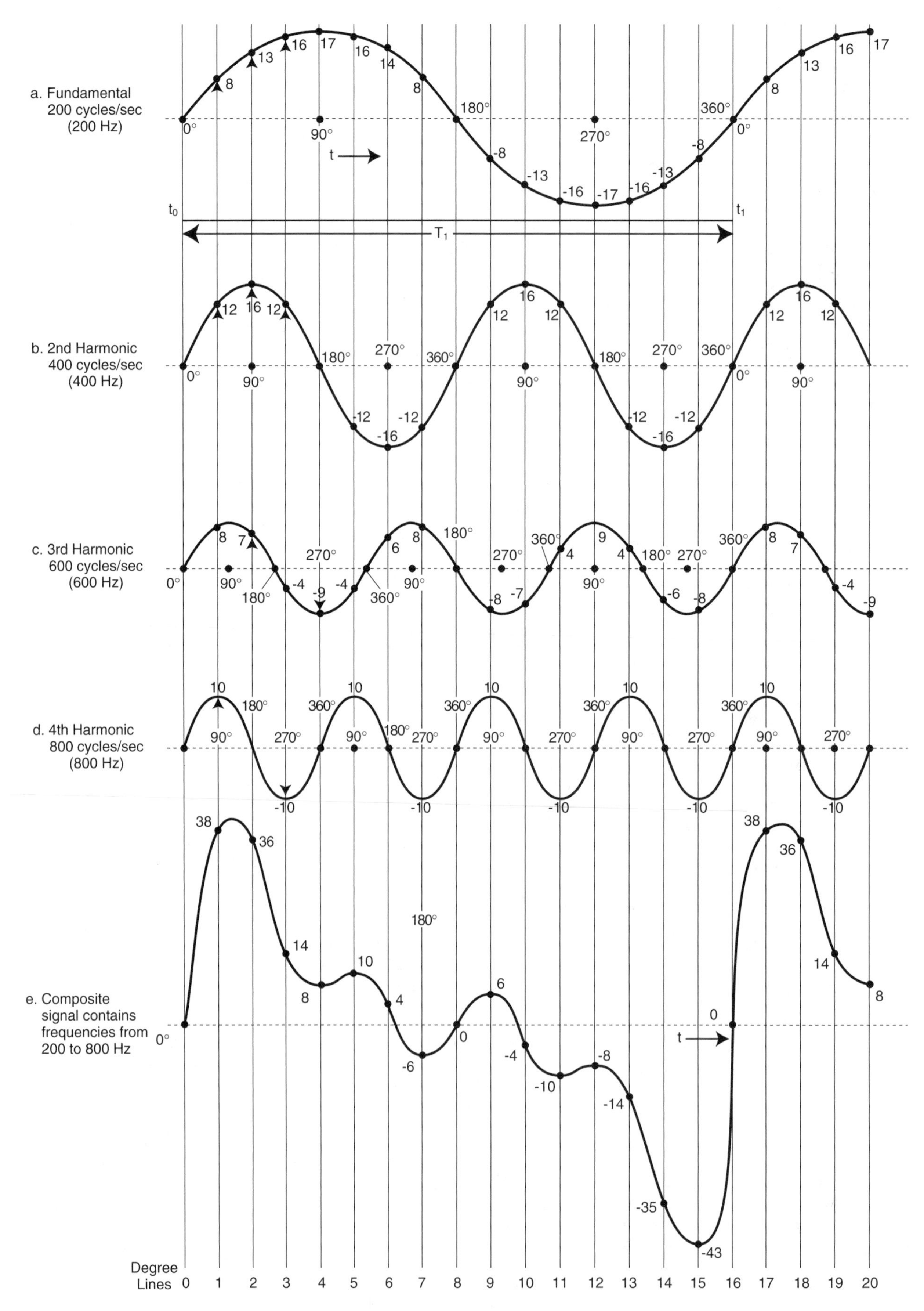


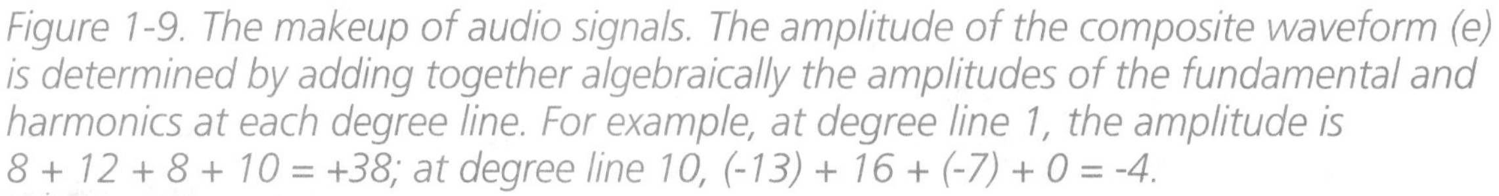

*Figure 1-9. The makeup of audio signals. The amplitude of the composite waveform (e) is determined by adding together algebraically the amplitudes of the fundamental and harmonics at each degree line. For example, at degree line 1, the amplitude is 8 + 12 + 8 + 10 = +38; at degree line 10, (-13) + 16 + (-7) + 0 = -4.*

### *Example 2. Harmonic Frequencies*

What is the frequency of the following signals?

A. The 3rd harmonic of 1000 Hz?
B. The 5th harmonic of 5000 Hz?
C. The 2nd harmonic of 200 Hz?
D. The fundamental of 12,000 Hz?

***Solution:***

The fundamental frequency is the actual frequency of the signal. Harmonics are multiples of the fundamental. Therefore:

A. The third harmonic of 1000 Hz is 3 x 1000 = 3000 Hz.
B. The fifth harmonic of 5000 Hz is 5 x 5000 = 25,000 Hz.
C. The second harmonic of 200 Hz is 2 x 200 = 400 Hz.
D. The fundamental of 12,000 Hz is 12,000 Hz.

## Bandwidth

Since audio frequencies can be anywhere from 10 Hz to 20,000 Hz, an information channel (transmission link) that handles audio frequencies must have a *bandwidth* of 20,000 Hz. In other words, since audio signals are going to be handled, the information channel of the communication system must have the capability to handle the full range of all of the frequencies of the signals that might occur in that information — in the case of audio, 10 Hz to 20,000 Hz. If the bandwidth of the transmission link is only 10,000 Hz, then certain audio signal frequencies may not be transferred and some of the original information would be lost. *Bandwidth is a specified range of frequencies, from a lower frequence ($f_1$) to a higher frequence ($f_2$), in which the signal response remains approximately the same — it is the difference between the highest and lowest frequencies necessary to accurately transfer information in a signal.* Communication systems can restrict the bandwidth for their use. For example, telephone systems only have an audio bandwidth of 3 kHz because voice signals require only frequencies from 80 Hz to 3 kHz for good communications.

## Video Signals

*Figure 1-10* shows the waveforms of a TV signal that transfers the information in a television picture. *Figure 1-10a* shows the expanded waveform, or the "field," of a TV video signal that uses an electron beam to paint the picture on a TV screen. Actually, the information shown not only provides the variable information for developing the contrast required for the picture, but also provides the timing to sweep the electron beam horizontally and vertically to paint the picture in correct correspondence to what the TV camera sees in the studio. The complete signal is called the composite TV video. Like a movie, full-motion video actually is a series of rapidly-changing still images. A single frame of information consists of 525 horizontal lines across the TV screen repeated at 30 frames per second, or 60 fields per second because there are two fields per frame. Because the human eye cannot detect changes at 30 frames per second, the rapid changes are seen as full-motion, smooth movements and scene changes on the TV screen.

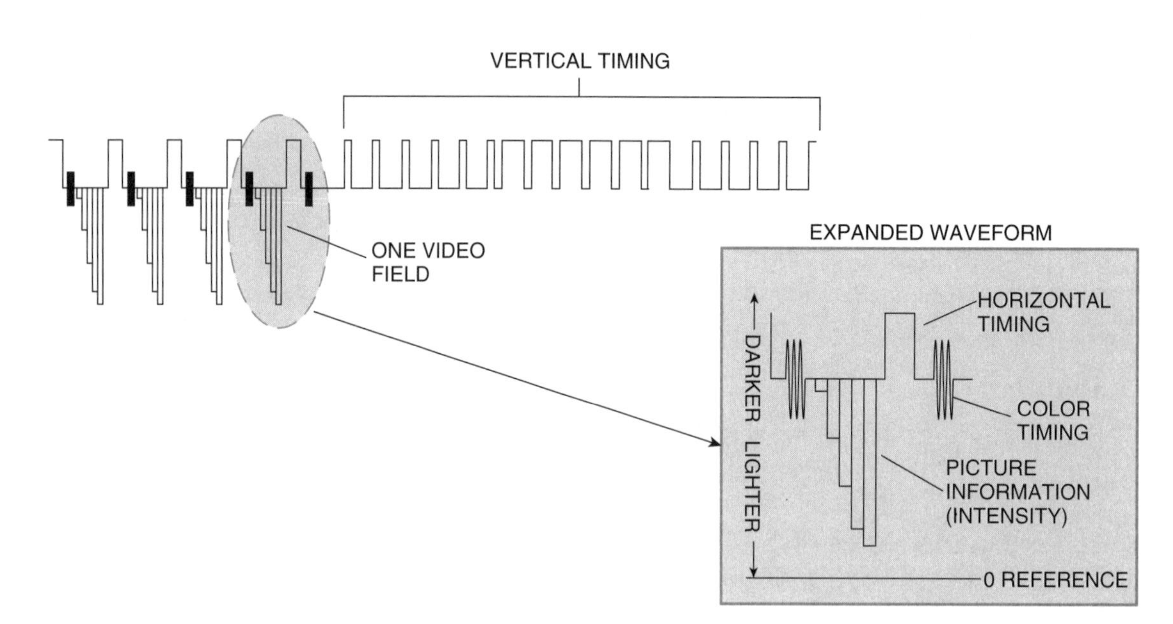


a. A TV Composite Video Signal
*Source: Using Video in Your Home, G. McComb. © 1989, Master Publishing, Inc., Lincolnwood, IL.*

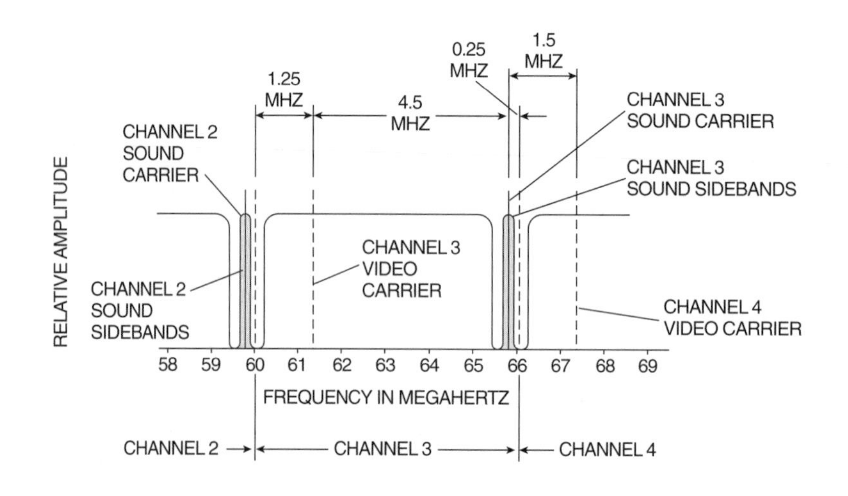


b. Bandwidth for a TV Channel
*Source: Antennas, A. Evans, K. Britain,. © 1988, Master Publishing, Inc., Lincolnwood, IL.*

*Figure 1-10. The video signal is composed of three major components: picture, timing, and color. Television sets "know" how to decode this stream of information into pictures that reproduce the changing scenes captured by the video camera.*

When compared to an audio signal, more information must be transferred each second to make-up the composite TV video signal used to create the full-motion picture on your TV screen. Thus, a larger bandwidth is required for TV than for audio for all the information to be transferred accurately. A TV signal can contain frequencies up to six million Hz — more than 300 times greater than audio frequencies; therefore, a system for the transfer of TV information must have a 6-MHz bandwidth. Our broadcast TV stations are licensed to operate on one of 68 channels. *Figure 1-10b* shows the 6-MHz bandwidth required for each TV channel. The frequencies shown are for Channel 3. The source transmission in a TV communication system must have a 6-MHz bandwidth for each TV channel that it transmits while, at the destination, the receiver must have the capability to tune-in all 68 channels, requiring tuning circuits that span 68 x 6 MHz in bandwidth. The information channel bandwidth of the receiver is 6 MHz.

### *Example 3. Channel Bandwidth*

If a video telephone needs a 3-MHz bandwidth, which of the communication systems listed would you need to transmit the signal accurately in real time?

System A. A standard telephone system with 3000-Hz bandwidth.
System B. A microwave system with 3-MHz bandwidth.
System C. A satellite system with 30-MHz bandwidth.

***Solution:***

The video telephone information would be transferred accurately in real time by both System B and System C, but you only need System B with its 3-MHz bandwidth. The 30-MHz bandwidth of System C has much more capability than required. System A, with its 3000-Hz bandwidth, would not work at all in transmitting the telephone video in real time.

## What is Phase and the Phase of a Signal?

Let's refer back to *Figure 1-9* where a sine-wave signal at the fundamental frequency and at the 2nd, 3rd and 4th harmonic frequencies were shown. All of the sine waves shown start at time $t_0$ and are rising in amplitude. In *Figure 1-8*, let's assume that vector A generates the fundamental sine wave. It would mean that all the vectors that rotate to generate the respective *harmonic* sine waves start from the same position on the zero axis as vector A. At $t_0$, all the waveforms are in the same time relationship. They are said to be "in phase" at time $t_0$. The natural question is, "What is out-of-phase?"

Let's demonstrate this in *Figure 1-11*, which is the same type of diagram as *Figure 1-8* but contains three vectors — A, B and C — instead of just A. The three vectors *maintain their position relative to one another* as they rotate. Vector A is on the

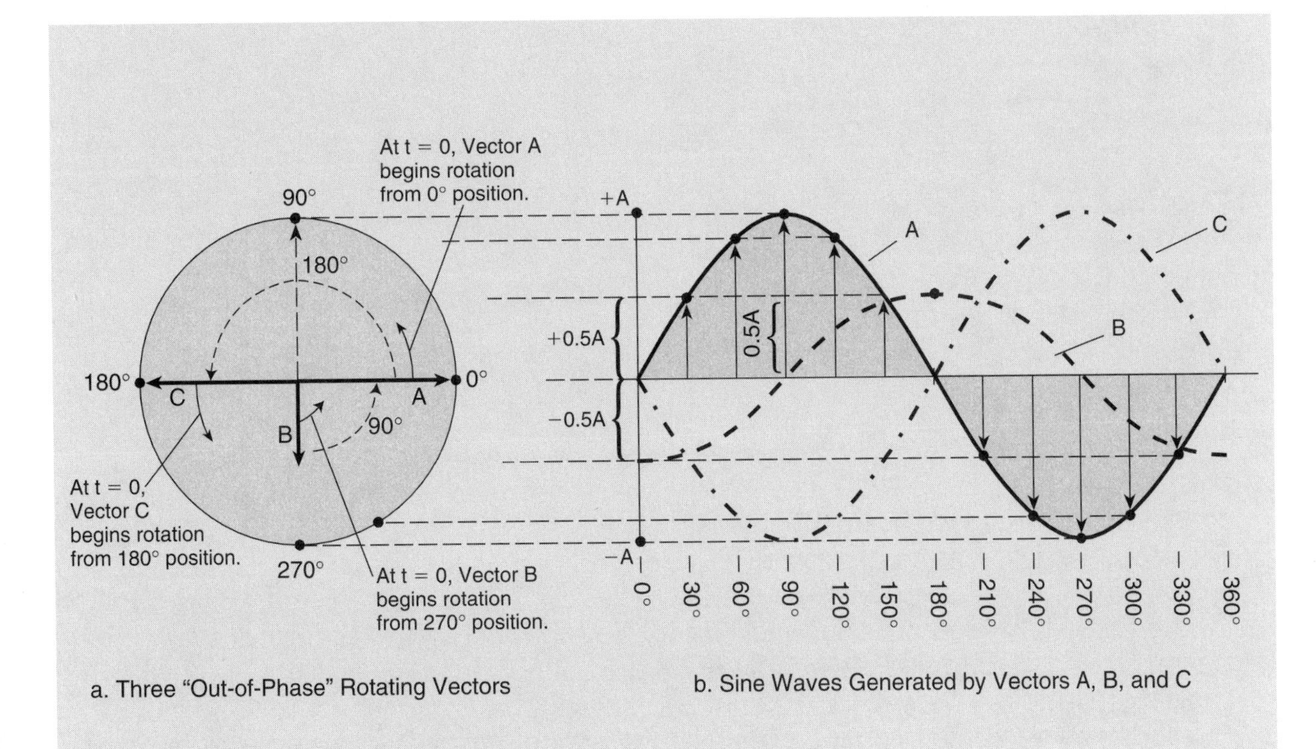


*Figure 1-11. Phase relationships of vectors A, B, and C.*
Source: *Basic Electronics,* G. McWhorter, A. Evans, ©1994, Master Publishing, Inc., Lincolnwood, IL.

0° axis when time starts at $t_0$. Vector B, which is one-half the amplitude of vector A, is positioned 90° behind vector A, pointing down at a 90° right angle in the 270° position. Vector C, which has the same amplitude as vector A, is positioned 180° behind vector A, or pointing in the exact opposite direction. The sine waves that these three vectors generate as they rotate are shown in *Figure 1-11b*. From $t_0$, vector A rotates through a positive amplitude first; vector B of one-half amplitude begins at its most negative amplitude and reduces toward zero; and vector C of the same amplitude as vector A rotates through a negative amplitude first. So there is a different time relationship between the waveforms — they are in different phase relationships. Vector B is said to be "90° out of phase" with vector A, and vector C is said to be "180° out of phase" with vector A. Phase, therefore, is *a measure of the difference in time relationships of signals*.

We now can construct the composite waveform of vectors A, B, and C as they rotate by adding the waveform amplitudes algebraically at each rotational angle $\theta$. One can easily see that because vector C is "180° out of phase" with vector A that it cancels vector A at every point, and we are left only with vector B's sine wave. Thus, the phase relationship of signals, along with the amplitude relationship of signals, determines what the final, composite signal looks like. If the waveforms in *Figure 1-9* had a different phase relationship, the output composite signal would look different because the algebraic adding of the amplitudes would be different and it would be different again if the respective waveform amplitudes were different as well from those in *Figure 1-9*.

### *Example 4. Signal Phase*

Draw a vector D that is 270° behind (lags A by 270°) vector A in *Figure 1-11* and plot the sine wave it generates as it rotates in a constant time relationship to vector A. Vector D is the same amplitude and frequency as Vector A.

***Solution:***

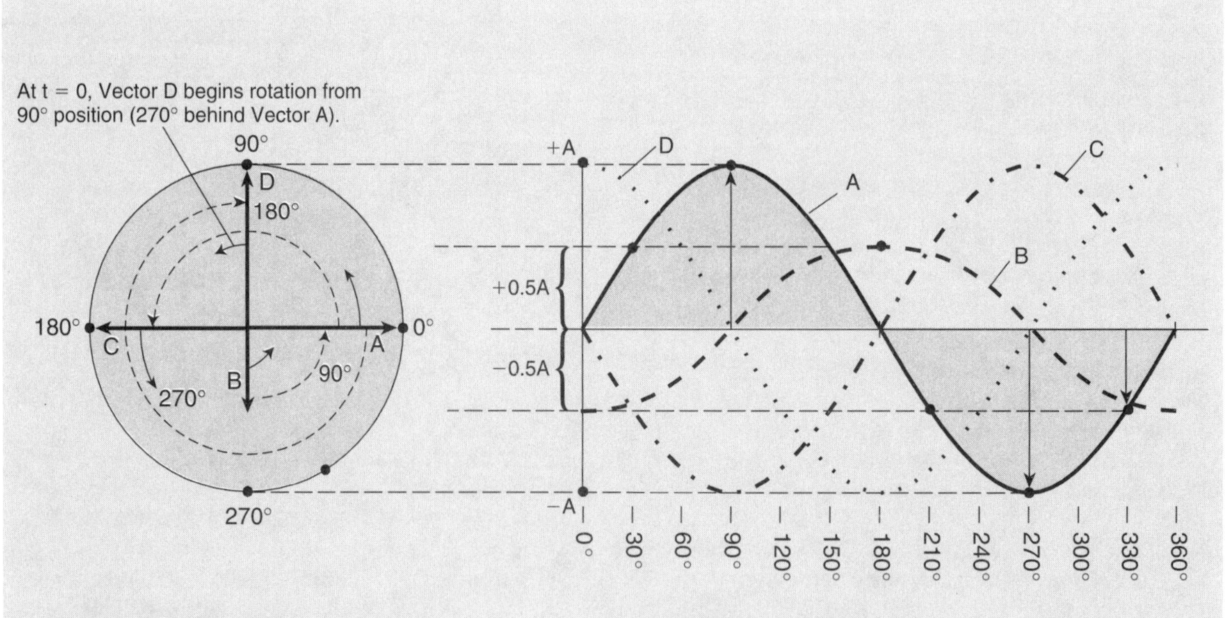


a. Adding Vector D 270° Behind Vector A

b. Sine Waves Generated by Vectors A, B, C and D

# Frequency, Period, Spectrum and Wavelength

## Frequency and Period

The frequency of the sine waves generated by vectors A, B and C are related directly to how fast the vectors rotate. As shown in *Figure 1-12b*, the time for one cycle, T, is called the *period* of the sine wave. The *frequency* of the sine wave is f = 1/T. The faster the vector rotates the smaller the time T and the higher the frequency. The smaller T is, the more cycles per second — the higher the frequency.

## Spectrum

The frequency of signals, especially electromagnetic radio waves, have been classified into what is called a frequency spectrum, as shown in *Figure 1-12a*. The spectrum has been divided into bands where the higher frequency is 10 times higher than the lower frequency. For example, the high-frequency or HF band extends from 3.0 MHz to 30 MHz; while the SHF band extends from 3 GHz to 30 GHz. A gigahertz (GHz) is 1000 megahertz. Since mega stands for million, that means the signal is vibrating at 1000 million cycles per second!

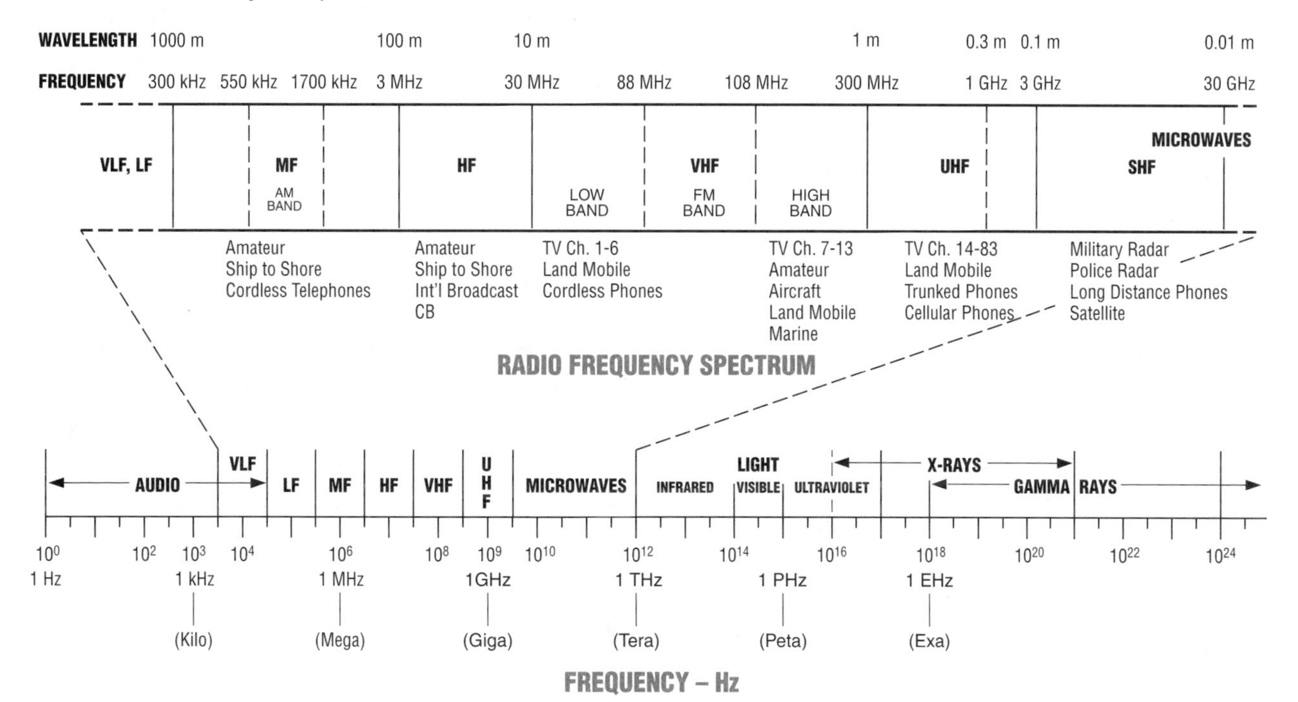


a. Frequency Spectrum
Source: *Mobile 2-Way Radio Communications,* G. West. © 1991, Master Publishing, Inc., Lincolnwood, IL.

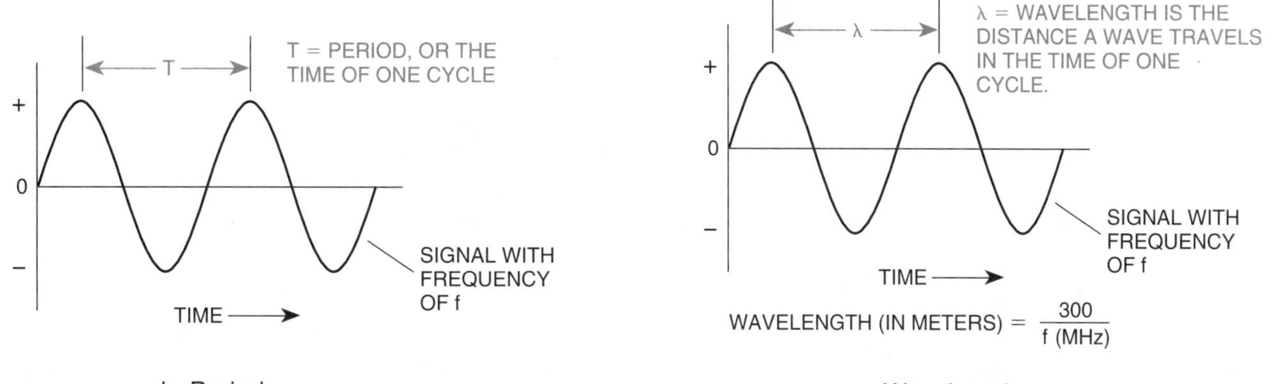


*Figure 1-12. Frequency spectrum, period, and wavelength.*

## Wavelength

The division of the frequency spectrum into bands occurs on submultiples or multiples of 300 MHz because the velocity of electromagnetic energy through space is 300,000,000 meters per second, which is 186,000 miles per second or the speed of light. As shown in *Figure 1-12c*, the distance that an electromagnetic wave travels through space in one cycle is the wavelength (λ) of the signal's frequency. Since wavelength is a distance, it is equal to the velocity of the signal times the amount of time for one cycle, or 300,000,000 x T. T is the period of the wave, or 1/f; therefore, the wavelength (λ) is 300,000,000/f.

$$\lambda \text{ (in meters)} = \frac{300{,}000{,}000}{f \text{ (in Hz)}}$$

If f is expressed in megahertz, then

$$\lambda \text{ (in meters)} = \frac{300}{f \text{ (in MHz)}}$$

### *Example 5. Calculating Frequency and Wavelength*

A. What is the frequency in Hz and MHz, and the wavelength in meters, of the signals with the following periods, T?

1. 0.01 second
2. 0.001 second
3. 0.0001 second
4. 0.00001 second

***Solution:***

Since f = 1/T, then

| f (in cycles) | *f (in MHz) | **λ (in meters) |
|---|---|---|
| 1.1/0.01 = 100 | 0.0001 | λ = 300/0.0001 = 3,000,000 |
| 2.1/0.001 = 1000 | 0.001 | λ = 300/0.001 = 300,000 |
| 3.1/0.0001 = 10,000 (10 kHz) | 0.01 | λ = 300/0.01 = 30,000 |
| 4.1/0.00001 = 100,000 (100 kHz) | 0.1 | λ = 300/0.1 = 3000 |

* Divide f in Hz by 1,000,000 to change to MHz.
** Divide 300,000,000 (the speed of electromagnetic waves) by f to determine wavelength in meters. When using MHz for frequency, divide 300 by f in MHz.

B. What is the wavelength in meters of an AM signal at 1000 kHz?
λ (in meters) = 300/f (in MHz) = 300/1 = 300 meters

## Importance of Wavelength

Wavelength is important to communication systems, especially when the transmission link is wireless electromagnetic radio waves. Antenna lengths for the transmitter and the receiver are much more efficient if they are submultiples or multiples of the wavelength. In addition, longer wavelength (lower frequency) signals tend to radiate in a much wider area over a longer distance, while shorter wavelength signals radiate for less distance and on line-of-sight paths.

As shown in *Figure 1-12a*, the AM (meaning *Amplitude Modulation*) broadcast band extends from 550,000 Hz (550 kHz) to 1,7000,000 Hz (1700 kHz), while the FM (meaning *Frequency Modulation*) broadcast band extends from 88,000,000 Hz (88 MHz) to 108,000,000 Hz (108 MHz), much higher frequencies. We stated previously that k stands for kilo or thousands, and M stands for mega or millions. As a result,

AM broadcast frequency wavelengths are much longer than the FM broadcast frequency wavelengths, and AM broadcast stations can be heard for much longer distances than FM broadcast stations. More detail on propagation is provided in Chapter 9.

Antennas for each of the bands should be at least one-quarter wavelength long. Therefore, AM antennas will be much longer than FM antennas. Other critical electrical lengths follow the same rule. For example, cellular telephones that operate at UHF frequencies have very short antennas. At microwave frequencies, components, circuit lead lengths, and circuits get so small that critical lengths are measured in centimeters or millimeters.

## How is Information Carried from Source to Destination?

In order to transfer information, we said we needed a transmitter (source), a receiver (destination), and a transmission link with communications channels in between the source and destination. Within the transmission link, information is transported at frequencies appropriate to the transmission link. There are typically five types of media used for transmission links. They are:

1. Wire — When you pick up a telephone receiver to talk to a friend, the information comes over two wires.
2. Cable — Cable TV comes into the home and to the TV via a cable. Yes, a cable has a wire in it, but it is a special combination of wire, insulation, and shielding so the signal loss is very small. It is a distinct type of transmission link.
3. Wireless — When there are no cables or wires, electromagnetic waves (radio) are used as a "wireless" transmission link.
4. Fiber Optics — Many new telephone and computer systems use light as the transmission link. Lasers generate the light and the information is carried within fiber-optic cables — cables manufactured to transmit light at low losses.
5. Sound — Pressure waves are generated in air, water, or other media for the transmission link.

The transmission link has within it the information to be transferred. In the case of physical media — wire, cable, fiber optics, air, or water — the physical media connect the transmitter and receiver directly. In the case of wireless, the transmission link is electromagnetic radio waves. In the same way as for the physical media, the usual practice (as in standard AM or FM broadcasting) is for the information to modulate a higher, single-frequency carrier with the information to be transferred. However, in wireless transmission links there are cases where the carrier signal is removed before transmission and put back at the receiver to recover the transferred information. Such a technique is called suppressed-carrier single-sideband transmission, and it will be covered in greater detail in following chapters.

## Modulation Techniques

### CW (Morse code)

If you were to turn on a shortwave radio receiver and scan across the dial, you would hear a wide variety of signals. Morse code signals from amateur radio operators, mobile and marine radio voice communications, and other signals can be heard at specific frequencies. As shown in *Figure 1-13a*, the continuous signal (CW for continuous wave) at a specific frequency is the *carrier* — the interruption of the carrier for different

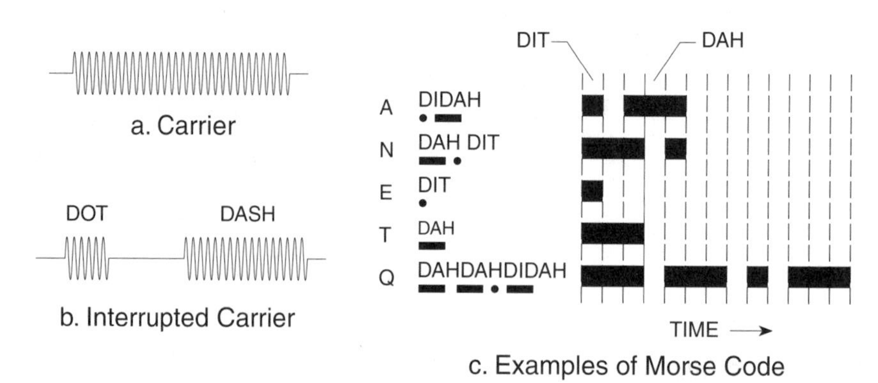

Figure 1-13. Morse code uses CW carrier.

periods of time (the dits and dahs of Morse code) carries the information. *Figure 1-13b* shows an example of how operators sending Morse code interrupt the carrier in order to produce codes for letters, numbers and other characters such as those shown in *Figure 1-13c*. Such a form of changing the carrier to transfer information is called *modulation*.

## AM and FM

*Figure 1-14* shows two of the most common forms of modulation. They are amplitude modulation (AM) and frequency modulation (FM). The information to be transferred appears to change the amplitude of the carrier corresponding to the input information for *amplitude modulation*. (In Chapter 5, we will show that the actual carrier amplitude does not change.) The information to be transferred changes the frequency of the carrier corresponding to the input information for *frequency modulation*. There is a third form of modulation that produces similar results to FM called *phase modulation* (PM). It changes the time (phase) relationship of the carrier signal to the original signal based on the input information.

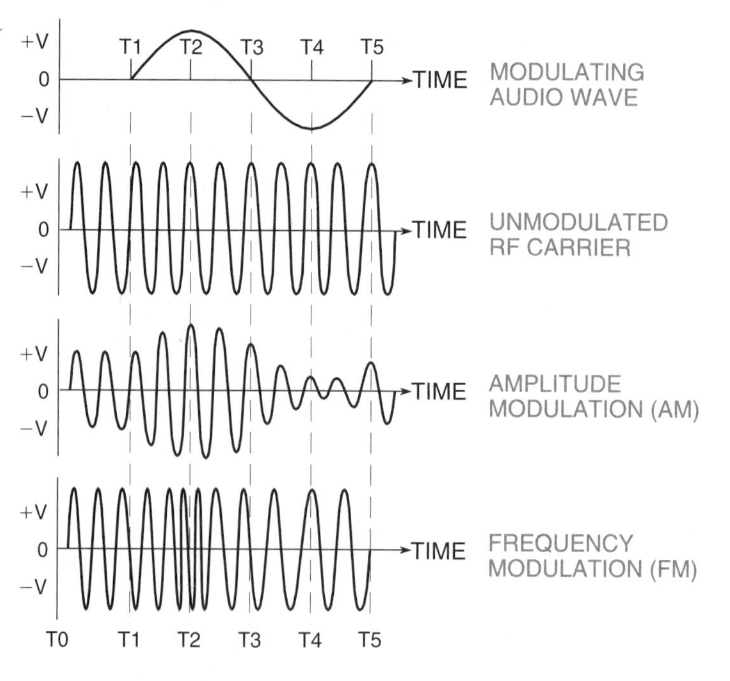


Figure 1-14. AM and FM modulation.

## Summary

Let's summarize what has been learned in our introductory Chapter. We've learned that:

1. Systems in which the information changes continuously and smoothly are called *analog.*
2. The *purpose* of a communication system is to *transfer information*.
3. Every communications system has three fundamental parts, a *transmitter*, the *transmission link*, and a *receiver*.
4. Analog signals are made up of signals containing one or more *frequencies*.
5. The more information contained in a signal to be transferred in a given amount of time the more channel *bandwidth* required to transfer the information accurately.
6. Communication systems that must handle a larger bandwidth (or many channels with a smaller bandwidth per channel) must operate at higher frequencies in order to provide the total *bandwidth* required.
7. *Wavelength* gets shorter as frequency increases.
8. Antennas and other critical electrical lengths get *shorter* as frequency increases.

In the next Chapter, we will begin looking at the analog functions used to create communications systems.

# Quiz for Chapter 1

1. What are the characteristics of an analog signal? An analog signal:
   a) Is turned on and off mechanically.
   b) Varies smoothly and continuously.
   c) Is made up of numeric codes.
   d) Moves erratically and is unstable.
2. How do digital and analog quantities differ?
   a) Digital quantities vary in discrete values, analog quantities are continuous.
   b) Analog quantities are made up of numeric codes; digital quantities are smooth and continuous.
   c) They are the same; there is no difference.
   d) None of the above.
3. What distinguishes a video signal from an audio signal?
   a) Video signals contain pictures; audio signals contain sound.
   b) Video signals contain more information than audio signals.
   c) Video signals for TV are higher in frequency than audio signals for radio.
   d) All of the above are correct.
4. What is meant by the frequency of a sine wave?
   a) The length of time to complete one cycle.
   b) The distance that a radio wave travels in one cycle.
   c) The number of cycles per second.
   d) The spectrum of the sine wave.
5. What is meant by the period of a sine wave?
   a) The length of time to complete one cycle.
   b) The distance that a radio wave travels in one cycle.
   c) The number of cycles per second.
   d) The spectrum of the sine wave.
6. What is meant by the wavelength of a sine wave?
   a) The length of time to complete one cycle.
   b) The distance that a radio wave travels in one cycle.
   c) The number of cycles per second.
   d) The spectrum of the sine wave.
7. What is meant by the term "frequency spectrum?"
   a) The combination of two frequencies.
   b) The classification of signals into bands, especially electromagnetic radio waves.
   c) The effect of sun spots on radio waves.
   d) The number of times a signal oscillates per minute.
8. What are the fundamental parts of a communications system?
   a) The transmitter, transmission link, and receiver.
   b) Picture, sound, and timing signal.
   c) Wire, cable, radio waves, fiber-optic cable.
   d) An AM/FM radio, TV set, and computer.
9. What is the frequency range of audio signals?
   a) 10 Hz to 20 kHz
   b) 1000 Hz to 20000 Hz
   c) 20 kHz to 1 MHz
   d) 88 MHz to 108 MHz
10. What is the frequency range of the standard FM broadcast band?
    a) 88 to 108 MHz.
    b) 3 to 350 GHz.
    c) 30 to 88 MHz and 108 to 300 MHz.
    d) 300 to 1700 kHz.
11. What is meant by the bandwidth of a signal?
    a) The channel number selected on the TV to view a station's programming.
    b) The station number used to tune your radio.
    c) The different type of signals used for AM, FM and TV stations.
    d) The range of frequencies necessary to transfer all of the information contained in a signal.
12. What are harmonics?
    a) Hand-held musical instruments.
    b) Secondary signals that are sub-multiples of the original signal.
    c) Frequencies that are multiples of the fundamental frequency.
    d) A way of tuning-in a radio station.
13. What is meant by AM, FM, PM, and CW?
    a) They are forms of modulating a carrier with a baseband signal.
    b) Amplitude modulation, frequency modulation, phase modulation, and continuous wave modulation.
    c) They are different broadcast bands for radio signals.
    d) a and b.

**Answers:**
1 b, 2 a, 3 d, 4 c, 5 a,
6 b, 7 b, 8 a, 9 a, 10 a,
11 d, 12 c, 13d

## Questions & Problems for Chapter 1

1. We selected four categories to classify types of information. What are they?

   a) ______________

   b) ______________

   c) ______________

   d) ______________

2. The three fundamental parts of a communications system are _______________, _______________, and _______________.

3. ___________ are used by engineers to represent quantities that vary in time, amplitude, and phase.

4. The name given to the vibrations per second of a guitar string is _______________ or _______________.

5. A range of frequencies from $f_{lo}$ to $f_{hi}$ is called _______________.

6. The waveform, important to communications systems signals, that is generated by a vector rotating in an X-Y rectangular coordinate system is called a _______________.

7. How much bandwidth is required for an audio signal? For a TV signal?

8. Calculate the period of signals with the following frequencies:

   a) 500 kHz

   b) 20 kHz

   c) 10 MHz

9. Calculate the following:

   a) The 3rd harmonic of a 2 kHz signal.

   b) The 5th harmonic of a 5 MHz signal.

   c) The 2nd harmonic of a 720 kHz signal.

   d) The fundamental of an 88 MHz signal.

10. Calculate the frequency of signals with the following periods:

    a) 0.001 second.

    b) 0.05 second.

    c) 0.00001 second.

11. Calculate the wavelengths of signals with the following frequencies:

    a) 600 kHz

    b) 100 MHz

    c) 250 MHz

## Questions & Problems for Chapter 1

12. If at a moment in time, three rotating vectors are at these positions: Vector A at 60°, Vector B at 120°, and Vector C at 90°. And the vectors maintain these relative positions as they rotate.

    a) What is the phase difference between Vector A and Vector B? Does Vector A lead or lag Vector B?

    b) What is the phase difference between Vector A and Vector C? Does Vector A lead or lag Vector C?

    c) What is the phase difference between Vector B and Vector C? Does Vector B lead or lag Vector C?

13. If the amplitude of a signal varies due to modulation, what type of modulation is being used?

14. If the frequency of a signal varies due to modulation, what type of modulation is being used?

15. If the phase of a signal varies due to modulation, what type of modulation is being used?

*(Answers on page 209.)*

# CHAPTER 2
# Analog System Functions



In Chapter 1, we established what a communication system is, what makes up a system, and became familiar with common terms used when describing the system. In this Chapter, we will explore the functions that are needed in analog systems, which are combined in a variety of ways to form communications systems. We will describe the basic concepts of the functions; actual details of the circuits that are built to provide the functions will be covered in later chapters. But before discussing the functions, let's talk a bit more about the *information* and the *transmission link* of a communication system.

## More on the Information Transferred

### Baseband Information

Recall that analog information, which varies continuously, usually contains signals of more than one frequency. In fact, information normally contains a *band of frequencies* (for audio — 20 Hz to 20 kHz; for video — up to 6 MHz). The information channel that is used to transfer the information must have a *bandwidth* equal to or greater than the band of frequencies that contains the information in order to transfer the information accurately. A signal with all of the frequencies that contain the information is called the *baseband signal*. Thus, an audio baseband signal might contain frequencies from 20Hz to 20 kHz (kilohertz). Recall that "kilo" means 1000 or, in mathematical terms, a multiplier of 1000. The band of frequencies for baseband signals can be specified for an information channel. For example, the telephone company's baseband signal, even though it is in the audio range, is specified to be only about 80 to 3000 Hz.

A *spectrum analyzer* is a piece of test equipment that detects the frequencies contained within a signal and displays the results on a video screen, as shown in *Figure 2-1*. The horizontal scale is calibrated in frequency, and the vertical scale in relative amplitude of the signal. *Figure 2-1a* is a spectrum analyzer output of an audio signal that contains five units of a 1 kHz signal, 10 units of a 5 kHz signal, and 12.5 units of a 9 kHz signal. An audio baseband signal with many frequencies present might look like *Figure 2-1b* because the baseband signal can contain frequencies from 20 Hz to 20 kHz.

### Information Channels

Communication systems depend on separating information into channels, where a channel must have at least the bandwidth of the baseband signal in order to transfer the baseband signal accurately. A well-known example is a TV set. The antenna of the TV set,

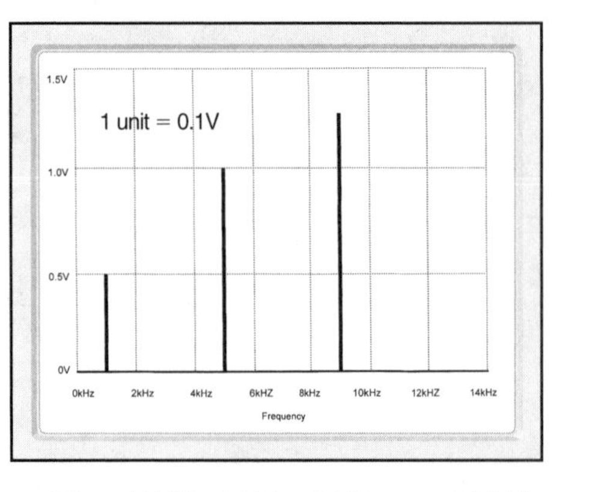


a. Signal With 1 kHz, 5 kHz, and 9 kHz

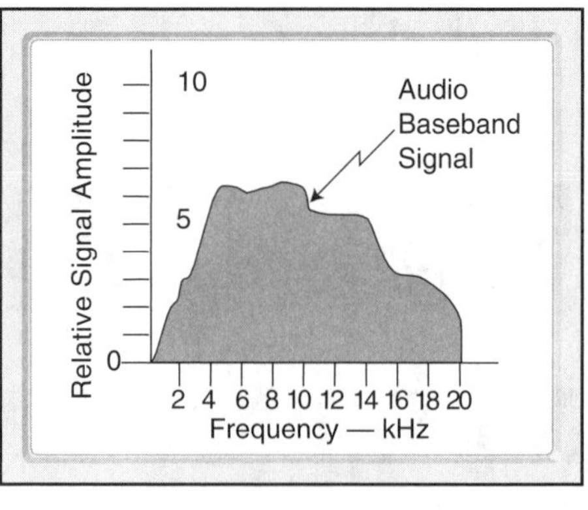


b. Baseband Signal.

*Figure 2-1. Spectrum analyzer output of audio signals.*

as shown in *Figure 2-2*, receives all of the "off-the-air" broadcast signals from many TV stations but the tuner selects only the signal of one specific TV channel to be passed on, processed, and displayed on the screen. Changing the tuner channel changes the information selected and presents a different TV station's program material on the screen.

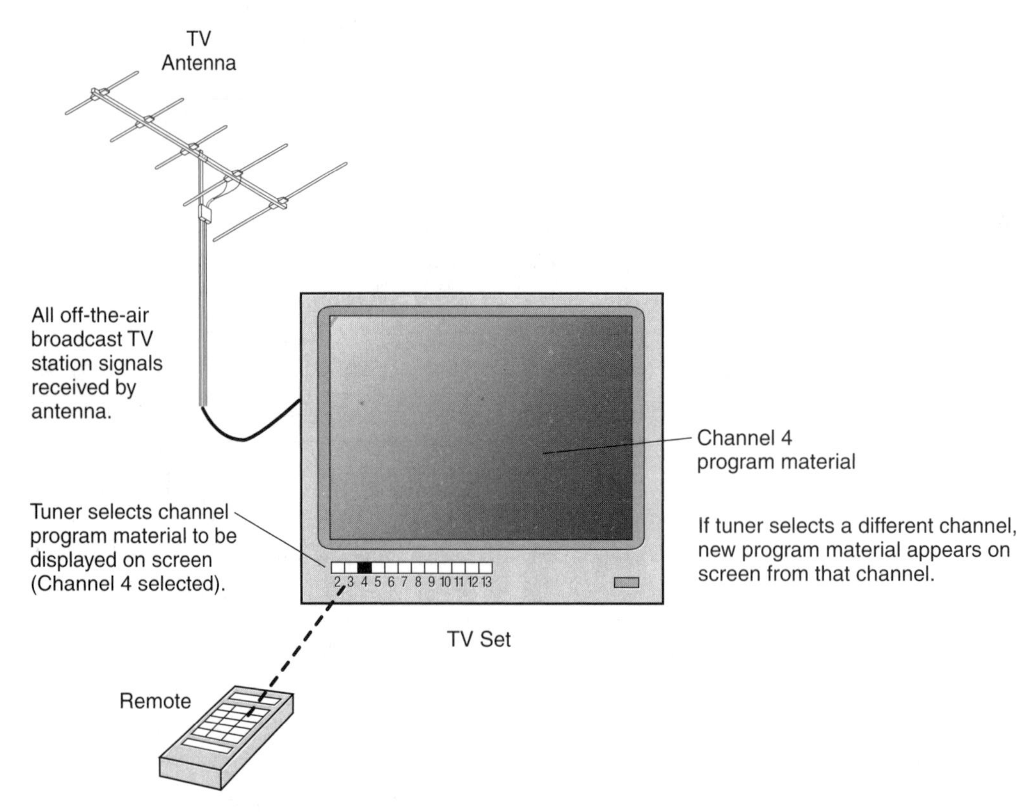

*Figure 2-2. Information channel selection.*

Such "channeling" and selection of channels to process only specific information is a prime way that communication systems are designed. The information channel, whose purpose is to transfer information as rapidly and accurately as possible, is created within the transmitter, within the transmission link, and within the receiver. All systems will have limitations placed on them by the characteristics of the information channel, but armed with an understanding of strengths and limitations of the communications channel, the system designer can determine the required performance of the transmitter, transmission link and receiver in the system. For many systems, the information channel specifications are standard and system components are manufactured to meet these established standards. An AM radio broadcast system is an example of such a system. We will see in later chapters how communication engineers and system designers take the information channel characteristics into account to set system specifications.

### Noise in Systems

As information is processed in the transmitter, transmission link, and receiver, unwanted signals are added to the original information signal by the electronic circuits in the equipment, by the power sources for the equipment, and by the environmental conditions of the transmission link, such as thunderstorms. These unwanted signals — when present with the original signal — are called *noise*. It is important that the noise be maintained at a low level compared to the received signal level so that the original signal can be transferred accurately. Therefore, the ratio of signal level to noise level (called the signal-to-noise — or S/N — ratio) has particular importance in communication systems and is a continuing concern of communication system design engineers.

## Transmission Link and Spectrum

The transmission link in a communication system is not really a function but a *means* to link transmitter and receiver. For example, the most common means of transferring information between humans is by sound. The transmitter is our voice; the transmission link is the variations in air pressure propagated through the air by our voice; and the receiver is our ear(s). The information transferred is whatever is said. Efficient transfer of sounds directly *through the air* over long distances is not practical because of the enormous system losses and wide dispersion of the signals.

Man, therefore, turned to other means to improve long distance communications. Earliest electronic communications used wires and cables to communicate. The most notable were the telegraph and telephone. Voltages were applied to one end of wires and cables to cause currents that were received at the other end, processed and interpreted to transfer information. Today, facsimile (fax) and digital data have been added. Radio, which came on the scene in the early 20th Century, began the era of "wireless" communications and advanced the technology significantly. It uses electromagnetic waves that propagate through "free space," are received, detected and interpreted at a distant point to transfer information.

Electromagnetic waves — including light — are a special combination of electric fields and magnetic fields at right angles to each other that propagate and move through free space, or for all practical purposes, air (because it behaves like free space). Early physicists thought that the electromagnetic waves needed a substance through which to propagate and named it "the ether." We now know that this is not true, but occasionally radio communications people still refer to the "ether" or "ether waves." Incidentally, electromagnetic waves do not propagate particularly well in water.

### Electromagnetic Spectrum

Unlike wires and cables, transmitted electromagnetic waves tend to spread out as they propagate and thus are available over fairly large areas. Since they propagate widely, radio waves are useful for broadcasting. AM and FM broadcasting, shortwave radio, VHF and UHF TV broadcasts are all electromagnetic waves that are covered by the term "wireless transmission link" in this book. However, as shown in *Figure 2-3*, the electromagnetic spectrum not only covers radio electromagnetic waves, but light waves, X-rays, and Gamma rays as well. In total, the spectrum covers frequencies from 1 Hz to $10^{24}$ Hz.

Light waves, both visible and invisible (infrared), are very useful for communications because they are very high in frequency compared to those used for radio. The higher the frequency of the wave the easier it is to focus the energy in the wave. This is the case in

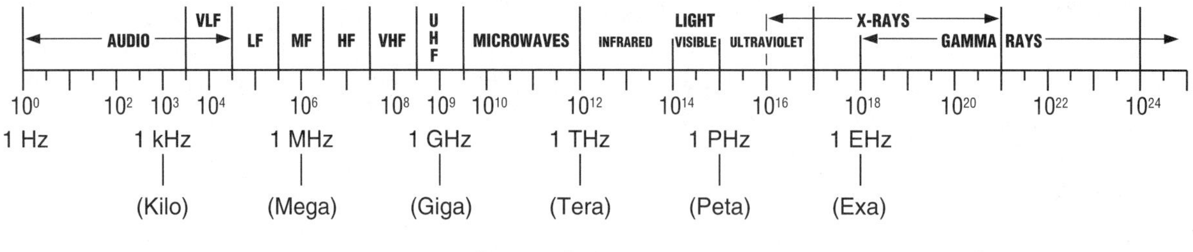


*Figure 2-3. Full electromagnetic spectrum.*
Source: Mobile 2-Way Radio Communications, G. West, ©1993, Master Publishing, Inc.

fiber optic cable where the light is focused or concentrated so that most of the light stays in the optic cable. Thus, the fiber optic cable behaves just like other electronic communications cable except that it can handle very high frequencies and has a very wide bandwidth. As a result, fiber optic cables have a large information capacity and can generally handle many channels of information. Infrared communications are used for short-range communications, notably your TV remote control, but also for nighttime military applications.

### Sound in Water

Water is a very useful transmission medium for sound. Sound propagates well in water as contrasted to electromagnetic waves. An example of systems designed to take advantage of this property are called sonar communication systems, and are the mainstay of certain types of submarine communications.

### Carrier

The term "carrier" is an old radio term. It implies that a radio wave is present "carrying" information. But it must not be thought of as something that "carries" information in the same way we might carry a book. The carrier can be thought of as a single-frequency continuous sine wave that sets the basic frequency of operation for the system. It is then modified by a process called modulation to create information at frequencies higher or lower than the carrier. It is these higher and lower sideband frequencies (which we will explain later) that "carry" the information. The carrier is very important to some systems but, as we will learn, some systems operate without the carrier being transmitted, and those systems have no energy transmitted at the carrier frequency.

## Modulation and Frequency Conversion

### Modulation

As stated in Chapter 1, modulation is the function of modifying a carrier with the information that is to be transferred so that the information is transmitted at frequencies close to the carrier. The carrier is a single-frequency signal, and the information is the baseband signal with its band of frequencies. The carrier is generally much higher in frequency than the baseband signal to allow for proper modulation. *Figure 2-4* shows the result of a 5-kHz signal modulating a 50-kHz carrier. A composite modulated signal is produced where the amplitude of the entire signal appears to change according to the 5-kHz modulating signal. This is amplitude modulation (AM) that we mentioned previously.

To understand the AM modulation function, look at *Figure 2-5*, which shows the amplitude of signal frequencies displayed on a spectrum-analyzer screen. If a single carrier signal at 200 kHz is transmitted, the spectrum analyzer screen would look like

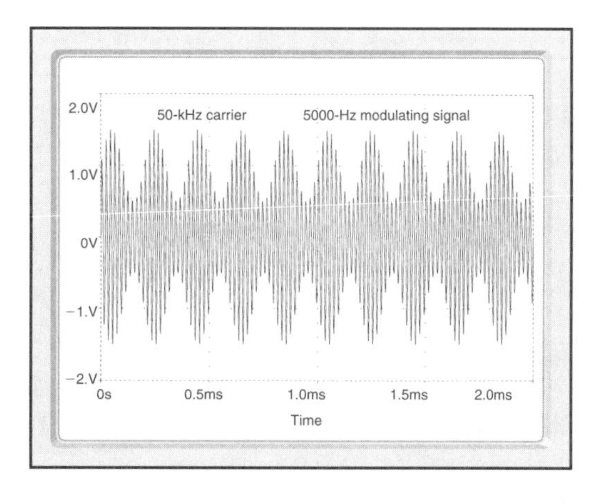


*Figure 2-4. A 50-kHz carrier amplitude modulated with a 5-kHz baseband signal.*

*Figure 2-5a*. The display of frequency is expanded so the screen only displays the frequency range from 180 kHz to 215 kHz. There is a single frequency signal at 200 kHz. *Figure 2-5b* is the spectrum analyzer display when a 5-kHz signal modulates the 200-kHz carrier. Note that a signal, $f_c$, appears for the carrier at 200 kHz as before, but two additional signals also appear. One at 195 kHz, or $f_c - f_m$ where $f_m$ is the 5-kHz modulating frequency; and another at 205 kHz or $f_c + f_m$.

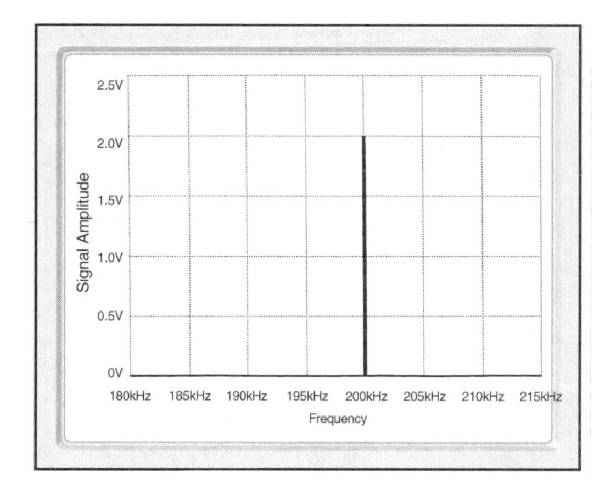


a. Single 200-kHz Carrier

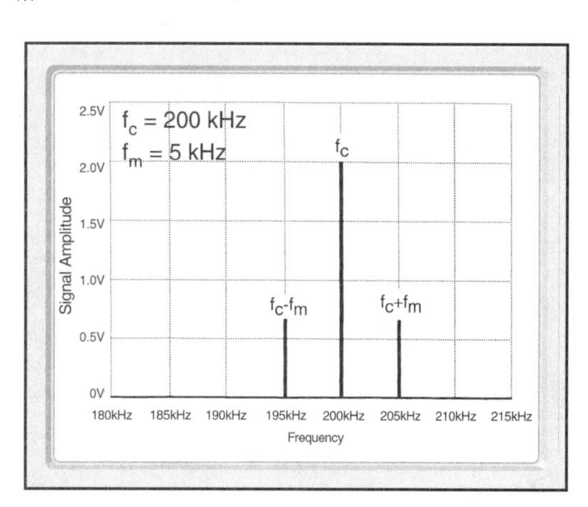


b. 5-kHz Signal Modulating 200-kHz Carrier

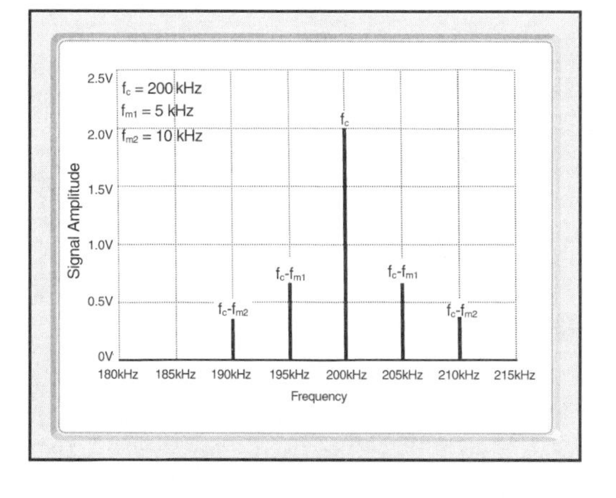


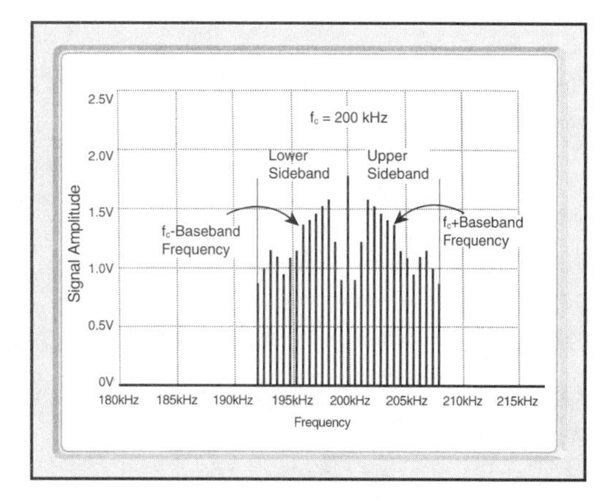

c. 5-kHz and 10-kHz Signals Modulating Carrier

d. Audio Baseband Signals Modulating Carrier

*Figure 2-5. Spectrum analyzer displays of modulation.*

*Figure 2-5c* shows the spectrum analyzer output when two frequencies, 5 kHz and 10 kHz, are present in the modulating signal. As a result, there is signal energy at the carrier $f_c$, at $f_c \pm 5$ kHz and at $f_c \pm 10$ kHz. Each of the modulation signals of 5 kHz and 10 kHz produce signal energy at the frequencies of $f_c \pm f_m$, where $f_m$ is 5 kHz in one case and 10 kHz in the other.

*Figure 2-5d* shows the result on the spectrum analyzer display when an audio baseband signal modulates the 200-kHz carrier. Each one of the frequencies in the baseband signal produces a signal at a frequency above and below the carrier frequency in correspondence to $f_c + f_m$ and $f_c - f_m$ such that the spectrum analyzer shows a whole band of frequency energy above and below the carrier. The largest deviation of $f_c + f_m$ and $f_c - f_m$ is generated by the maximum frequency in the baseband signal. In the example of *Figure 2-4d* it is 20 kHz, the maximum frequency in an audio baseband signal. Note that there is frequency energy indicated over the complete band. The baseband signal energy has been translated to the frequency position around the carrier frequency. The band of signal energy below the carrier is called the *lower sideband*, and the band of signal energy above the carrier is called the *upper sideband*. AM modulating a carrier with a baseband signal results in signal energy in the carrier and in the upper and lower sidebands.

## Three Types of Modulation

Recall that there are three types of modulation:

1. Amplitude Modulation (AM) — where the *amplitude* of a carrier appears to vary according to the variations in amplitude of the modulation signal (See *Figure 2-4*).
2. Frequency Modulation (FM) — a form of angle modulation where the *frequency* of a carrier is varied according to the variations in amplitude of the modulating signal (See *Figure 1-14*).
3. Phase Modulation (PM) — a form of angle modulation where the *time relationship (phase)* of the carrier is varied instantaneously according to the variations in amplitude of the modulating signal (results similar to *Figure 1-14*). Because the phase of the signal is changing, apparent frequency changes result.

In each modulation, upper and lower sidebands are produced around the carrier. The frequency spread, ($f_c + f_m$) or ($f_c - f_m$ ), produces a total bandwidth of $2f_m$ and depends on the highest frequency in the modulating signal ($f_m$). AM modulation has only one set of sidebands and only with AM modulation is the carrier at a constant amplitude. With angle modulation, multiple upper and lower sidebands may be produced from a single modulating frequency, and the carrier amplitude varies.

## Frequency Conversion

Note that in all modulation types, the baseband signal is converted to frequencies around the carrier frequency. In the illustrations shown in *Figure 2-4* and *2-5*, the information is converted to upper ($f_c + f_m$) and lower ($f_c - f_m$) sidebands around the carrier frequency — 50 kHz in *Figure 2-4,* and 200 kHz in *Figure 2-5*.

## Mixing — A Cousin of Modulation

In analog systems, there is a function called "mixing." It is a frequency-conversion cousin of modulation. Two signals of different frequencies are multiplied together and the sum and difference frequencies are produced. The purpose of mixing is to move (translate) the information signal to a different position in the frequency spectrum. Moving the information signal to a different frequency does not affect the information.

To demonstrate this function, let's use the carrier signal generated by a local oscillator modulated with a baseband audio signal, as shown in *Figure 2-5d*. It is the information input signal in the block diagram of *Figure 2-6*. It has a carrier of

$f_c$ = 200 kHz plus upper ($f_c + f_m$) and lower ($f_c - f_m$) sidebands, where $f_m$ stands for all the frequencies in the baseband signal. The local oscillator input, $f_{lo}$, is a single-frequency, 1000-kHz signal that usually has a constant, larger amplitude than the information signal. The output signals that result from the mixing are at a frequency of $f_{lo} + f_c$ and $f_{lo} - f_c$ — the sum and difference of the two carriers of the signals that are inputted.The output used in *Figure 2-6* is the difference signal, $f_{lo} - f_c$, centered at 800 kHz (1000 - 200). The modulation of the original information signal at 200 kHz is maintained for the signal at 800 kHz. In other words, the information signal has been moved in position on the frequency spectrum as shown in the spectrum analyzer display of *Figure 2-6b*. Instead of the information signal with its modulation being centered at 200 kHz, it is now centered at 800 kHz. The *mixer* circuit combined with the local oscillator circuit has been called a *converter* because the combination *converts* the information signal to a new position in the frequency spectrum.

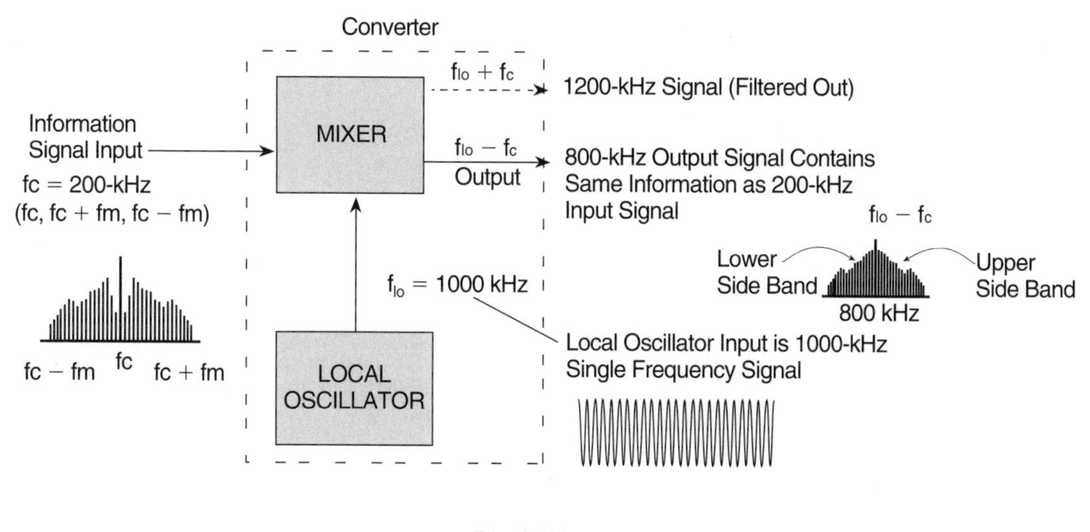


a. Block Diagram

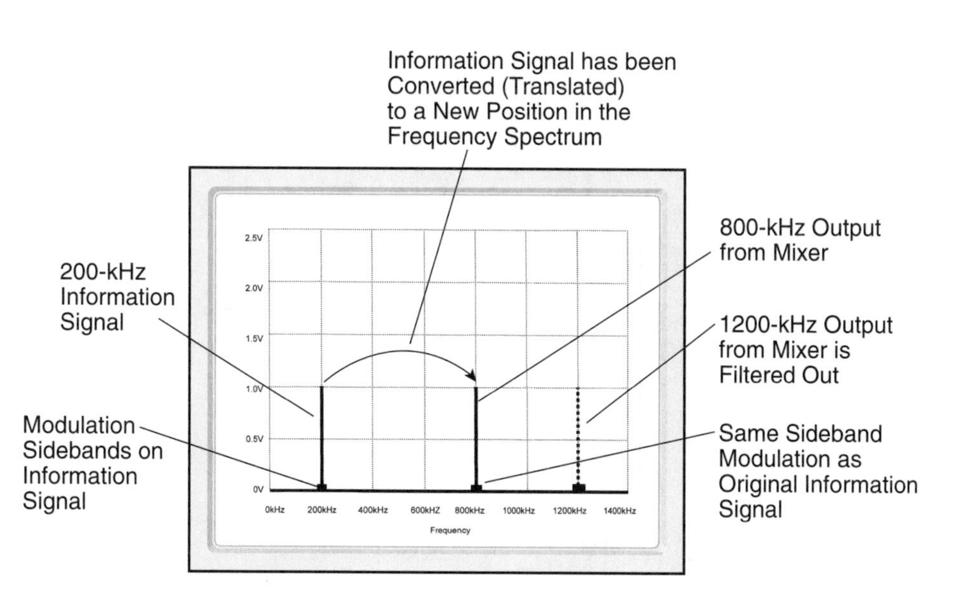


b. Spectrum Analyzer Display Showing Conversion of Signal Frequency to New Spectrum Position

*Figure 2-6. Mixer function.*

### *Example 1: Calculating Frequency Conversion*

Modulation and mixing both cause a shift in the location of a signal to a new position in the frequency spectrum. Calculate the new frequencies that result from modulating a 2-MHz signal with a 50-kHz signal; and mixing a 720-kHz signal with a 450-kHz signal.

***Solution:***

Modulation and mixing signals result in the multiplication of the signals $f_1 \times f_2$, which results in signals at two new frequencies, $f_1 + f_2$ and $f_1 - f_2$.

A. For modulation: $f_1 = 2$ MHz and $f_2 = 50$ kHz, therefore
$f_1 + f_2 = 2 \text{ MHz} + 50 \text{ kHz} = 2.05 \text{ MHz}$
$f_1 - f_2 = 2 \text{ MHz} - 50 \text{ kHz} = 1.95 \text{ MHz}$

B. For mixing: $f_1 = 720$ kHz and $f_2 = 450$ kHz, therefore
$f_1 + f_2 = 720 \text{ kHz} + 450 \text{ kHz} = 1170 \text{ kHz}$
$f_1 - f_2 = 720 \text{ kHz} - 450 \text{ kHz} = 270 \text{ kHz}$

## Detection

Detection reverses the process of modulation. It is the recovery at the receiver of the original information that was modulated onto the carrier at the transmitter. The information usually is a lower-frequency signal modulated onto the higher-frequency carrier, as was shown in *Figure 1-14* and *Figure 2-4*. Therefore, detection is the recovery of the baseband signal by translating the sidebands centered around the carrier frequency back to the original baseband frequency positions in the spectrum. Note that this downward frequency translation is the reverse of what happens in modulation.

*Figure 2-7* is a block diagram that summarizes the AM detection function. The input signal is the same AM signal shown in *Figure 2-4*. The 5-kHz information signal modulates a 50-kHz carrier. The output of the detector is the original 5-kHz information. The high-frequency 50-kHz carrier has been removed, leaving only the information that modulated the carrier. It may be obvious already, but all receivers have detectors because the receiver is the final destination of the transferred information, and the output of the receiver is the original baseband information.

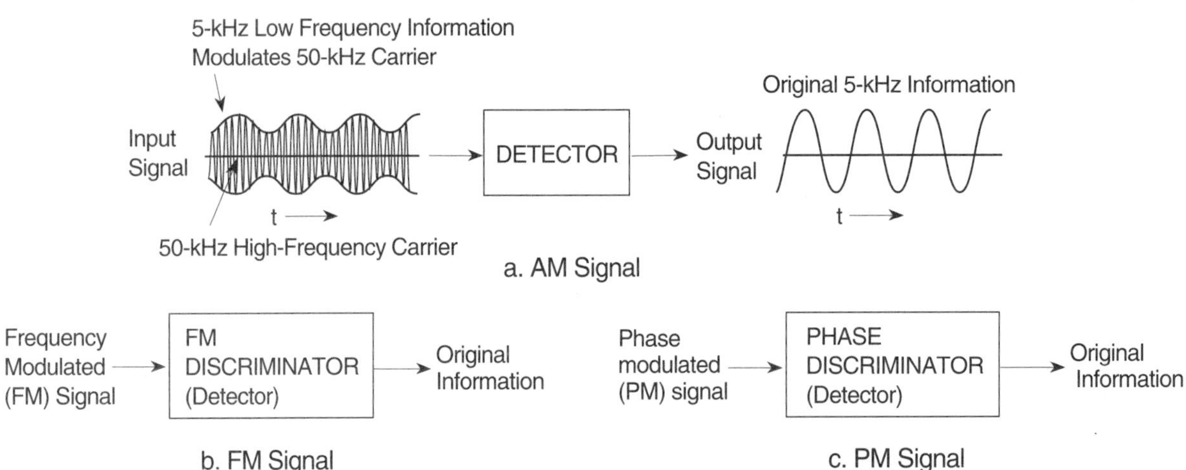


*Figure 2-7. Detector function.*

*Figure 2-7b* shows the same detector function for FM signals. Here the detector is called an *FM discriminator. Figure 2-7c* shows the same detector function for phase modulation (PM) signals. Here the detector is called a *phase discriminator*.

## Amplification

In almost every analog system there is a need to increase the amplitude of a voltage signal, or increase the amount of current supplied to a load, or increase the power level of a signal. All of these are commonly called changes in level. Changes in level have been given the name "gain." The different gains are defined in *Figure 2-8*.

### Review of DC and AC Power

Before proceeding, there is a need to review what is meant by gain in analog systems. Let's examine *Figure 2-8* closer. *Figure 2-8a* is a dc circuit. A constant voltage, $V_{IN}$, is applied to the circuit with an input resistance of $R_{IN}$, producing an output current, $I_O$, into a load resistor, $R_{LOAD}$, for an output voltage of $V_O$. The input power, $P_{IN}$, equals $V_{IN} \times I_{IN}$; the output power, $P_O$, equals $V_O \times I_O$; and the power gain equals $P_O / P_{IN}$.

### RMS Values

Even though the amplitude of an ac sine-wave voltage is alternating from positive to negative as time varies, there is a voltage value in an ac circuit, such as the one shown in *Figure 2-8b,* that is equivalent to the dc voltage value shown in *Figure 2-8a*. If the impedance ($Z_{IN}$) of *Figure 2-8b* is a resistance equal to $R_{IN}$ of *Figure 2-8a*, then a value of input voltage of $V_{in\,(rms)} = V_{IN}$ will provide *the same amount* of power to the ac circuit to heat the resistance $R_{IN}$ as that supplied to the dc circuit of *Figure 2-8a*. $V_{in(rms)}$ is called the RMS (for *root-mean-square*) value of an ac sine wave and, as shown in *Figure 2-8c*, is equal to 0.707 $V_{pk}$, where $V_{pk}$ is the peak value of the sine wave. Thus, $V_{rms} = 0.707\ V_{pk}$. In like fashion, the RMS value of a current sine wave is $I_{rms} = 0.707\ I_{pk}$. Since power is equal to voltage times current, $V_{rms} \times I_{rms} = P_{rms}$, and the power gain is $\frac{P_{o(rms)}}{P_{in(rms)}}$.

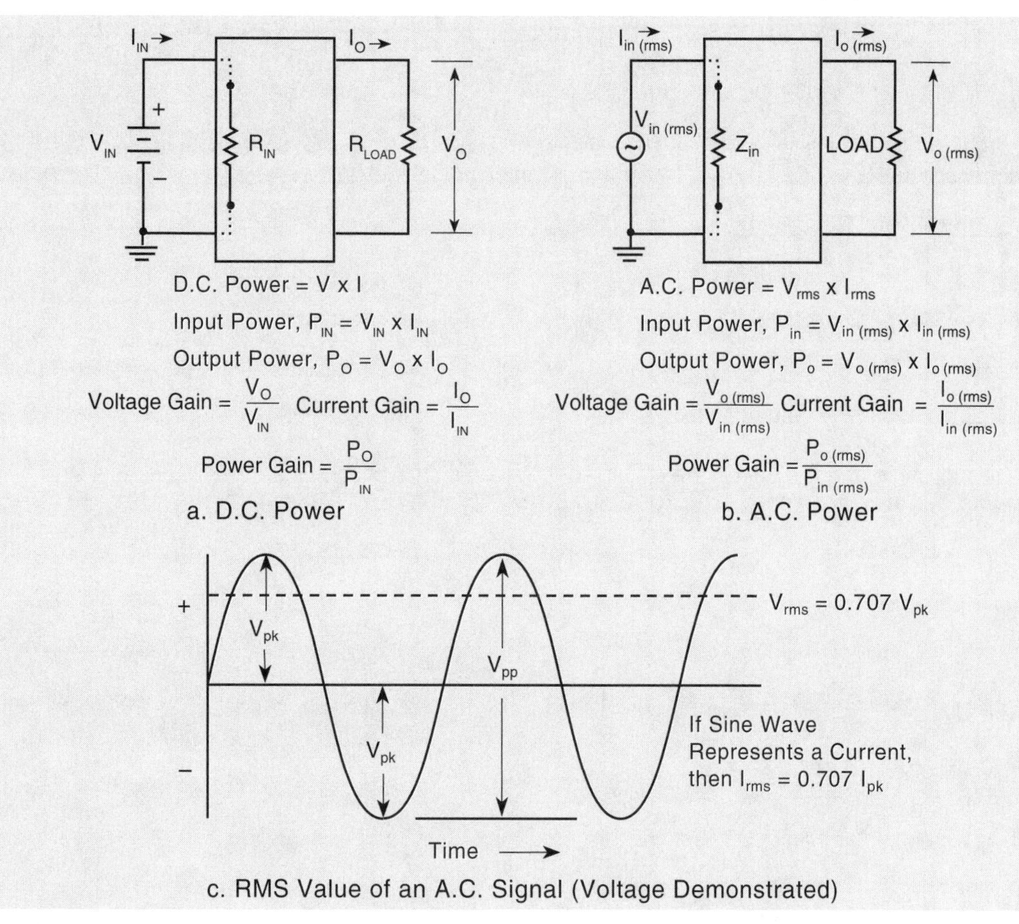


*Figure 2-8. Comparison of gain and power levels in DC and AC circuits.*

Therefore, voltages and currents in ac circuits (or in analog circuits) usually are measured as RMS values, and ac power = $V_{rms} \times I_{rms}$. In all of these calculations, the voltage and current are considered to be in phase. If they are not, the difference in phase must be taken into account and the calculations are much more difficult. In the discussions in this book, unless told otherwise, the voltages and currents are considered to be in phase.

### *Example 2: Calculating RMS Values*

The RMS (Root-Mean-Square) of an ac sine-wave voltage dissipates the same amount of heat in a resistor as an equivalent dc value. Calculate the RMS values for a sine-wave voltage with a peak value of 9 volts and a sine-wave current with a peak value of 15 amps. If these values are measured across a resistor in a circuit, use the results to calculate the RMS power dissipated in the resistor.

***Solution:***

$V_{rms} = 0.707\ V_{pk} = 9.0\text{ V} \times 0.707 = 6.363\text{ volts}$

$I_{rms} = 0.707\ I_{pk} = 15\text{ A} \times 0.707 = 10.605\text{ amps}$

$P_{rms} = V_{rms} \times I_{rms} = 6.363\text{ V} \times 10.606\text{ A} = 67.48\text{ watts}$

## Examples of Voltage, Current and Power Gain

*Figure 2-9* shows examples of voltage, current and power gain in an analog ac circuit. In the calculations, all voltage, current and power are RMS values. Further examples are shown in *Example 3.* AC circuits may have large voltage gain, but because $Z_{LOAD}$ is often large, the output current is small, and thus the power gain is small. This is the case for circuit A of *Example 3.* If $Z_{LOAD}$ is small, as in circuit B of *Example 3,* but voltage gain is small, power gain is still small. Combining large $V_O$ and large $I_O$ as in circuit C of *Example 3* provides large power gain.

$V_{in}$ is a sine-wave voltage generated by a vector of amplitude X rotating through an angle Θ. It is expressed as $V_{in} = X \sin\Theta$. Since X is the peak amplitude of the sine wave, $V_{pk}$ of $V_{in}$ = X. Therefore,

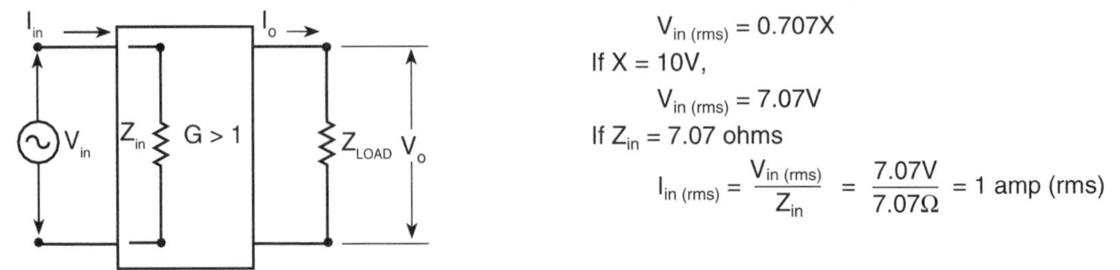


In this example, there is gain in the circuit, so $V_o$ will be an ac voltage larger than $V_{in}$. $V_o$ is a sine wave that can be expressed as $V_o = Y \sin\Theta$. For $V_o$, $V_{pk}$ of $V_o$ = Y. Therefore,

$V_{o\,(rms)} = 0.707Y$

If Y = 20V,

$V_{o\,(rms)} = 20 \times 0.707 = 14.14\text{V}$

If $Z_{load}$ = 7.07 ohms,

$$I_{o\,(rms)} = \frac{V_{o\,(rms)}}{Z_{load}} = \frac{14.14\text{V}}{7.07\Omega} = 2\text{ amps (rms)}$$

$$\text{Voltage Gain} = \frac{V_{o\,(rms)}}{V_{in\,(rms)}} = \frac{14.14\text{V}}{7.07\text{V}} = 2$$

$$\text{Current Gain} = \frac{I_{o\,(rms)}}{I_{in\,(rms)}} = \frac{2}{1} = 2$$

$$\text{Power Gain} = \frac{P_o}{P_{in}} = \frac{V_{o\,(rms)} \times I_{o\,(rms)}}{V_{in\,(rms)} \times I_{in\,(rms)}} = \frac{14.14 \times 2}{7.07 \times 1} = \frac{28.28}{7.07} = 4$$

*Figure 2-9. Examples of voltage, current, and power gain in AC circuits. (All voltages and currents are in phase.).*

### *Example 3: Calculating Power Gain*

Using the circuit block diagram at *Figure 2-9*, calculate the power gain in the following circuit variations. All voltage and currents are RMS values:

| | | | | |
|---|---|---|---|---|
| Circuit A: | $V_{in} = 1V$ | $I_{in} = 0.1A$ | $V_o = 50V$ | $Z_{load} = 5k\Omega$ |
| Circuit B: | $V_{in} = 1V$ | $I_{in} = 0.1A$ | $V_o = 2V$ | $Z_{load} = 2\Omega$ |
| Circuit C: | $V_{in} = 1V$ | $I_{in} = 0.1A$ | $V_o = 30V$ | $Z_{load} = 4\Omega$ |

***Solution:***

Circuit A:

$P_{in} = V_{in} \times I_{in} = 1 \times 0.1 = 0.1W$

$I_o = \frac{V_o}{Z_{load}} = \frac{50V}{5000\Omega} = 0.01A$

$P_o = V_o \times I_o = 50 \times 0.01 = 0.5W$

$\text{Power Gain} = \frac{P_o}{P_{in}} = \frac{0.5}{0.1} = 5$

$\text{Voltage gain} = \frac{50}{1} = 50$

$\text{Current gain} = \frac{0.01}{0.1} = 0.1$

Circuit B:

$P_{in} = V_{in} \times I_{in} = 1 \times 0.1 = 0.1W$

$I_o = \frac{V_o}{Z_{load}} = \frac{2V}{2\Omega} = 1A$

$P_o = V_o \times I_o = 2 \times 1 = 2W$

$\text{Power Gain} = \frac{P_o}{P_{in}} = \frac{2}{0.1} = 20$

$\text{Voltage gain} = \frac{2}{1} = 2$

$\text{Current gain} = \frac{1}{0.1} = 10$

Circuit C:

$P_{in} = V_{in} \times I_{in} = 1 \times 0.1 = 0.1W$

$I_o = \frac{V_o}{Z_{load}} = \frac{30V}{4\Omega} = 7.5A$

$P_o = V_o \times I_o = 30 \times 7.5 = 225W$

$\text{Power Gain} = \frac{P_o}{P_{in}} = \frac{225}{0.1} = 2{,}250$

$\text{Voltage gain} = \frac{30}{1} = 30$

$\text{Current gain} = \frac{7.5}{0.1} = 75$

Note in the circuits that when the output voltage is large and the load impedance is large, there is a large voltage gain but a small power gain. While when the output voltage is small with a small load impedance, the current gain is reasonably large, but the power gain is still relatively small. Only when the voltage gain and the current gain are large is the power gain very large.

## Power Amplifiers

Analog power amplifiers are circuits designed specifically to provide large power gains because they must drive components with low-valued impedances. They are classified as "large-signal" amplifiers. Some have larger voltage swings than others.The active devices used must deliver large amounts of power to a load, but also must dissipate a lot of power within the devices. In many cases, circuits with semiconductor active devices can supply large currents but cannot handle large voltage amplitudes. Semiconductors are used as the active devices up to about the 1,000-watt (1-kW) level, but vacuum tubes usually are used at power levels above that because of the amount of power that must be dissipated as heat within the device to provide the power gain. *Figure 2-10* shows three large power amplifier tubes use for transmitters. Small-signal devices can handle the dissipation with or without cooling, but large-signal devices may require water cooling or even special cooling systems.

a. VHF TV Linear Amplifier

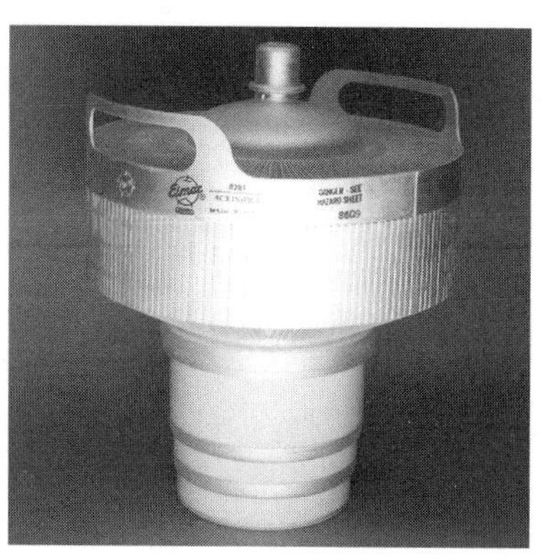

b. RF Amplifier

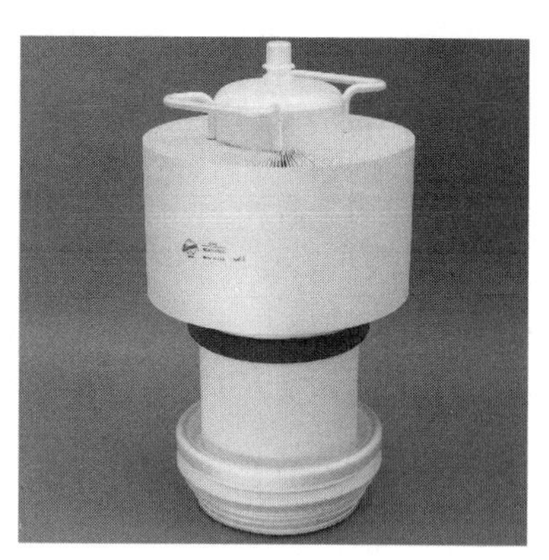

c. 50 to 150 kW Power Amplifier

*Figure 2-10. Vacuum tubes are still used for large power output in 1-kW and up circuits.* *(Photos courtesy of CPI Eimac Division).*

## The "Small-Signal" Amplifier

There are many cases in analog circuits where either the voltage or current levels must be increased, but no large power gain is required. Such analog circuit amplifiers are known as "small-signal" amplifiers because signal levels are in microvolts to millivolts and current levels are in microamperes to milliamperes. Remember, "micro" means a multiplier of one one-millionth, or $1 \times 10^{-6}$ (0.000001) and "milli" means a multiplier of one one-thousandth, or $1 \times 10^{-3}$ (0.001). Semiconductor devices, both bipolar and field-effect transistors, handle the small-signal amplification function well because they operate at low power levels, which means low power dissipation.

## Active Devices

Components in analog circuits are classified as "active" or "passive." Active devices in circuits provide gain. Passive devices, such as resistors, inductors, capacitors and transformers, provide no power gain. Passive devices can change voltage and current levels up or down, but do not increase the power level of a given signal. They are used to set voltages or currents around which signals cause variations so that semiconductors (active devices) have the proper fixed, "no-signal" or "dc" operating point. Setting the no-signal operating point around which the small ac signals vary is called "biasing" the circuit.

The example circuit of *Figure 2-11* uses an NPN transistor as the active device that typically provides current gains from 50 to 200. The resistors (passive devices) set the "no-signal" operating point such that the collector voltage to ground is about one-half the supply voltage, $V_{CC}$, and the steady-state current into the base, $I_B$, around which the signal current, $i_b$, varies, is small compared to the bias current, $I_{BIAS}$. It is important to understand that active devices do not "create" power, but simply have the ability to take power in one form, such as a dc power supply, and use it to increase the amount of power in the signal that is being amplified.

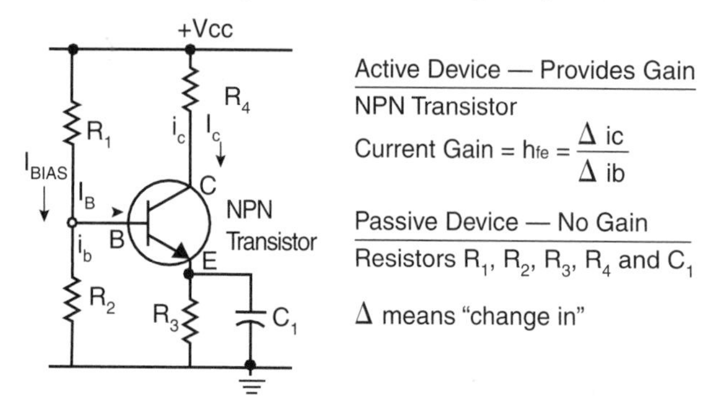


A Note on Signal Nomenclature: DC bias currents and voltages are normally identified with upper-case letters and subscripts. AC currents and voltages are normally identified with lower-case subscripts. However, there is quite a variance among publishers.

*Figure 2-11. Small-signal amplifier with an NPN transistor as the active device and resistors and a capacitor as passive devices that set the "no-signal" (DC) operating point.*

It is very important that small-signal amplifiers used in circuits contribute very little noise so that a high signal-to-noise ratio is maintained. Some noise may be contributed by passive devices, but the major contribution will come from the active device itself and may vary with the device's operating point. In many cases, the output is not exactly the same as the input; it is distorted or the amplifier cannot respond fast enough to inputs that change very rapidly . Both of these conditions — called distortion and poor transient response — contribute to amplification errors and reduce the quality of the signal.

## Integrated Circuit Amplifiers

In many cases, especially for small-signal amplifiers, semiconductor manufacturers provide amplifiers in integrated circuit (IC) form. All the active devices, and a majority of the passive devices, are located inside the IC package, which contains a small chip of semiconductor material with all the components, devices, and interconnections on or within the chip of semiconductor material. The package is environmentally sealed with all the necessary external connections brought out on pins to connect to the circuit in a given application. External components may be required to set the gain of the amplifier, adjust its frequency response, or stabilize the circuit. In Chapter10, we will explain how ICs are manufactured.

### *Operational Amplifiers (Op-Amps)*

One very important integrated circuit amplifier that has somewhat universal use for low-frequency amplifiers is the operational amplifier or, as it is called in the trade, "op-amp." *Figure 2-12a* shows typical IC operational amplifier packages. The schematic of *Figure 2-12b* shows the characteristics of an ideal op-amp. Operational amplifiers like these are used to provide accurately-controlled voltage and current gains, tailored frequency response, level comparisons, and control circuit functions.

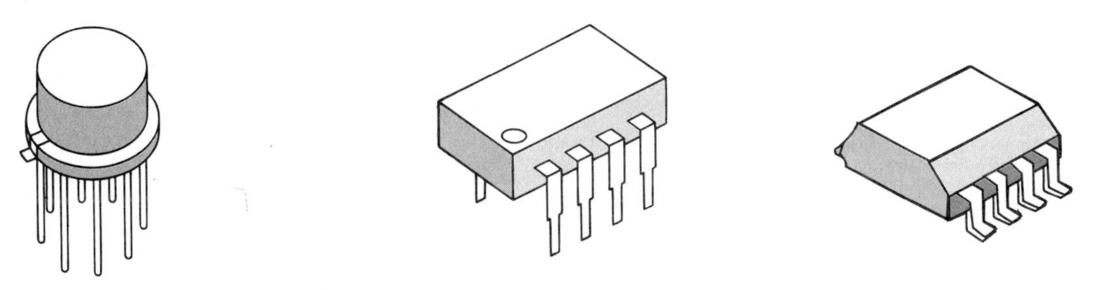

a. Typical IC Packages for Op-Amps—Can, 8-Pin In-Line Plastic Package, and Package for Surface Mounting

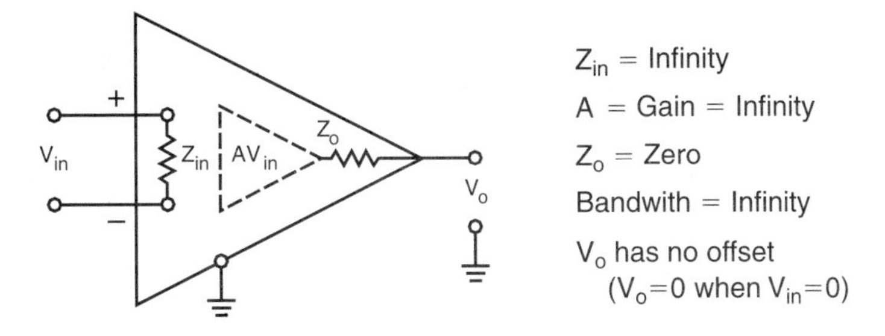


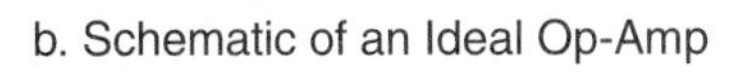

b. Schematic of an Ideal Op-Amp

*Figure 2-12. Integrated circuit (IC) operational amplifier (Op-Amp).*

## Oscillators and Oscillation

A function used many times over and over in analog systems is an *oscillator*. It is an electronic circuit designed to produce a signal with a constant frequency and amplitude. Sometimes the electronic circuit can vary the frequency, but in most cases the oscillator circuit produces an output signal at a specific, constant frequency.

To understand how an oscillator produces a signal at a constant frequency, it must be understood that an oscillator is a special kind of *feedback* amplifier. Look at *Figure 2-13*, which shows an oscillator in block diagram form, but a special one that has a switch, S, in the feedback path. Waveforms identify the signals at points A, B and C.

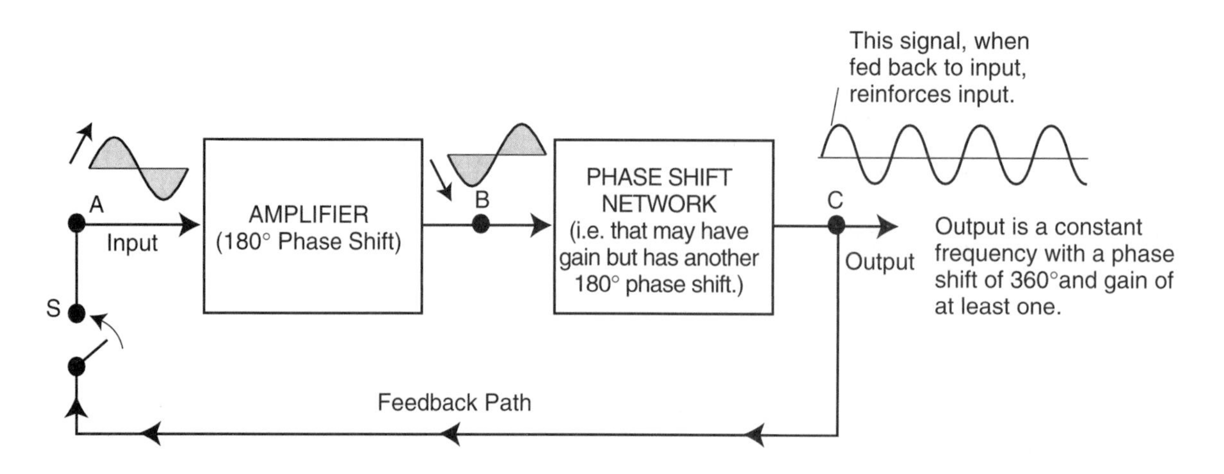


*Figure 2-13. An oscillator outputs a signal of constant frequency.*

The switch, S, is open so the feedback path is open. Look at the beginning of the signal at A. It is increasing positively, and is amplified in the first block to give an output at B. Note that through the first block there is a 180° phase shift. In other words, the output at B is increasing negatively while the signal at A is increasing positively. The second block may or may not amplify the signal; but at a minimum it contains circuitry that changes the phase of the signal by another 180°. The output at C, because of the additional 180° phase shift from B, is going positive as the signal at A is going positive.

With the feedback path open, the oscillator is just an amplifier whose output is in phase with its input. Now let's close S. Hopefully, it is easy to see that the signal fed back from output C to input A aids or reinforces the signal at A. As a result, the circuit is now an oscillator that oscillates at the set frequency *where the phase shift is 360° and the gain is at least one*. So an oscillator is really an amplifier whose output is fed back to the input so that the input is reinforced to allow the circuit to maintain itself without external signal energy, except from the power supply. Note that the second block in *Figure 2-13* may or may not have gain, but it must have enough phase shift (180° if the first block has 180° phase shift) to complete the full 360° phase shift and the full circuit amplification must be at least one so that the feedback signal reinforces the input.

Many engineers who design amplifier circuits find that, in some of their layouts, a portion of the output signal unintentionally feeds back to the input and they have designed an oscillator rather than an amplifier. The circuit will oscillate at the frequency where the phase shift is 360° and the gain is at least one.

## Transmitter

A transmitter is at the beginning of a communication system. It is a unit that converts the information to be transferred from baseband form into a form that can be transferred over the transmission link. *It prepares the information for transmission*.

A transmitter can be as simple a unit as the microphone of a PA system, or a more complete unit such as a telephone with a microphone for sound pick-up and the keypad to convert the press of keys into multiple tones sent over the telephone lines to ring the party who is to receive the information. Or a transmitter can be a unit that is a combination of many of the functions discussed above that prepares the signal for broadcast (voice and music for radio, or picture and sound for TV). The point is that the transmitter function can vary in scope depending on the communication system. It may contain only one of the basic functions, or it may be a combination of basic functions that are subfunctions of the total transmitter.

In this book, we will be concentrating on the later because our major emphasis is on communication systems that use radio (electromagnetic waves) as the transmission link. *Figure 2-14* demonstrates how the subfunctions we have been discussing are combined to build a radio transmitter. Amplification, both small-signal and power, oscillation and modulation are all contained in the radio transmitter. But also note the signal conditioner, commonly called a transducer. This is a function we haven't talked directly about before, but it is a major subfunction of a transmitter.

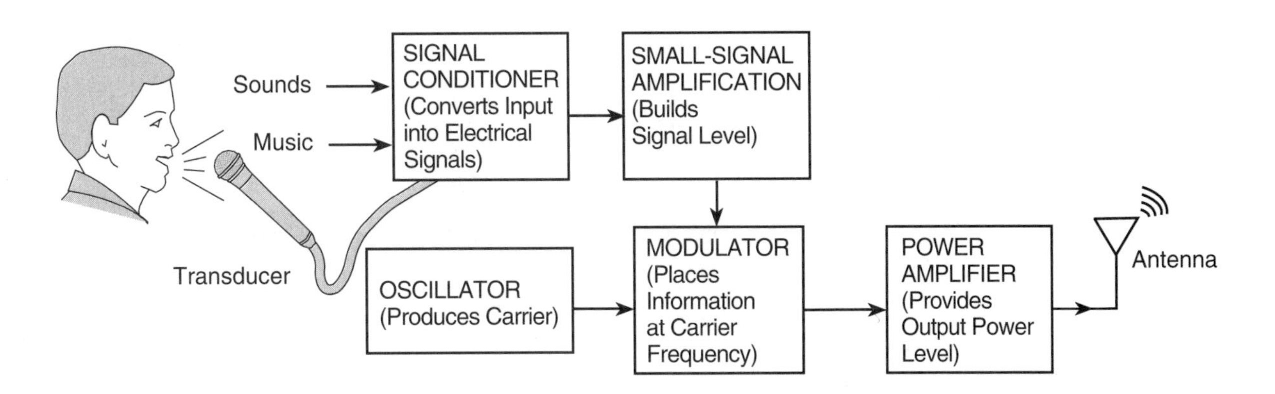


*Figure 2-14. Radio broadcast transmitter showing how basic subfunctions are combined to provide overall transmitter function.*

## Transducers

A microphone converts sound into electrical signals, a tape head converts music stored on a magnetic tape into electrical signals, a compact disk player converts music stored on the surface of a CD into electrical signals, and a turntable converts music stored as impressions on a vinyl disc into electrical signals. These are all examples of converting non-electrical energy into electrical signals. There are reverse functions as well — converting electrical signals into non-electrical energy. Such functions are commonly performed by transducers. Transducers take the information, in many cases in a form understood by humans, and convert it into electrical signals to be processed by electronic circuits, or take the output electrical signals and convert them into non-electrical energy in a form useful to humans.

## Receiver

A receiver is at the end of a communication system. It is a unit that recovers the information that has been transferred and restores it (in most cases) into a form similar (or in many cases, exactly) to the original. Again, it can be as simple as the ear piece in your telephone, or as complicated as your TV set. It may be built with only one of the basic functions, or it may be a combination of subfunctions built to provide the overall function.

For our emphasis on communication systems that use radio waves as the transmission link, the overall receiver usually is a combination of subfunctions, as shown in *Figure 2-15*. Amplification, signal selection, signal detection and a transducer are the subfunctions contained in the receiver. The transducer — in this case, a speaker — converts the electrical signals into sounds and music.

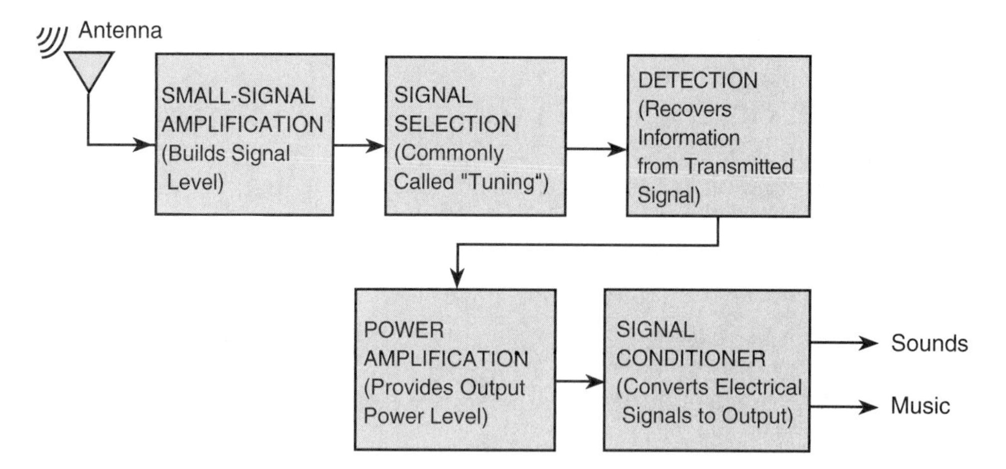


*Figure 2-15. Radio broadcast receiver showing how basic subfunctions are combined to provide overall receiver function.*

## Signal Control

There are four main types of signal control in analog systems:

1. ON-OFF — Signals are turned ON or OFF according to a timed pattern, according to a code, or they may be turned ON or OFF randomly.
2. Gain — Signal levels are maintained constant through a system or subsystem by varying the gain of circuits.
3. Frequency — Signals are selected by their frequency.
4. Time Relationships — Signals are selected by their position in a time frame.

### On-Off

The prime example of controlling the transfer of information by turning it ON and OFF is the transmission of Morse code. We showed the example in *Figure 1-13*. It was used in one of the earliest communication systems — the telegraph — and is still used today by amateur radio, ship communications, and emergency radio operators. The complete Morse code is shown in *Figure 2-16*. The longer ON transmission is a dash (called a "dah") and the shorter transmission is a dot (called a "dit"). Combinations of dits and dahs produce codes for each character in the alphabet, numbers and other special characters to allow the transfer of information by combining the codes into messages.

| | | | |
|---|---|---|---|
| A | • — | N | — • |
| B | — • • • | O | — — — |
| C | — • — • | P | • — — • |
| D | — • • | Q | — — • — |
| E | • | R | • — • |
| F | • • — • | S | • • • |
| G | — — • | T | — |
| H | • • • • | U | • • — |
| I | • • | V | • • • — |
| J | • — — — | W | • — — |
| K | — • — | X | — • • — |
| L | • — • • | Y | — • — — |
| M | — — | Z | — — • • |
| 1 | • — — — — | 6 | — • • • • |
| 2 | • • — — — | 7 | — — • • • |
| 3 | • • • — — | 8 | — — — • • |
| 4 | • • • • — | 9 | — — — — • |
| 5 | • • • • • | 0 | — — — — — |

a. International Morse Code

| | | |
|---|---|---|
| $\overline{AR}$ | (end of message) | • — • — • |
| K | Invitation to transmit (go ahead) | — • — |
| $\overline{SK}$ | End of work | • • • — • — |
| $\overline{SOS}$ | International distress call | • • • — — — • • • |
| V | Test letter (V) | • • • — |
| R | Recieved, OK | • — • |
| $\overline{BT}$ | Break or Pause | — • • • — |
| $\overline{DN}$ | Slant Bar | — • • — • |
| $\overline{KN}$ | Back to You Only | — • — — • |
| Period | | • — • — • — |
| Comma | | — — • • — — |
| Question mark | | • • — — • • |

b. Special Signals and Punctuation

*Figure 2-16. Morse code (CW) uses On-Off signals in codes to send messages.*

## Gain

Control is provided in analog circuits by varying the gain of circuits. Thus, as shown in *Figure 2-17*, the signal levels through a chain of amplifiers are held constant because a signal, G, controls the amplifier gain. This is called automatic gain control (AGC). If the input signal is at a lower level than desired, the level-sensing circuit increases the gain to increase the output level. If the output level is too high, the level-sensing circuit decreases the gain.

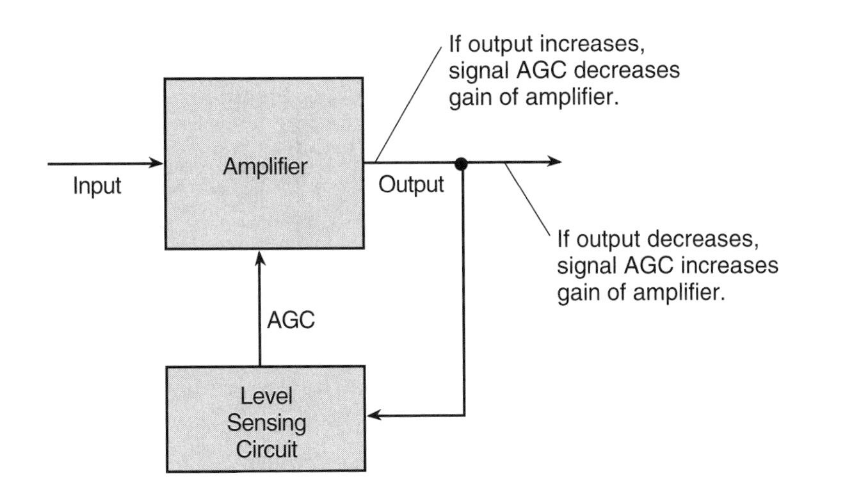


*Figure 2-17. Automatic gain control keeps signal level constant.*

## Frequency Selection

Possibly the most used signal control is frequency selection. Signals are selected to be passed on and processed based upon their frequency. Examples are shown in *Figure 2-18. Figure 2-18a* is a block diagram of a frequency selection circuit — the common name is filter or filter network. Three different types are shown in *Figure 2-18b, c* and *d*. In *Figure 2-18b*, called a *low-pass* filter, all signals with frequencies lower than the cutoff frequency, $f_c$, are passed through the filter. Any signals with frequencies higher than $f_c$ are blocked from passing. In *Figure 2-18c*, called a *high-pass* filter, all signals with frequencies higher than the cutoff frequency, $f_c$, are passed through the filter. Any signals with frequencies lower than $f_c$ will be blocked. *Figure 2-18d* combines the low-pass and high-pass into a *band-pass* filter. Any signals with frequencies in the pass band between $f_L$ and $f_H$ will be passed on, while signals with frequencies lower than $f_L$ and higher than $f_H$ will be blocked and not processed. In many communication systems, filters and amplifiers are combined to create *"tuned-circuit"* amplifiers. Such filter-amplifier combinations work just like a band-pass filter with amplification included. Any one or all of the filter examples are used over and over with different frequency pass bands to implement many different types of communication systems.

### *Frequency Response*

While we are talking about frequency selection, there is an important term that needs to be understood. The term is *frequency response*. Frequency response has been defined as follows:

> *A plot of how a circuit or device responds to different frequency inputs. Usually refers to the amplitude response at the output as a constant-amplitude signal with varying frequency is applied to the input.*[1]

[1] *Technology Dictionary*, © Copyright 1987, Master Publishing, Inc.

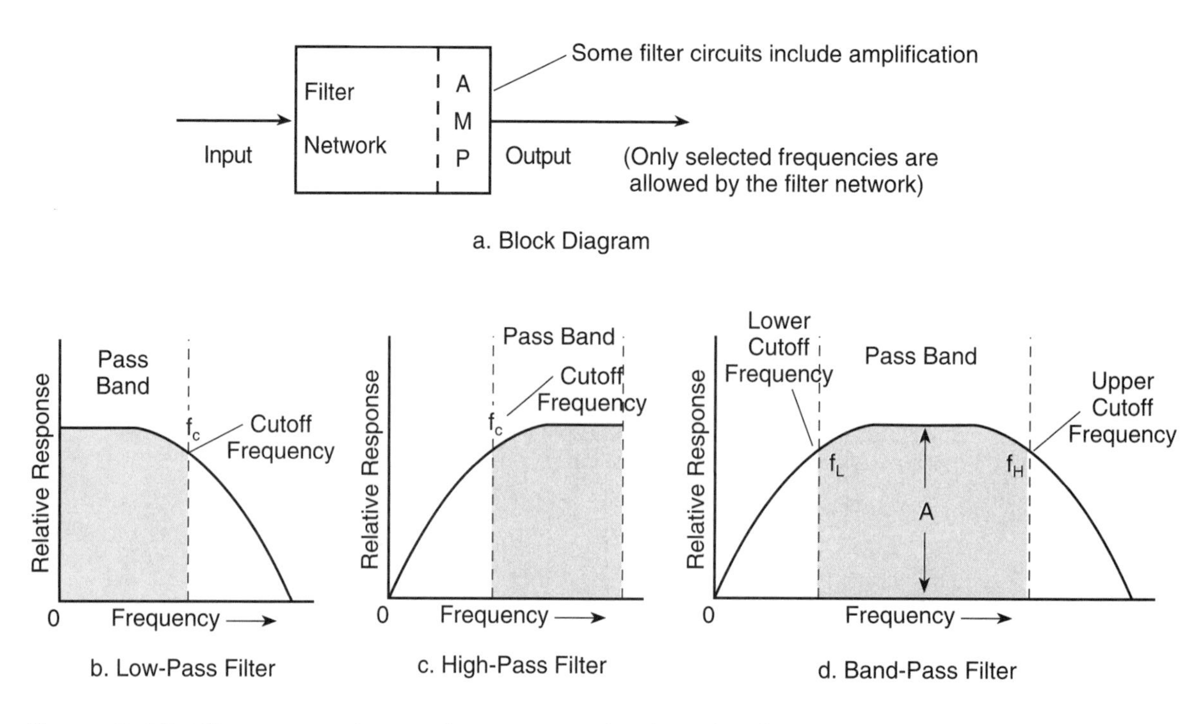


*Figure 2-18. Filter networks are frequency selection circuits.*

Look at *Figure 2-18d*. A is the amplitude response of the circuit's output signal reacting to an input signal that has a constant amplitude but is varied in frequency from zero to greater than $f_H$. It is almost constant as input frequencies vary from $f_L$ to $f_H$. The amplitude is plotted on the vertical axis and the frequency is plotted on the horizontal axis. The response is obtained by applying a constant-amplitude signal to the input, varying the frequency from zero to frequencies higher than $f_H$, and recording the amplitude of the output at regular steps of frequency across the band. Generally, the flatter or steadier the amplitude of the output signal as the frequency is varied the better for an audio amplifier used for stereo systems. Designers like the flat response so it can be adjusted with special filters called "equalizers." However, every system is different, and system designers will tailor the frequency-response characteristics to meet the requirements of the application. Therefore, when a circuit's frequency response is mentioned or shown, it means how the circuit responds to input signals at different frequencies. Frequency response is a very important specification that is relied on in the design of communication systems.

## Time Relationships

The selection of signals also can be controlled by placing a number of signals in a specific time relationship to one another and then selecting them based upon their position in time. *Figure 2-19* is an example. Signals are input into an electronic circuit, called a *multiplexer*, which takes the signals and places them in time slots one after the other and outputs them to the transmission link. Signal #1 goes in time slot #1, signal #2 goes in time slot #2, and signal #3 goes in time slot #3. The signals stay in this time relationship until they arrive at the demultiplexer. At the demultiplexer, which is synchronized (controlled to react at a specific time) to the multiplexer with the synchronizing signal included with the other signals, the signals are picked from their respective time slots and sent to the correct output. If one is interested only in signal #2, the demultiplexer can be instructed to select only signal #2. With this technique, many signals can be placed in many time slots and transmitted in sequence. Specific signals can then be selected for processing by selecting the correct time slot. This time selection process is know technically as *Time Division Multiplexing* (TDM.)

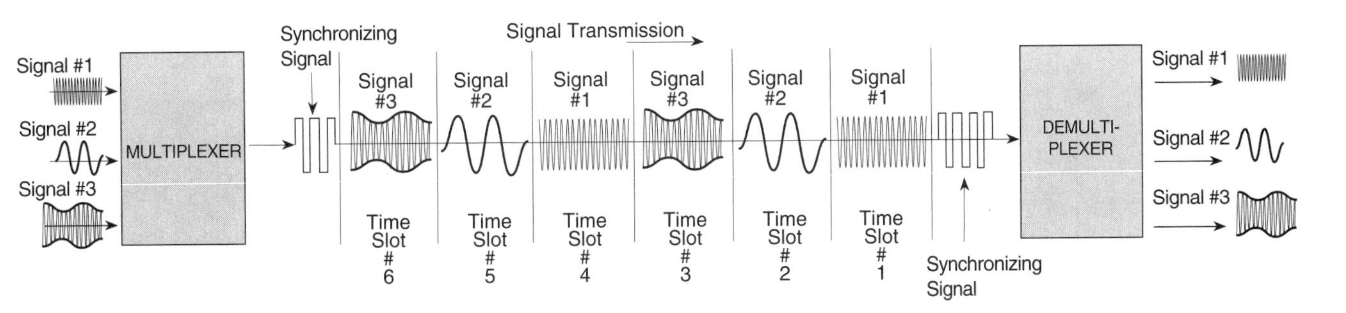


*Figure 2-19. Time multiplexing and demultiplexing.*

There also is a very similar system based on frequency slots where signals are placed in particular frequency pass bands. It is called *Frequency Division Multiplexing* (FDM) and has been used for a long time in the telephone industry to place many different frequency signals onto a single carrier. At the receiving end, the signals are detected by the frequency selection techniques that were discussed.

## Power Supply

One of the most common functions in analog systems is that of a power supply. Its job is to supply a constant voltage or voltages with the capability to provide current at prescribed levels to all of the electronic circuits in the total system. The power supply output voltage is "regulated," which means that the voltage output is maintained within system design limits even though the input power, the output current requirements, or the power requirements of internal components may change. The usual input power to the power supply is the 120-volt ac, 60-Hz power supplied by the power company. Power supplies that maintain very closely-regulated voltage tolerances are said to be "stiff." Generally, the stiffer the power supply for an analog system the fewer circuit problems that will occur because of power supply voltage variations. System technicians have learned over the years to check the power supply first if there is a system problem.

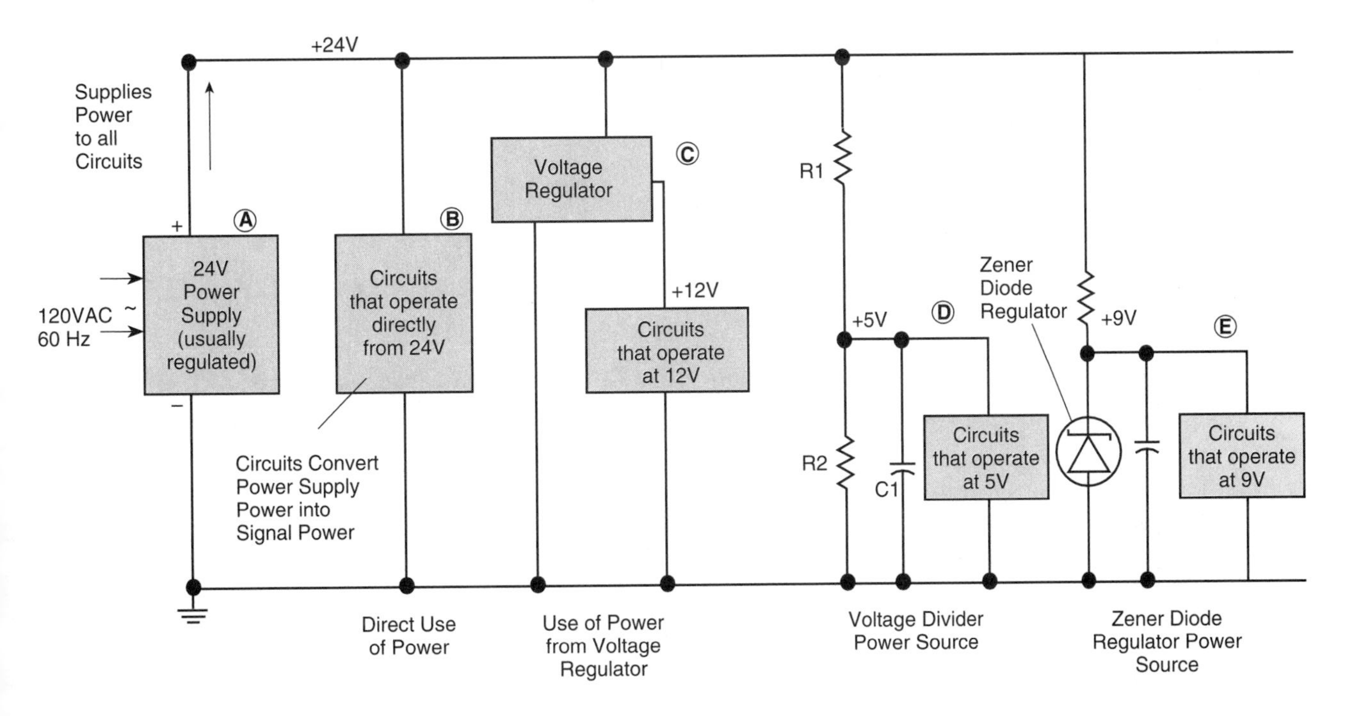


*Figure 2-20. Regulated power supply provides power to the system in a variety of ways.*

*Figure 2-20* shows different ways the main power supply voltage, A, may be distributed to the electronic circuits in the system. For example, at B the regulated voltage is used directly by many of the circuits. At C, another voltage regulator is inserted in the power supply line to step down the voltage for circuits that operate at a lower voltage, but which need close regulation because of the current load. In D, the lower voltage circuits can operate just from a voltage that is stepped down with a resistor divider. While in E, the voltage supplied to the circuits is regulated with a special semiconductor device called a Zener diode. As a varying current is needed by the circuits at E, the Zener diode gives up current while maintaining its voltage.

In general, analog circuits operate from higher voltages than digital circuits. Also, many analog circuits operate from a power supply that outputs both a positive and a negative voltage. The signal output thus swings from positive to negative around a common ground. And one last point bears repeating — the electronic circuits supplied by the power supply use power, they do not create it. They use the power from the power supply and convert it into different types of signal power.

## Powering Portable Equipment

Portable equipment substitutes batteries for the power supply shown in *Figure 2-20*. Batteries supply a constant (until they wear down) dc voltage. They generally do not need the regulation of an electronic power supply, but circuits that operate using batteries for their power supply must be designed to operate from a beginning full-charge voltage to a lower minimum voltage at which point the battery is considered discharged.

## Summary

This Chapter has covered the basic functions found in analog system — modulation and frequency conversion (including modulators, mixers, and detectors), amplifiers (including amplification and gain), and oscillators. We showed how these functions are combined to create transmitters and receivers. We began by looking at information channels. And we ended by discussing the use of power supplies to provide power to the systems.

The following Chapters will detail electronic circuits that perform the functions, and then show how these circuits are combined into communication systems.

First, however, we want to provide you with a "refresher" on some fundamental concepts critical to your understanding of the rest of the material in the book.

# Quiz for Chapter 2

1. What is a transmission link ?
   a) The means used to tie the transmitter to the receiver.
   b) The physical media, such as a wire or cable, linking the transmitter and receiver.
   c) In radio, the electromagnetic wave used to send information from transmitter to receiver.
   d) All of the above.
2. What is the importance of noise in a communication system?
   a) It helps improve the quality of the signal as it is sent from transmitter to receiver.
   b) It makes it easier to modulate a baseband signal to a higher frequency.
   c) It causes errors in transmission of information.
   d) It is of no concern in communications.
3. What is a carrier?
   a) A single-frequency sine wave that sets the basic frequency of operation for the system.
   b) A control signal used at the transmitter that has no information.
   c) A control signal used at the receiver that has no information.
   d) The modulating information.
4. What is the purpose of modulation?
   a) To improve the tonal quality of audio signals.
   b) To encode the analog signal for transfer from transmitter to receiver.
   c) To move the baseband signal to a higher frequency for easy transmission over the selected transmission link.
   d) All of the above.
5. What are the three basic types of modulation?
   a) Drum, trumpet, and string modulation.
   b) Amplitude, frequency, and phase modulation.
   c) Sky-wave, ground-wave, and sound modulation.
   d) Filter network, transistor, and diode modulation.
6. What is detection?
   a) The process of recovering the baseband signal at the receiver.
   b) The reverse process of modulation.
   c) The process of translating the sideband signals back to the original baseband frequencies.
   d) All of the above.
7. What is the mixing function?
   a) The combining of circuits to blend ac and dc electricity.
   b) The process of multiplying two signals together to move one of the signals to a new position in the frequency spectrum.
   c) A method used to create sideband signals for transmission over the Internet.
   d) The use of diodes and transistors to create transmitter circuits.
8. What is meant by system gain?
   a) The ability of a circuit to provide an increase in the voltage, current, or power level of a signal.
   b) The ability of a circuit to provide noise to the system.
   c) The increase of the power supply voltage to the circuit in question.
   d) An unexpected increase in volume of the signal output.
9. What is the difference between small-signal and power amplifiers?
   a) Small-signal amplifiers are used to amplify large voltage signals; power amplifiers amplify microvolt signals.
   b) Power levels from small-signal amplifiers usually will be less than 1 watt; while those from power amplifiers greater than 1 watt.
   c) Small-signal amplifiers increase signal levels below 1650 kHz, while power amplifiers are used to increase signal levels above 1650 kHz.
   d) Small-signal amplifiers are only used inside electronic devices, while power amplifiers are used outside electronic devices.
10. What is an oscillator?
    a) It is an electronic circuit designed to output a signal with a constant frequency and amplitude.
    b) It is special type of feedback amplifier where a portion of the signal output is fed back to the signal input.
    c) A circuit that has a 360° phase shift within gain of one from input to output, resulting in an output signal with constant frequency and amplitude.
    d) All of the above.

**Answers:**
1 d, 2 c, 3 a, 4 c, 5 b, 6 d, 7 b, 8 a, 9 b, 10 d

## Questions & Problems for Chapter 2

1. A _________ is a piece of test equipment that detects the frequencies contained within a signal and displays their amplitude against frequency on a video screen.
2. The _________-to-_________ ratio should be maintained at a low level compared to the original signal level so that the information can be transferred accurately.
3. Electromagnetic waves, including light, are a special combination of _________ fields and ______ fields at right angles to each other that propagate and move through free space.
4. Calculate the frequencies of the sidebands for a 1-MHz carrier AM modulated with a 2-kHz signal.
5. If a 20-kHz signal is mixed with a 100-kHz signal, what are the resultant frequencies?
6. What is the RMS value of a sine wave with a peak value of 100 volts? Of 500 volts?
7. If the ac input voltage is 5.0 volts and the input current is 0.1 amperes, and the output voltage is 40 volts and output current is 1.0 amperes, what is the power gain through the amplifier?
8. What are the four main types of signal control?
9. A ____________ filter allows a selected range of frequencies to be passed on for further signal processing.
10. An ____________ circuit maintains a constant output from an amplifier.

*(Answers on page 209.)*

# CHAPTER 3
# A Refresher

There are a number of fundamental concepts that will be critical to your understanding of the detailed discussion explaining the operation of electronic circuits and systems that follow in the remaining chapters of the book. This chapter provides a "refresher" about these topics to assure that all readers have the same level of understanding as we move forward through our explanation of basic communications electronics. The subjects covered in this Chapter are:

- Ohm's Law.
- Decibels.
- Passive Devices — the characteristics of resistors, inductors, and capacitors.
- One-way Valve for Current — the characteristics of diodes.
- Active Devices — the characteristics of NPN and PNP bipolar transistors; and of N-Channel and P-Channel FETs.
- Coupling — the meaning and use of dc, ac, transformer, and optical coupling.

Many readers may be knowledgeable about these subjects since most of them were covered in *Basic Electronics*[1]. Readers who have a thorough understanding of this material may choose to skip past this Chapter and go on to Chapter 4 "Amplifiers & Oscillators."

## Ohm's Law

The law of electricity most used in electronic circuit design is Ohm's Law. It is named in honor of Georg Simon Ohm, who formulated the relationship between voltage, current and resistance in the 19th Century. Ohm's Law states:

*"The current in an electrical circuit is directly proportional to the voltage applied to the circuit, and inversely proportional to the resistance."* In equation form, Ohm's Law is:

$I = \frac{E}{R}$ where: I is current in **amperes**

$E = I \times R$ E is voltage in **volts**

$R = \frac{E}{I}$ R is resistance in **ohms**

### Simple Aid for Using Ohm's Law

A simple aid for remembering Ohm's law is shown in *Figure 3-1a*. Just cover the letter in the circle that you want to find and read the equation formed by the remaining letters.

[1] *Basic Electronics*, G. McWhorter and A. J. Evans, ©1994, Master Publishing, Inc., Lincolnwood, IL

When the current is unknown, but the voltage and resistance are known, the basic equation to be solved for I is found by using the aid of *Figure 3-1b*. The result is:

$$I = \frac{E}{R}$$

Similarly, knowing the current and resistance, the voltage can be calculated by using the equation shown in *Figure 3-1c*:

$$E = IR$$

*Figure 3-1d* shows the aids when the current and resistance are in values other than amperes and ohms, respectively.

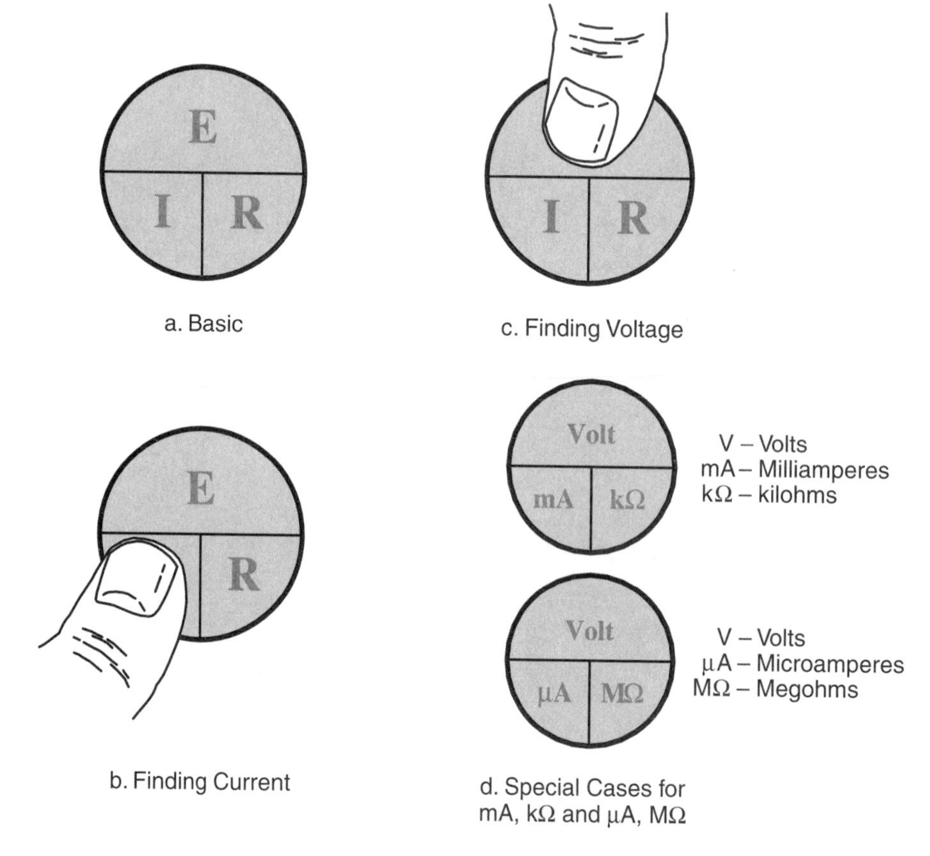


*Figure 3-1 Ohm's law circle.*
*Source: Basic Electronics, G. McWhorter and A.J. Evans, ©1996, Master Publishing, Inc., Lincolnwood, IL.*

## Decibel — A Quantity to Describe Gain

Almost all technicians and engineers, as well as marketing personnel in communications systems, use the term "decibel." The decibel (abbreviated dB) is one-tenth of a bel. It is a standard unit for expressing the ratio between output power and input power or, in special cases, output voltage and input voltage. (It also is used to express differences in sound levels (power levels) in audio systems.) The decibel is expressed as:

Power

$$dB = 10 \log_{10} \frac{P_{out}}{P_{in}}$$

Voltage

$$dB = 20 \log_{10} \frac{V_{out}}{V_{in}} \text{ when } R_{in} = R_{out}$$

To review its use, what power ratio is represented by 20 dB?

**Remember:** The logarithm of a number is the *exponent* to which the base of the logarithm must be raised in order to arrive at the number, e.g. $10^x = Y$, therefore, $x = \log_{10} Y$. Also recall that ∴ means "Therefore."

$$20 = 10 \log_{10} \frac{P_{out}}{P_{in}}$$

$$2 = \log_{10} \frac{P_{out}}{P_{in}}$$

$$\therefore 10^2 = \frac{P_{out}}{P_{in}}$$

∴ $P_{OUT}$ is 100 times $P_{in}$

What voltage ratio is represented by 60 dB?

$$dB = 20 \log_{10} \frac{V_{out}}{V_{in}}$$

$$60 = 20 \log_{10} \frac{V_{out}}{V_{in}}$$

$$3 = \log_{10} \frac{V_{out}}{V_{in}}$$

$$\therefore 10^3 = \frac{V_{out}}{V_{in}}$$

∴ $V_{out}$ is 1000 times $V_{in}$

There is a caution in using $dB = 20 \log_{10} V_{out} / V_{in}$. It is assumed that $V_{out}$ and $V_{in}$ are across the same value of resistance. If this condition is not met, $dB = 10 \log_{10} P_{out} / P_{in}$ must be used. The following reference table shows the equivalent power and voltage ratios to various decibels.

| dB | $P_{out}/P_{in}$ | $V_{out}/V_{in}$ |
|---|---|---|
| 3 | 2 | 1.4 |
| 6 | 4 | 2 |
| 10 | 10 | 3 |
| 20 | 100 | 10 |
| 30 | 1000 | 31.6 |
| 40 | 10,000 | 100 |
| 50 | $10^5$ | 316.2 |
| 60 | $10^6$ | 1000 |

Table 3-1.

The advantage of using dB units is that they can be added and subtracted directly to obtain the final result. For example, if two amplifier stages each have 10 dB of power gain, the total power gain is 20 dB. 10 dB of power gain is a ratio of 10. The gains of two amplifiers cascaded multiply; therefore, 10 x 10 = 100 power gain for the two stages. 20 dB of power gain is a ratio of 100.

## Passive Devices

All resistors, inductors and capacitors have impedance (symbol Z) because they "impede" (resist) current in electronic circuits. Impedance is measured or expressed in ohms (symbol Ω ). The schematic symbols for these passive devices are shown in *Figure 3-2.*

1. Resistors impede (resist) current equally well in dc or ac circuits. Unless a resistor has inductance or capacitance, its impedance is $Z = R \angle 0°$ , just the resistance with a zero phase angle. Because the phase angle is zero, it is left off and Z is just equal to R (Z=R).

2. Inductors are coils of wire that have inductance (symbol L), a capability to store energy in a magnetic field surrounding the coil when there is current through the coil. The stored energy opposes changes in the existing current through the coil. The opposition to the changing current is called inductive reactance (symbol $X_L$). The impedance of an inductor is made up of the coil's resistance and inductive reactance added as vectors at right angles to each other. Its value is $Z_L = \sqrt{R^2 + X_L^2}$ with a phase angle, θ, whose $\tan \theta = X_L / R$. When R = 0, $Z_L = X_L \angle 90°$ . Inductors with zero or very small resistance are a short circuit (zero impedance) to dc current, but increase their inductive reactance as frequency increases according to the expression $X_L = 2\pi fL$, where f is frequency in Hz and L is inductance in henries. As frequency increases, the inductive reactance of an inductor increases. At 10 MHz (10,000,000 cycles), even a small inductance (0.1 millihenry) with little or no resistance has 6,280 ohms (Ω) of impedance.

3. Capacitors are made from two metal plates separated by an insulator that have capacitance (symbol C), a capability to store a charge in an electrostatic field. The stored energy opposes changes to the existing voltage across the capacitor. The opposition to the changing voltage is called capacitive reactance (symbol $X_C$). The impedance of a capacitor is made up of the capacitor's resistance and capacitive reactance added as vectors at right angles to one another. Its value is $Z_C = \sqrt{R^2 + X_C^2}$ with a phase angle (θ) whose $\tan \theta = -X_C / R$. The minus sign on $X_C$ means that the right triangle leg is in the opposite direction from $X_L$. The R for capacitors is the dc resistance of the leads and the metal plates, which is very small. Therefore, $Z = X_C \angle -90°$ . Capacitors are open circuits (infinite impedance) to dc current, but decrease their capacitive reactance as frequency increases according to the expression $X_C = 1 / 2\pi fC$, where f is frequency in Hz and C is capacitance in farads. As frequency increases, the capacitive reactance of a capacitor decreases. At 10 MHz, even a fairly large capacitor (0.1 microfarad) has only about 0.2 ohms (Ω) impedance.

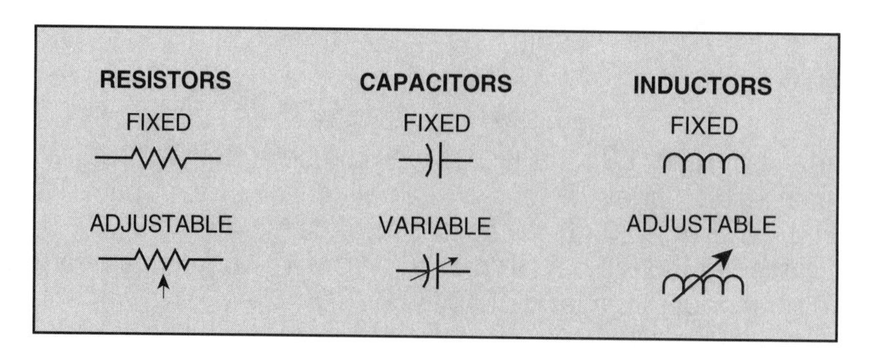

*Figure 3-2. Schematic symbols for resistors, capacitors, and inductors.*

### *Example 1. Impedances of R, L and C*

A. What is the impedance of a 10,000-ohm resistor that has no inductance or capacitance, first to current in a dc circuit and second to current in a circuit powered by 60 VAC?

B. What is the impedance of a 1 millihenry inductor with zero resistance at 1000 Hz, 1 MHz, and 1000 MHz?

C. What is the impedance of a 1 microfarad capacitor with zero resistance at 1000 Hz, 1 MHz, and 1000 MHz?

***Solution:***

A. The impedance of a resistor is the same for a dc or an ac circuit — 10,000 Ω. Since it has no inductance or capacitance, resistance is the only component and there is no phase angle.

$$Z = 10{,}000\ \Omega \angle 0^\circ = 10{,}000\Omega$$

B. With R = 0, $Z = X_L \angle 90^\circ$.

Use the equation $X_L = 2\pi fL$. Remember that π is a constant of 3.14.

1. For 1000 Hz: $X_L = 6.28 \times 1 \times 10^3 \times 1 \times 10^{-3} = 6.28$ ohms $\angle 90^\circ$

2. For 1 MHz: $X_L = 6.28 \times 1 \times 10^6 \times 1 \times 10^{-3} = 6.28 \times 10^3 =$ 6280 ohms = 6.28 kilohms $\angle 90^\circ$

3. For 1000 MHz: $X_L = 6.28 \times 1 \times 10^9 \times 1 \times 10^{-3} = 6.28 \times 10^6$ ohms = 6.28 Megohms $\angle 90^\circ$

C. With R = 0, $Z = X_C \angle -90^\circ$

Use the equation $X_C = \dfrac{1}{2\pi fC}$. Remember that π is a constant of 3.14.

1. For 1000 Hz: $X_C = \dfrac{1}{6.28 \times (1 \times 10^3) \times (1 \times 10^{-6})} =$

$$\frac{1}{(6.28 \times 10^{-3})} = 0.159 \times 10^3 = 159 \text{ ohms } \angle -90^\circ$$

2. For 1 MHz: $X_C = \dfrac{1}{6.28 \times (1 \times 10^6) \times (1 \times 10^{-6})} =$

$$\frac{1}{6.28} = 0.159 \text{ ohms } \angle -90^\circ$$

3. For 1000 MHz: $X_C = \dfrac{1}{6.28 \times (1 \times 10^9) \times (1 \times 10^{-6})} =$

$$\frac{1}{(6.28 \times 10^3)} = 0.159 \times 10^{-3} = 0.159 \text{ milliohms } \angle -90^\circ$$

## The Diode — A One-Way Valve for Current

A diode is a semiconductor chip (usually silicon) with a PN junction. The P material is the anode; the N material is the cathode. For a silicon diode, as shown in *Figure 3-3*, conventional current will flow easily from the anode to the cathode when the voltage at the anode is 0.7 V more positive than the cathode. If the anode is less than 0.7 V more positive than the cathode, no current will flow. Therefore, the diode is a one-way valve for current. This characteristic is used extensively in electronic circuits, including rectifiers in power supplies, as well as in detection and mixing circuits. The 0.7 V differential in a silicon junction is used extensively as a relatively constant voltage in amplifier circuits.

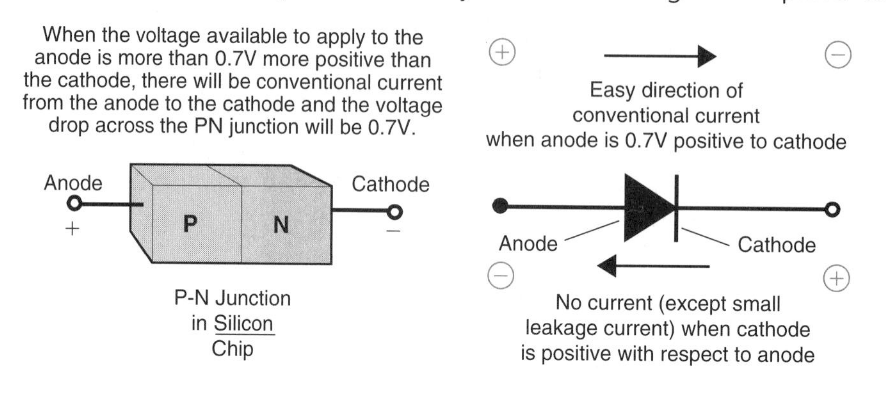


*Figure 3-3 A diode conducts current only in one direction.*

## Active Devices

In Chapter 2, you learned that *electronic devices that provide gain are called active devices*. The most important active device for electronic circuits is a transistor. There are two types of transistors — bipolar and field-effect transistors.

### Bipolar Transistors

A bipolar transistor is a combination of two junctions of semiconductor material built into a semiconductor chip (usually silicon). There are two types of bipolar transistors — PNP and NPN — and their junction structures are shown in *Figure 3-4a* and *3-4d*. For a transistor that is producing gain, the emitter-base junction is a forward-biased diode and the collector-base junction is a reverse-biased diode. The diode equivalents of a PNP and an NPN transistor are shown in *Figure 3-4b* and e, respectively. The collector-base junction of a bipolar transistor is distinctly different from a reverse-biased diode. Normally, a reverse-biased diode conducts no current, except for a very small leakage current. The reverse-biased collector-base junction of a bipolar transistor conducts collector current that is *controlled* by the current into the base at the base-emitter junction. The ratio of the collector current to the base current is $h_{FE}$. Under normal operation, the collector current is greater than the base current; so there is a current gain and $h_{FE}$ is a number greater than one — typically 50 to 200. However, there are special cases of operation or manufacture where $h_{FE}$ is less than one.

#### *NPN*

The normal active device operation is shown in *Figure 3-5*. An NPN silicon transistor P base is 0.7 V more positive than its N emitter, and the N collector is several more volts more positive than the emitter, as shown in *Figure 3-5a*. The emitter current, $I_E$, is the sum of the base current, $I_B$, and the collector current, $I_C$. The current gain of the transistor under any dc operating condition is $h_{FE}$, the ratio of $I_C$ to $I_B$. $h_{FE}$ current gains of 50 to 200 are common in modern day silicon transistors. $h_{FE}$ is actually called "the common-emitter" current gain because the emitter is common in the circuit. In Chapter 4, other circuit connections will be discussed.

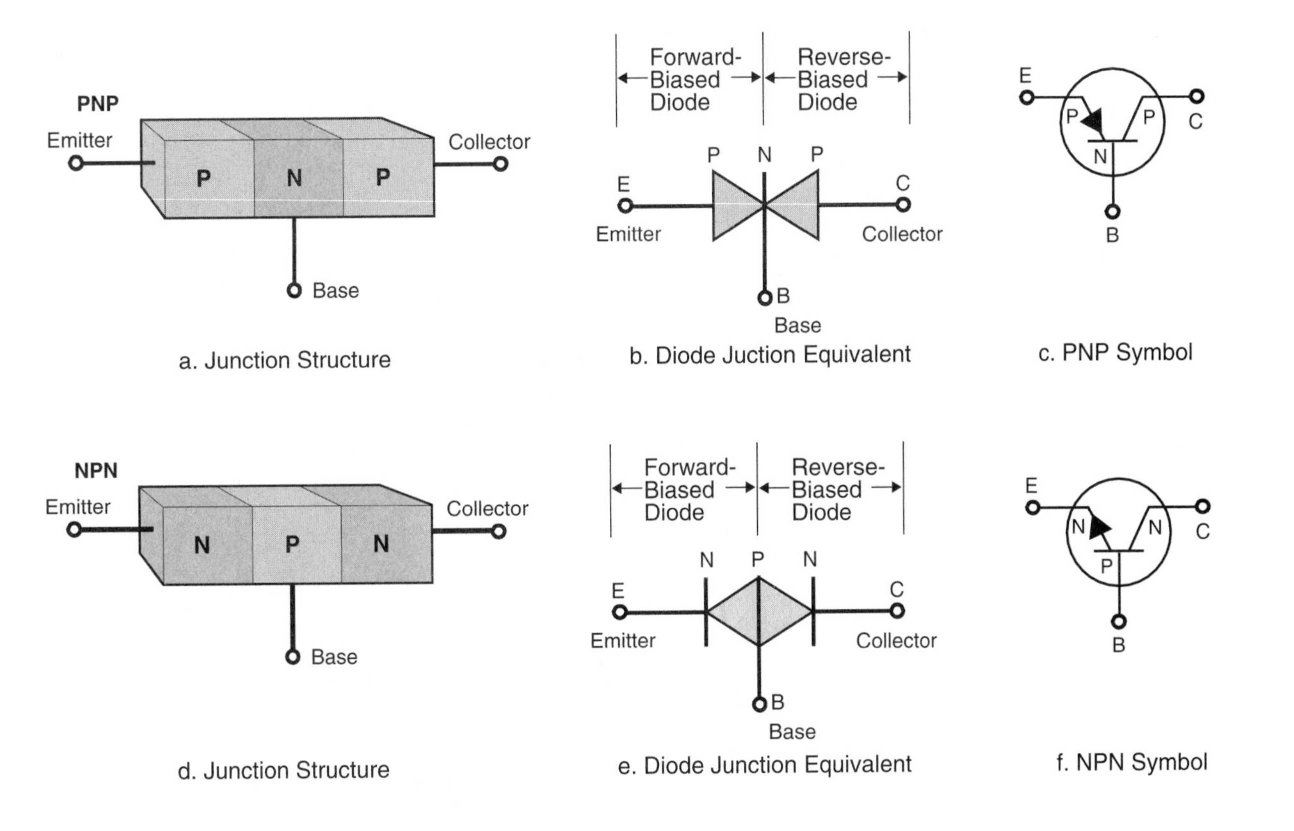


*Figure 3-4. Bipolar transistors — their construction and symbols.*

## *PNP*

A silicon PNP transistor base is 0.7 V *negative* with respect to its emitter in order to have the P emitter more positive than the N base. The P collector is several volts *negative* from the emitter to keep the collector-base junction reverse biased. As shown in *Figure 3-5b*, the same current equations apply and the current gain, $h_{FE}$, is the same. The major difference is in the polarity of the voltages for operation. For the NPN common-emitter operation the base and collector voltages are *positive* with respect to the emitter; while for the PNP the voltages are *negative*.

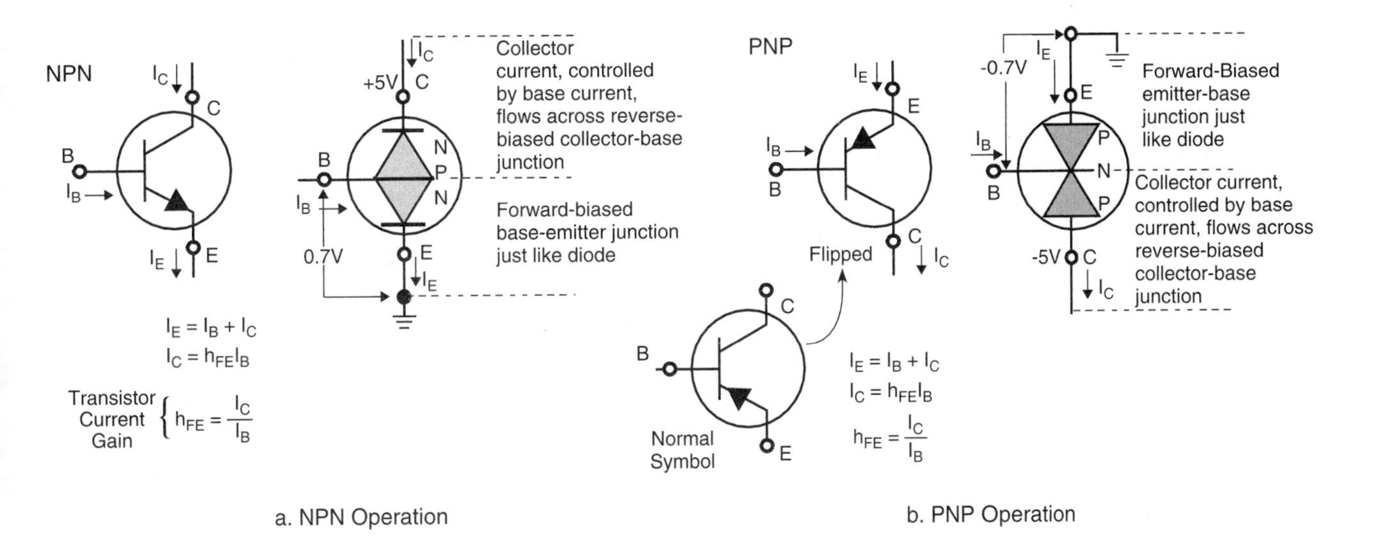


*Figure 3-5. Bipolar transistor operation.*

## Field-Effect Transistors (FETs)

Unlike bipolar transistors, which depend on current into the base to control collector current, field-effect transistor current between *source* and *drain* is controlled by a *voltage* on a *gate*. Look at the basic structure of an N-channel MOSFET (Metal-Oxide Semiconductor Field-Effect Transistor), as shown in *Figure 3-6a*. Heavily-doped N semiconductor material forms *source* and *drain* regions in a P semiconductor material substrate. The region between the *source* and *drain* is the *gate* region, where a thin layer of oxide insulates the P semiconductor substrate underneath from a metal plate that is deposited over the thin oxide. A thick oxide layer over the *source* and *drain* regions insulates metal connection pads from the substrate. Holes in this thick oxide layer allow the metal pads to contact the *source* and *drain*.

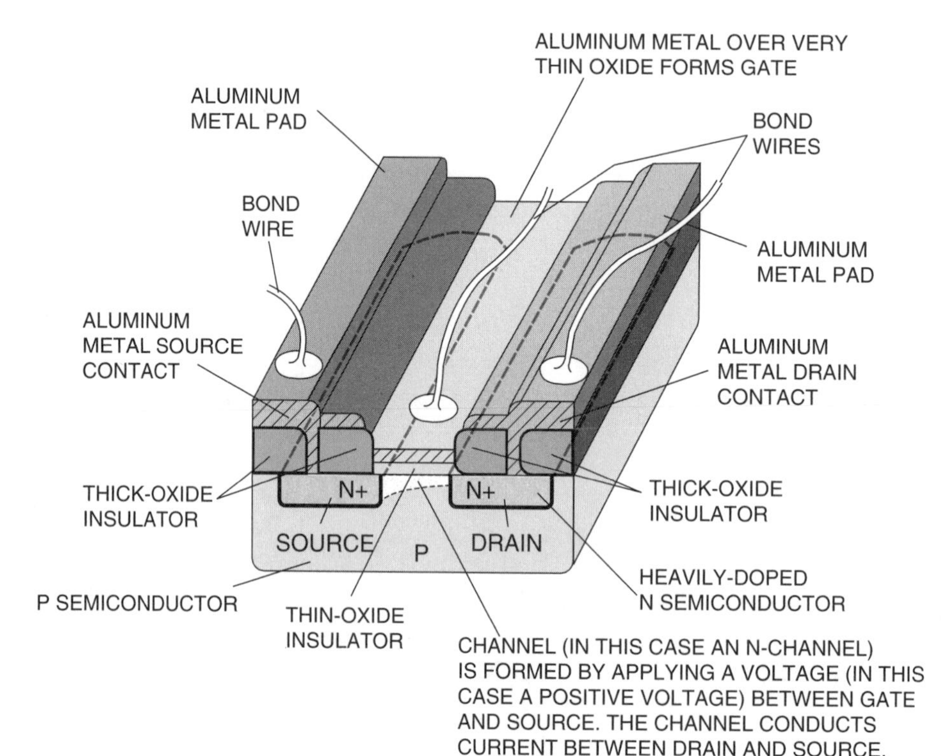


a. Pictorial of Construction (N-Channel Enhancement)

P-CHANNEL | N-CHANNEL

DEPLETION MODE – GATE BIAS TURNS OFF

A. (P-channel): + G OFF; S D; + → –

B. (N-channel): – G OFF; S D; – ← +

S – SOURCE
G – GATE
D – DRAIN

ENHANCEMENT MODE – GATE BIAS TURNS ON

C. (P-channel): – G ON; S D; + → –

D. (N-channel): + G ON; S D; – ← +

b. Schematic Symbols of MOSFETs

SOURCE-DRAIN CURRENT
GATE VOLTAGE CONTROLS CURRENT IN A CURVED NON-LINEAR WAY
GATE-SOURCE VOLTAGE (+ OR –)

c. Characteristic Curve of Field-Effect Transistor

*Figure 3-6. MOS (Metal-Oxide-Semiconductor) Field-Effect Transistor.*

## N-Channel Operation

Symbol D of the schematic symbol diagrams of *Figure 3-6b* represents an N-channel enhancement-mode MOSFET. It indicates that a positive voltage is applied to the *drain* of an N-channel MOSFET with respect to the *source*. When no voltage is applied to the *gate* with respect to the *source*, no current flows from *drain* to *source*. However, applying a positive voltage to the *gate* with respect to the *source* produces a channel underneath the *gate* in the P-semiconductor substrate. This channel conducts current between *drain* and *source*. The characteristic curve that shows *drain*-to-*source* current plotted against *gate*-to-*source* control voltage is shown in *Figure 3-6c*. Therefore, in field-effect transistors we have a voltage (between *gate* and *source*) controlling current between *drain* and *source*. The ratio of the *drain*-to-*source current* change to the *gate*-to-*source voltage* change that caused it is called the *transconductance* (abbreviated **gm**) of the field-effect transistor.

## Four Common Types

*Figure 3-6b* shows that there are four common types of MOSFETs: P-channel depletion and enhancement mode devices, and N-channel depletion and enhancement mode devices. P-channel devices have the semiconductor materials just reversed from the N-channel materials shown in *Figure 3-6a*. The *drain* and *source* are P semiconductor material and the substrate is N semiconductor material. In an enhancement-mode MOSFET, current is produced and increased as an increasing voltage is applied between the *gate* and *source*. In a depletion-mode MOSFET, there already is current from *drain* to *source* when there is no *gate* to *source* voltage. Applying a *gate* voltage reduces (depletes) the current. Voltage polarities are reversed when using P-channel field-effect transistors from those used for N-channel. The *source* is positive with respect to the *drain* and the *gate* voltage is negative with respect to the *source*. Since the *gate* is insulated from the substrate, there is a very high impedance from *gate* to *source* for field-effect transistors. The *drain* to substrate and *source* to substrate junctions are just the same as any other semiconductor diode junction, and for proper operation they must be kept reverse biased.

## Biasing

Biasing has already been mentioned and stated to be "setting the no-signal operating point around which the small ac signals vary." In the "Amplifiers and Oscillators" chapter that follows, the different classes of amplifiers will be discussed. Each depends on a different operating point when there is no signal.

To make sure we understand what the "operating point" is, let's look at *Figure 3-7a*, which is a plot of the characteristic curves of an NPN common-emitter transistor. If a base current, $I_B$, of 0.02 mA is caused to flow in the base-emitter junction and the collector-emitter voltage, $V_{CE}$, is varied from zero to 20 volts, the heavy-line characteristic curve will be drawn. Increasing the base current in 0.02 mA steps and varying $V_{CE}$ the same way again will draw the other curves shown. To make sure that the transistor used in an amplifier reproduces a signal at the output exactly as the signal at the input, with only minimal distortion, the operating point is chosen at **A**. The NPN transistor has an $h_{FE} = 100$, so the input signal that varies the base current produces a collector-current signal output, shown in *Figure 3-7b*, that is 100 times the base current signal. Point **B** is the $V_{CE}$ of the transistor when $I_C = 0$ and point **C** is the maximum current the transistor could draw if $V_{CE} = 0$ (it never gets there because of what is called a saturation condition at point **D**). The line from **C** to **B** is called the "load-line" and describes what the $V_{CE}$ will be with a given resistor load in the collector as the base current is varied. So the operating point determines where the transistor needs to be biased when there is no signal applied.

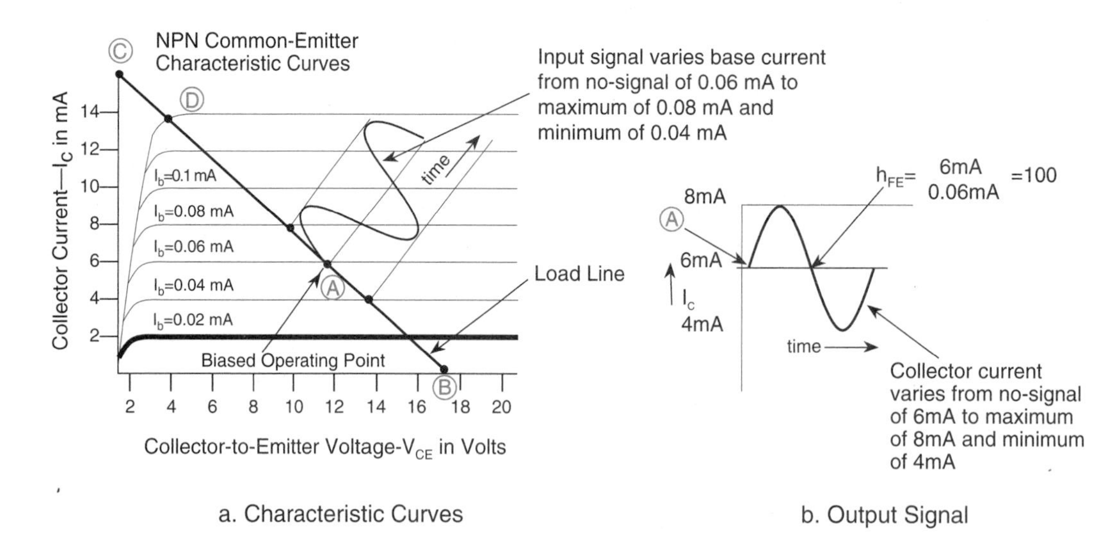


*Figure 3-7. Biasing sets "No-Signal" operating point*

Point **A** can be moved up or down the load line from point **B** to point **D** by changing the biasing that is chosen. Input signals vary the circuit operation up and down from the no-signal operating point to produce an output signal.

## Coupling

A circuit that performs a function usually is called a *stage*, such as an amplifier stage, detector stage, etc. Signals must be coupled from stage to stage, and there are four different types of coupling used in electronic circuits — dc, ac, transformer, and optical.

### DC Coupling

DC coupling means that the dc voltages used feed directly between stages as shown in *Figure 3-8a*. Any voltage at the collector of stage #1 is coupled directly to the base of the second stage and must be considered when setting the operating point of stage #2.

### AC Coupling

AC coupling using a capacitor is shown in *Figure 3-8b*. Note that the operating point dc voltages do not transfer from stage #1 to stage #2. Thus, the operating point of stage #2 can be set independent of stage #1. However, because the impedance of a capacitor is infinity at zero frequency (a capacitor is an open circuit to dc current), ac signals with low frequencies will not be coupled through the capacitor the same as high-frequency signals will be. The response of the ac gain through the stages plotted against frequency is as shown.

AC coupling also occurs when using an inductor for coupling the signal between stages, as shown in *Figure 3-8c*. The ac gain response is reversed from that of the capacitor. Now, dc and low-frequency signals are amplified fully but high-frequency signals are attenuated, as shown by the response curve. Because an inductor is a short to a dc current, stage #2's operating point is determined by the bias voltages of stage #1. Thus, more correctly, this coupling might be described as both dc and ac coupling.

### Transformer Coupling

Transformer coupling, shown in *Figure 3-8d*, depends on the coupling of a changing magnetic flux between the primary and secondary of the transformer. Thus, the dc bias voltages of stage #1 are isolated from stage #2. Time changing or ac signals are required to produce the changing flux for the transfer of a signal.

## Optical Coupling

In optical coupling, signal coupling is achieved using light waves, as shown in *Figure 3-8e*. A light-emitting diode (LED) varies its light output as current in the collector of stage #1 varies. The variations in light are picked up by a photo transistor in stage #2. The coupling provides dc isolation so that stage #1 bias voltages are isolated from stage #2 biasing. The light sensitivities of the LED and photo transistor are matched for maximum coupling efficiency.

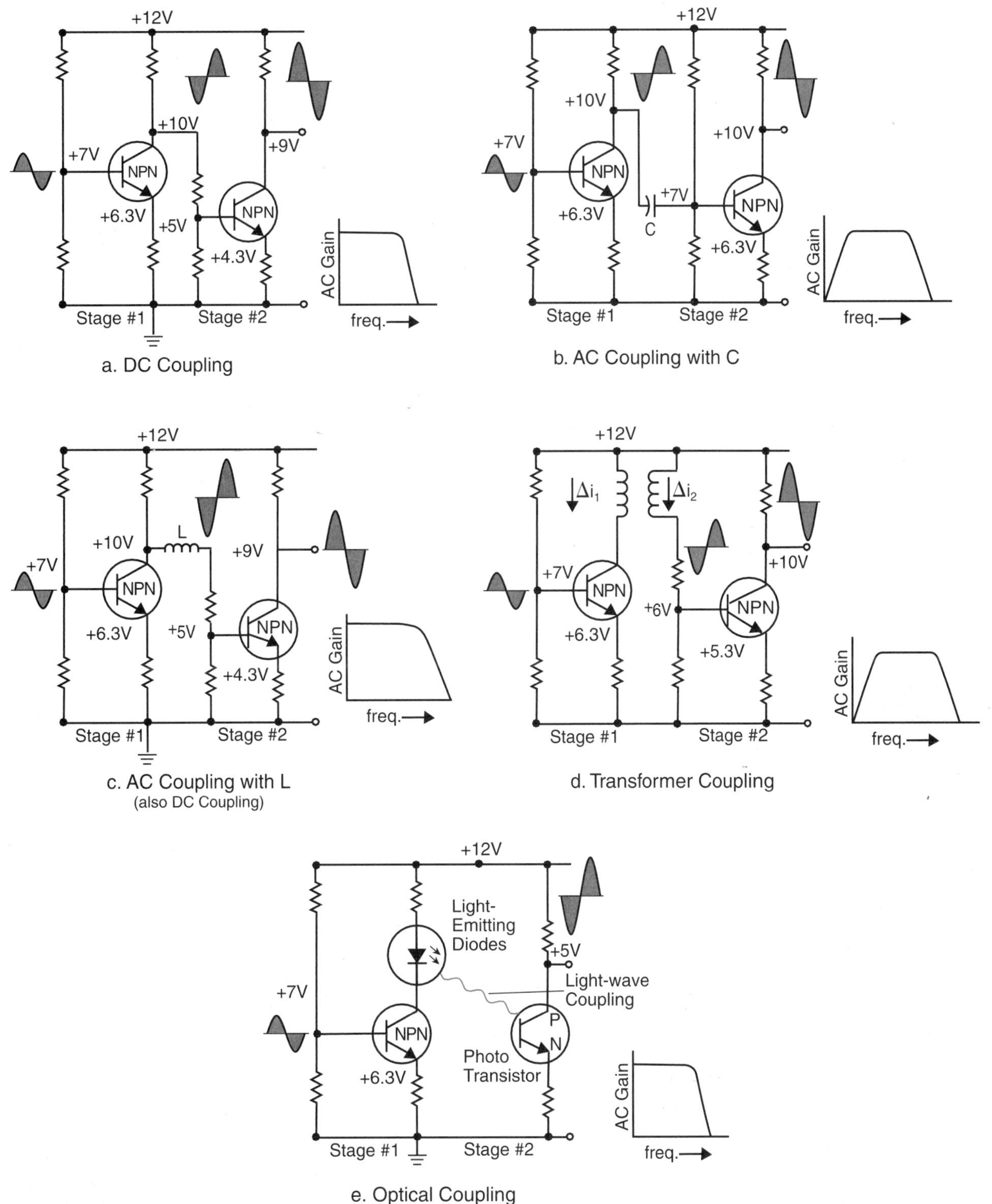


*Figure 3-8 Five types of coupling.*

## Summary

Now that you have reviewed Ohm's Law and practiced your math for decibel calculations, and refreshed your knowledge about the fundamental concepts of diodes, passive and active devices, and coupling, we are ready to study the operation of various circuits that perform functions used in communications electronics. We will begin by studying amplifier and oscillator circuits.

# Quiz for Chapter 3

1. Ohm's Law states: "The ________ in an electrical circuit is directly proportional to the __________ applied to the circuit, and inversely proportional to the __________."
   a) Current, voltage, resistance.
   b) Resistance, current, voltage.
   c) Voltage, resistance, current.
   d) Current, resistance, voltage.
2. The decibel (dB) is a standard unit for expressing the ratio between:
   a) Output power and input power.
   b) Output voltage and input voltage.
   c) Both a and b.
   d) None of the above.
3. Which of the following are passive devices?
   a) Resistors, transformers, and MOSFETs.
   b) Capacitors, inductors, and resistors.
   c) Capacitors, bipolar transistors, and inductors.
   d) Inductors, resistors, and FETs.
4. Describe the action of a diode.
   a) It is a one-way valve for current.
   b) It reverses the direction of current flow.
   c) It is a current divider.
   d) It amplifies ac signals.
5. For an NPN transistor, the base normally is:
   a) Connected to the common ground.
   b) Less positive than the emitter.
   c) Connected to the power source.
   d) More positive than the emitter.
6. For a PNP transistor, the base normally is:
   a) Connected to the diode bridge.
   b) Negative with respect to the emitter.
   c) Connected to the power supply capacitor.
   d) More positive than the emitter.
7. What is the difference between field effect transistors and bipolar transistors?
   a) FETs normally have an input forward-biased diode; bipolar transistors do not.
   b) FETs use a voltage on a gate to control source to drain current; bipolar transistors depend upon current into the base to control collector current.
   c) Bipolar transistors operate in the depletion mode; FETs in the saturation mode.
   d) FET transistors have $h_{FE}$ current gain; bipolar transistors do not.
8. What is meant by "biasing" with regard to transistor amplifiers?
   a) Deciding whether the amplifier should be used for ac or dc.
   b) Setting an amplifier circuit using an NPN transistor to operate in the depletion mode.
   c) Setting the no-signal operating point around which the small ac signals vary.
   d) Calculating the dB gain of the amplifier stage.
9. What are the four types of coupling used in electronic circuits?
   a) Resistor, diode, capacitor, and transistor.
   b) Transformer, bridge rectifier, filter, and flux.
   c) NPN, PNP, MOSFET, and GASFET.
   d) Optical, ac, transformer, and dc.
10. Which forms of coupling provide isolation of dc bias voltages from stage to stage?
    a) Dc and ac.
    b) Transformer, ac and optical.
    c) Optical and dc.
    d) Transformer and inductive.

**Answers:**
1 a, 2 c, 3 b, 4 a, 5 d, 6 b, 7 b, 8 c, 9 d, 10 b.

## Questions & Problems for Chapter 3

1. What is the voltage across a 50 ohm resistor if 1.5 amperes is the current through the resistor?
2. What is the resistance in ohms if there is 500 volts applied to the circuit and the current is 200 amps?
3. What is the current if there is 125 volts and a resistance of 75 ohms?
4. If an amplifier has a voltage gain of 40db, what is the ratio of $V_{out}$ to $V_{in}$?
5. What is the gain in dB if a system has 2 watts in and 2000 watts out?
6. If a voltage amplifier has a 100V output when the input is 1V, what is the amplifier gain in dB?
7. What is the impedance of a 5μH inductor with no resistance at 2MHz?
8. What is the impedance of a .01uf capacitor with no resistance at 2MHz?
9. If the inductor of problem 6 and the capacitor in problem 7 are connected in series in a circuit operating at 2MHz, what will be the resultant and impedance?
10. A diode will conduct current when the ______ is 0.7 V more positive than the ______.
11. What devices are used for optical coupling?
12. If an NPN transistor amplifier has an input base current of 0.01mA and the collector current is 1.8mA, what is the current gain provided by the transistor?
13. For normal operation, the base of a PNP is __________ with respect to its emitter, and the collector is __________ with respect to its emitter.
14. The current from Drain to Source of a FET is 5ma when 1.25 volts is applied between the gate and the source. What is the gm?
15. A __________ mode FET has current between source and drain when the gate is shorted to the source.

*(Answers on page 209.)*

# CHAPTER 4

# Amplifiers & Oscillators



Amplifiers and oscillators are major building blocks in communications systems. Amplifiers, in particular, are fundamental to communications systems, as well as to analog systems in general. This chapter focuses on the basic functions of amplification and oscillation, discusses the active devices required to build these circuits, describes their use in performing the functions, and explores the circuit operation of common amplifiers and oscillators.

Amplifier and oscillator circuits are referred to as active circuits since their operation is based on active devices. They are distinguished from passive circuits by their ability to provide power gain, which is essential to most electronic systems. This need for power gain is most evident in wireless communications systems where the signals received may be in the micro-microwatt (one millionth of a millionth) range. Such signal power is too weak to process without amplification. Power gain is required to raise the signal level so that it is at a high enough level to drive the output circuits. Keep in mind that all amplifiers, their power supplies, and associated processing circuits will add some noise to the signal.

As mentioned in Chapter 2, oscillators are the "first cousin" of amplifiers. Oscillators produce ac waveforms of specific amplitudes and frequencies by converting power from a dc power source. We will look at how oscillators work and how they are used after we study amplifiers.

## Small-Signal Amplifiers

Small-signal amplifiers are used to increase the voltage amplitude and/or current amplitude of a signal, and to provide some increase in the power of a signal. The signal may be audio or video, either by itself or modulated up to higher frequencies. As mentioned earlier, small-signal amplifiers always act at low signal levels — voltage levels of microvolts to millivolts, current levels of microamperes to milliamperes, and power levels of a few hundred milliwatts or less. Seldom are the small-signal transistor collector voltages as great as 20 volts, and power supply voltages usually are less than a total of 40 volts. Transistor bias currents usually are a few milliamperes, but occasionally may approach 100-milliampere levels.

Transistors can be used in virtually all small-signal amplifiers, either built into an integrated circuit (IC) or used as a discrete part in a circuit. Depending upon the particular circuit application, NPN or PNP bipolar junction transistors (BJTs) or a variety of field-effect transistors (FETs) are used. The most common FETs are the junction (JFET) and the metal-oxide (MOSFET) types, either N-channel or P-channel, and either enhancement or depletion mode. Most devices are fabricated from silicon but, occasionally in communication circuits, gallium arsenide FETs (GASFETs) are used for very-high frequency and low-noise operation. The choice of the type of device to use is a design consideration based upon the circuit application.

## Common-Emitter Stage

An example of a small-signal bipolar transistor amplifier is shown in *Figure 4-1*. It is built using an NPN 2N2222 bipolar transistor. Such an amplifier stage is called a common-emitter amplifier because the emitter lead is common to both the input and output of the circuit. Remember, the word "stage" means a single amplifying circuit. The dc voltages, called bias voltages, between the base, collector, and emitter of the transistor and ground are shown on the schematic in *Figure 4-1*. The biasing method for this particular circuit is called voltage-divider bias. The voltages are typical, but will vary depending on the design for a specific application.

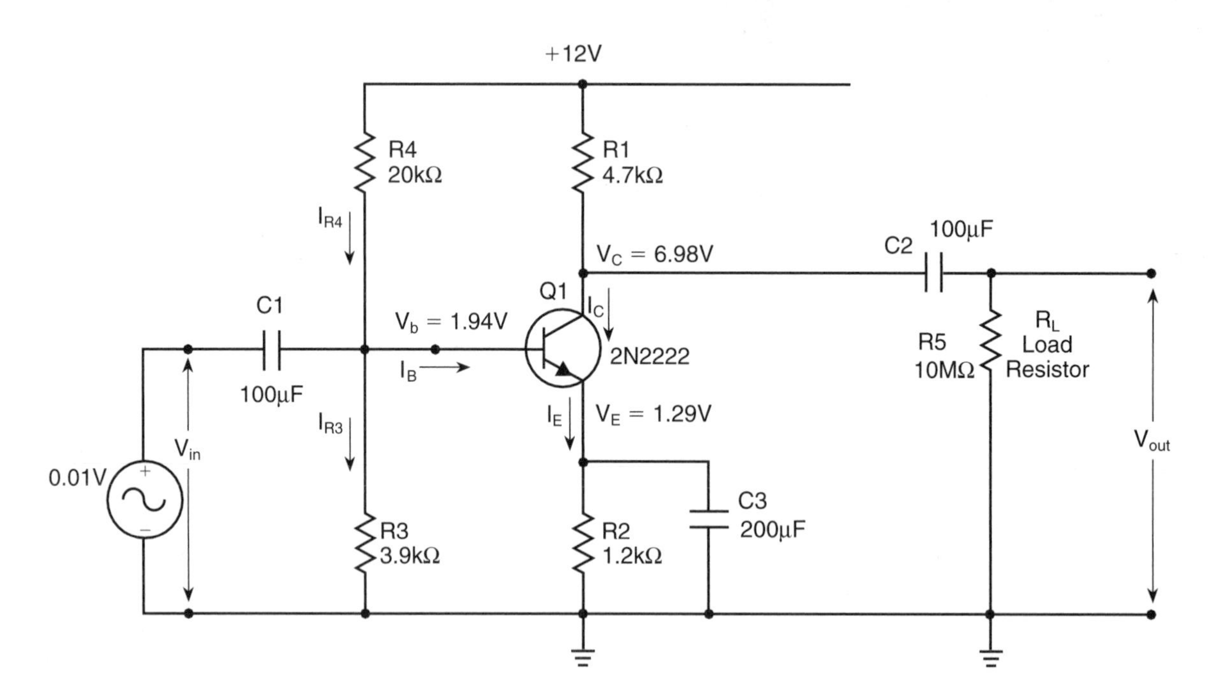


*Figure 4-1. Small-signal amplifier using NPN 2N2222 as the active device.*

As explained in Chapter 3, biasing refers to setting the dc operating point of the transistor in the "middle" of its operating range so that the ac signal can swing equally in both the positive and negative direction from the operating point and thus reproduce the input signal accurately and with minimal distortion at the amplifier output. The transistor operating point for the small-signal amplifier of *Figure 4-1* is placed on the load line at a point similar to point **A** of *Figure 4-15*, as done previously in *Figure 3-7*.

The amplifier circuit in *Figure 4-1* operates as follows: The signal to be amplified, $V_{in}$, is AC coupled through C1, and produces a small current change in the base of Q1. Because of the transistor current gain, there is a larger collector current change through the collector load resistor, R1. The collector current change is multiplied by the 4.7 kΩ resistor to produce a large voltage, coupled through C2, across R5, the load resistor. The net effect is that the stage acts as a voltage amplifier. Small input voltages (0.01 V) are amplified from 50 to 200 times.

The resistor in the emitter lead going to ground provides negative feedback to the input signal from dc to the frequency where C3 bypasses all ac signals. The net effect is to stabilize the operating point. If the input signal increases, the emitter current increases, which increases the voltage across the emitter resistor. The emitter voltage increase is in opposition to the input voltage increase, thus providing the negative feedback. In analog amplifier circuits, an emitter resistor should always be used to provide negative feedback to stabilize the operating point from changes due to temperature or the drift of parameters.

## *Rules for Common-Emitter Stage Biasing (Approximate Equations)*

The rules for biasing common-emitter stage amplifiers shown below are approximate and may vary from circuit to circuit. They are used as general guidelines.

1. Collector-to-ground voltage should be approximately one-half the power supply voltage.
2. Emitter-to-ground voltage should be 10% to 15% of the power supply voltage.
3. Base-to-ground voltage will be approximately 0.7 volts greater than the emitter-to-ground voltage.

With $I_{BIAS} >> I_B$, $h_{FE}$ greater than 50, and $I_C$ no more than 30-40 mA, the equations below will give approximate dc values in a circuit like that in *Figure 4-1* in which resistor values are set:

$$V_B = \frac{R3}{(R3 + R4)} \times V_{CC}$$

$$V_E = V_B - 0.7V$$

$$I_E = \frac{V_E}{R_E} = \frac{V_E}{R2}$$

$$I_C \cong I_E$$

$$V_C = V_{CC} - (I_C \times R_C) = V_{CC} - (I_C \times R1)$$

Current gain is specified for dc signals (or for large signals) as $h_{FE}$, but small-signal amplifier designers often use $\beta$ (beta) or $h_{fe}$, which is the ac current gain. Beta is the ratio of a small-signal change in collector current to a small-signal change in base current, or $\beta = i_c / i_b$. The small-signal parameters are noted by lower case identifications. Note that neither $\beta$ nor $h_{fe}$ appear in the approximation equations. Thus, this biasing scheme is referred to as "beta independent."

### *Example 1: Calculating NPN Common-Emitter Amplifier Voltages and Currents*

Use *Figure 4-1* and the approximate equations to calculate $V_B$, $V_E$, $I_C$ and $V_C$.

***Solution:***

$$V_B = \frac{3900}{3900 + 20000} \times 12V = 1.96V$$

$$V_E = 1.96 - 0.7 = 1.26 \text{ V}$$

$$I_C \cong I_E = \frac{1.26V}{1200\Omega} = 1.05 \text{ mA}$$

$$V_C = 12 - (0.00105 \times 4700) = 7.07 \text{ V}$$

These values check very well with the computer-generated values noted on *Figure 4-1*.

The amplifier circuit has a voltage gain of about 45 dB (181 raw gain) and a bandwidth of 7 MHz when calculated by a computer program. The voltage gain is expressed in decibels (dB). The approximate equation, in terms of the circuit elements, for the *ac gain* of this circuit is:

$$A_V = \frac{R_L \times I_E}{0.026}$$ where $R_L$ = total load resistance (R1 in parallel with R5) in **ohms,**
$I_E$ = dc emitter current in **mA,** and
$I_C \cong I_E$

The equation results because the voltage gain of the amplifier is essentially the voltage gain of the transistor itself. The voltage gain of the transistor, $A_V$, can be expressed as:

$$A_V = \frac{\Delta V_C \text{ (Change in Collector Voltage)}}{\Delta V_{BE} \text{ (Change in Base-to-Emitter Voltage)}}$$

$\Delta V_{BE}$ can be approximated by the emitter current, $I_E$, flowing through an emitter resistance $r'_e$ (termed $r_e$ prime.) Therefore, $\Delta V_{BE} = r'_e \times I_E$

Since $\Delta V_C = R_L \times I_C = R_L \times I_E$ (because $I_C \cong I_E$), $A_V = \frac{R_L \times I_E}{r'_e \times I_E}$

An internal voltage of 0.026V exists across a base-emitter junction in a BJT transistor, therefore, $r'_e$ can be approximated by $r'_e = \frac{0.026V}{I_E}$, a small resistance equal to the voltage across the junction divided by the current through the junction.

Substituting for $r'_e$, the transistor voltage gain is:

$$A_V = \frac{R_L \times I_E}{\frac{0.026}{I_E} \times I_E} = \frac{R_L \times I_E}{0.026}$$

### *Example 2: Calculating the approximate gain of a small-signal amplifier*

Use the voltages and resistor values from *Figure 4-1* to calculate the amplifier gain. Since R5 is very large compared to R1, R1 in parallel with R5 will be approximately R1.

***Solution:***

Since $R_E$ = R2 = 1.2k$\Omega$

$$I_E = \frac{V_E}{R_E} = \frac{1.29}{1200} = 1.07 \text{ mA}$$

Since $R_L$ = R1 = 4700 ohms,

$$A_V = \frac{4700 \text{ X } 0.00107}{0.026} = 193 = 45.7\text{dB}$$

This value compares well with the computer-calculated gain of 181. More complete equations exist, but this approximation is good for most common-emitter amplifiers. This equation is based on the physics of BJT transistors and amplifier characteristics. Amplifiers like this are used where high gain of all types is required at moderate impedance levels.

## *Designing a voltage-divider biased NPN common-emitter amplifier stage*

Some readers may be interested in a more detailed explanation of the step-by-step design of the small-signal amplifier circuit shown in *Figure 4-1*. Here is an example of designing a voltage-divider biased NPN common-emitter amplifier stage. Use the rules for common-emitter stage biasing to approximate the operating point and *Figure 4-1* for component identification.

Given: $V_{CC} = +12V$ $\quad V_B = +2V$
$V_C = +7V$ $\quad h_{FE} \geq 50$
$I_C \cong 1mA$

1. Solve for R1 first.

$$V_{CC} - (I_C \times R1) = V_C$$
$$12V - 1mA \times R1 = 7V$$
$$12V - 7V = 1mA \times R1$$
$$\frac{5V}{1mA} = R1 = 5\ k\Omega$$

2. Choose 4.7 kΩ because it is a standard resistor value, readily available.

$$\therefore I_C = \frac{5V}{4.7\ k\Omega} = 1.06mA$$

3. Since $h_{FE} \geq 50$ (Note: ≥ means "equal to or greater than")

$$I_B = \frac{I_C}{50} = \frac{1.06}{50} = 0.021mA\ (\text{maximum})$$

4. $I_E = I_C + I_B = 1.06 + 0.021 = 1.081mA$
5. $V_E = V_{BE} - 0.7\ V = 2V - 0.7V = 1.3V$
6. $R_E = R2 = \frac{V_E}{I_E} = \frac{1.3V}{1.081mA} = 1.2\ k\Omega$

A first design approximation is to assume the voltage divider current, $I_{R3}$, is at least 10 times greater than base current. To make the bias base voltage even more stable, this design assumes $I_{R3}$ is about one-half of collector current, or 0.5mA.

7. $\therefore R3 = \frac{2V}{0.5mA} = 4\ k\Omega$

Again, choose a standard resistor value of 3.9 kΩ.

8. $\therefore I_{R3} = \frac{2V}{3.9\ k\Omega} = 0.513mA$
9. $I_{R4} = I_{R3} + I_b = 0.513\ mA + 0.021\ mA = 0.534mA$
10. $\therefore R4 = \frac{(V_{CC} - V_B)}{I_{R4}} = \frac{12V - 2V}{0.534mA} = \frac{10V}{0.534mA} = 18.7\ k\Omega$

This time, choose 20 kΩ as standard value.

$\therefore R4 = 20\ k\Omega \qquad R3 = 3.9\ k\Omega$

11. Since $I_{BIAS} >> I_b$, then

$$V_B = \frac{R3}{R3 + R4} \times V_{CC} = \frac{3.9\ k\Omega}{3.9\ k\Omega + 20\ k\Omega} \times 12V = 1.96V$$

## Common-Emitter Stage Frequency Response

*Figure 4-2* shows a computer-generated plot of the frequency response of the common-emitter amplifier we have been studying in *Figure 4-1*. Recall that frequency response is a graph of amplifier gain, $A_V$ in dB, versus frequency in Hz. The amplifier circuit has a voltage gain of about 45 dB (181 raw gain) and a bandwidth of 7 MHz. The voltage gain (the ratio of the output voltage to the input voltage of an amplifier) is expressed in decibels (dB). It is calculated as shown in Chapter 3. The gain of 45 dB is the mid-band gain between the frequencies $f_L$ and $f_H$ where $A_V$ is 3 dB less than its mid-band value. As shown, the bandwidth between $f_L$ and $f_H$ is approximately 7 MHz. Capacitors C1 and C2 provide ac coupling for the input and output, respectively, and block dc coupling between stages. C3, connected from the emitter to ground, is called a bypassing capacitor. At frequencies where the capacitive reactance ($X_C$) of C3 is less than the emitter resistor, R2, the capacitor bypasses the ac signal and removes R2 from the ac circuit. This eliminates the negative feedback that occurs at dc and increases the ac gain of the circuit. Capacitors C1, C2 and C3 affect the low-frequency response shown in *Figure 4-2* and, as a rule of thumb, the larger the capacitors the better the low-frequency response. The emitter capacitor C1 generally has the greatest affect on low-frequency response.

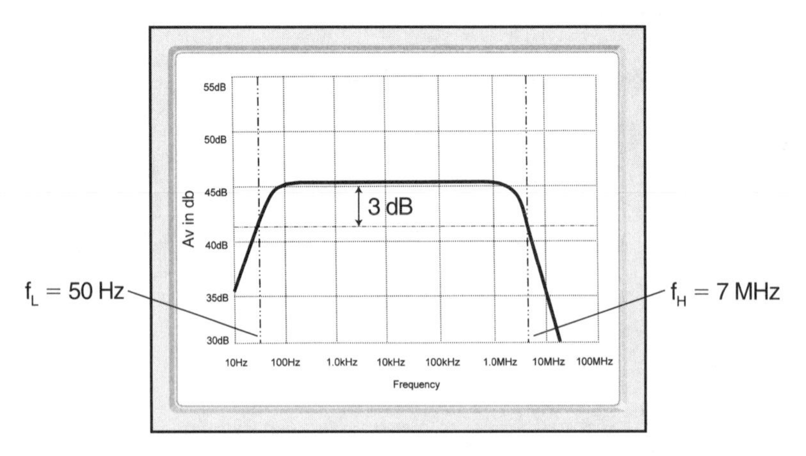

*Figure 4-2. Plot of frequency response of the common-emitter amplifier of Figure 4-1.*

# Other Common Amplifier Stages

There are two other amplifier configurations possible. They are referred to as common-collector and common-base circuits.

## Common Collector

The common-collector amplifier is called an emitter follower because signals on the base are followed at the emitter. It has no voltage gain ($A_V$ = 1), but does have current gain along with performing an impedance change. The emitter follower is used to isolate amplifier stages because it has a high input impedance and a low output impedance. In such cases, referred to as impedance matching, or impedance transformation, the amplifier acts to change the impedance levels within the system circuitry. Usually, this configuration just modifies the impedance to some acceptable level rather than achieving an exact impedance match.

## Common Base

The common-base amplifier circuit occasionally is used in radio-frequency circuits and has about the same gain as a common-emitter amplifier. The common-base amplifier is hard to drive since it normally exhibits a very low input impedance. This is because the driving circuit essentially is looking into a base-emitter diode junction.

*Figure 4-3* shows the schematics, with dc bias voltage values, for a common-collector and a common-base circuit. Even though the common-collector circuit has its collector at +12 V, the collector is still at ground for ac signals. Both of these circuits are working amplifiers and may be built and tested to help one understand more fully the operation of common-collector and common-base amplifiers.

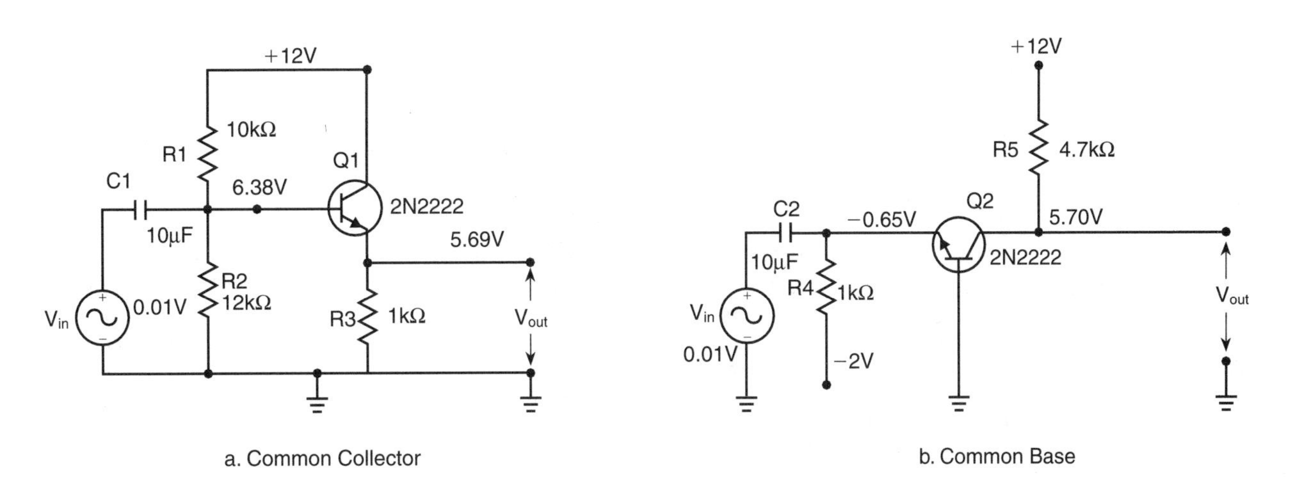


*Figure 4-3. Common-collector and common-base amplifier circuit schematics.*

## Cascaded Stages for More Gain

If more gain is required, transistor amplifiers may be connected together (cascaded) to provide more gain than is available from one stage alone. Most systems needing amplification will have multiple stages of gain and even have different types of amplifiers, including emitter-followers. *Figure 4-4* shows an example of two coupled stages. Amplifiers of this type can have voltage gains well into the thousands. When designing cascaded amplifiers, the designer must check the last stage in the chain to make sure it is not over-driven, which would cause distortion.

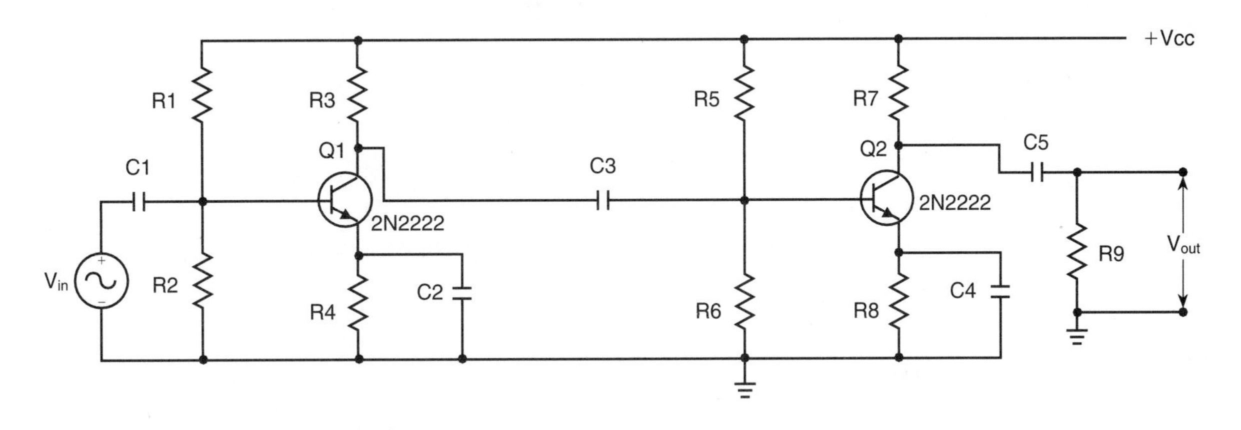


*Figure 4-4. Two-stage cascaded amplifier.*

## FETs in Small-Signal Amplifiers

As discussed previously, field-effect transistors are different than bipolar junction transistors in that they amplify voltage changes at the input into current changes at the output. That is, they depend on a voltage change between the gate and source to produce a current change between drain and source. They have a very large resistance looking into the gate of the device, which is a very useful property. *Figure 4-5* shows an N-channel depletion-mode small-signal amplifier. The JFET is biased at a drain current of

6.23 mA using a supply voltage of +12V. The source-to-ground capacitor, C1, bypasses the negative feedback that stabilizes the dc operating point. At this operating point, the JFET has a transconductance, gm, of 4,000 micromhos, now called *microsiemens.* Its gain is $A_V$ = –gm x $R_L$, the transconductance times the load resistor. Here are the advantages and disadvantages of FET amplifiers compared to BJT amplifiers for an application:

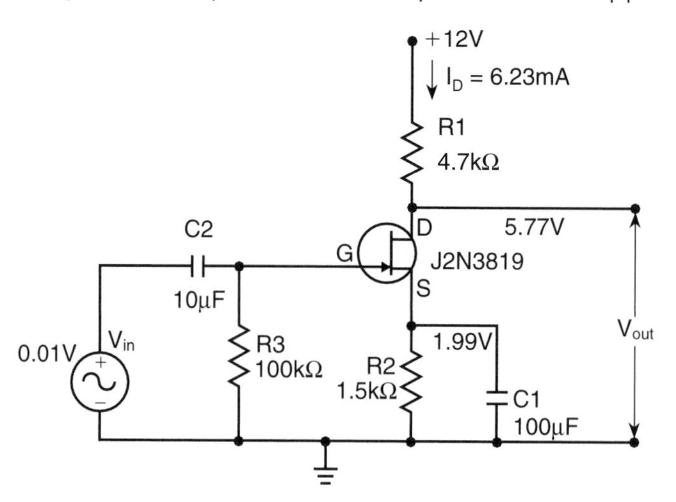


*Figure 4-5. Small-signal JFET amplifier.*

**Advantages**
- High input impedances
- Voltage inputs desirable for some circuits
- Very good in tuned circuits

**Disadvantages**
- Low gain with resistive loads
- FETs in small-signal circuits are difficult to bias because there tends to be a wide variation in device parameters. Thus, FET small-signal amplifiers are usually tuned amplifiers

### *Example 3. Calculating Voltage Gain of FET Small-Signal Amplifier*

Using the circuit of *Figure 4-5,* calculate the voltage gain of the JFET amplifier, and find the value of $V_O$.

***Solution:***

Since $R_L$ = R1 = 4700 ohms, and gm = 4000 microsiemens, the voltage gain is:

$A_V = -gm\ R_L$

$A_V = 4000 \times 10^{-6} \times 4.7 \times 10^{+3}$

$= 4.0 \times 4.7 = -18.8$ (the minus sign indicates a 180° phase shift)

Since $V_{in}$ = 0.01V (10 millivolts)

$V_O = A_V \times V_{in}$

$= 18.8 \times 0.01 = 0.188$ (188 millivolts)

This is typical of this type of JFET amplifier. The Siemen is a unit of conductance (mhos) and is the inverse of resistance. In this case, gm is a transconductance or the ratio of output $I_{ac}$ to input $V_{ac}$, expressed in (siemens).

## Amplifiers for High Frequency

Looking at the frequency response of the common-emitter amplifier shown in *Figure 4-2*, we see that it loses gain at frequencies higher than a few megahertz — 7 MHz in this case. Therefore, the usefulness of the common-emitter amplifier in higher-frequency circuits, such as those for radio and TV systems, is limited and another type of circuit is needed. So, how do we get amplifiers to operate higher in frequency?

There are special circuit designs that can handle a wide bandwidth. They are called wide-band amplifiers, or video amplifiers. Many times, the very wide-band nature of this type of amplifier makes it inappropriate for radio frequency (RF) work because, even though the frequencies are very high, the bandwidth need not be as wide. For this reason another type of small-signal amplifier is used. It is called the tuned amplifier.

## The Tuned Circuit and Resonance

Let's recall the way that the impedance of a capacitor and the impedance of an inductor vary with frequency. The capacitor has infinite impedance (it is an open circuit) at zero frequency and its impedance decreases as frequency is increased. The inductor is just the opposite — it has zero impedance (it is a short circuit if its resistance is zero) at zero frequency and its impedance increases with frequency. In *Example 1* of Chapter 3, we calculated the variation of capacitive reactance and inductive reactance with frequency. The results are plotted in *Figure 4-6* on log-log graph paper. The solid lines are for an inductance of one millihenry and a capacitance of one microfarad. The dotted lines are for an inductance of one henry and a capacitance of 0.001 microfarads. On these graphs of reactance in ohms vs. frequency in Hz, the reactance plots as a straight line. Note that there is a special point called *resonance*. It is the point where inductive reactance equals capacitive reactance. The frequency at which resonance occurs is called the *resonant frequency*, 5.0 kHz for points D and B. Since $X_L = 2\pi fL$ and $X_C = 1/2\pi fL$, $X_L = X_C$ results in the resonant frequency as $f_r = 1/2\pi\sqrt{LC}$. The derivation of $f_r$ and an example calculation are shown in *Figure 4-7c*.

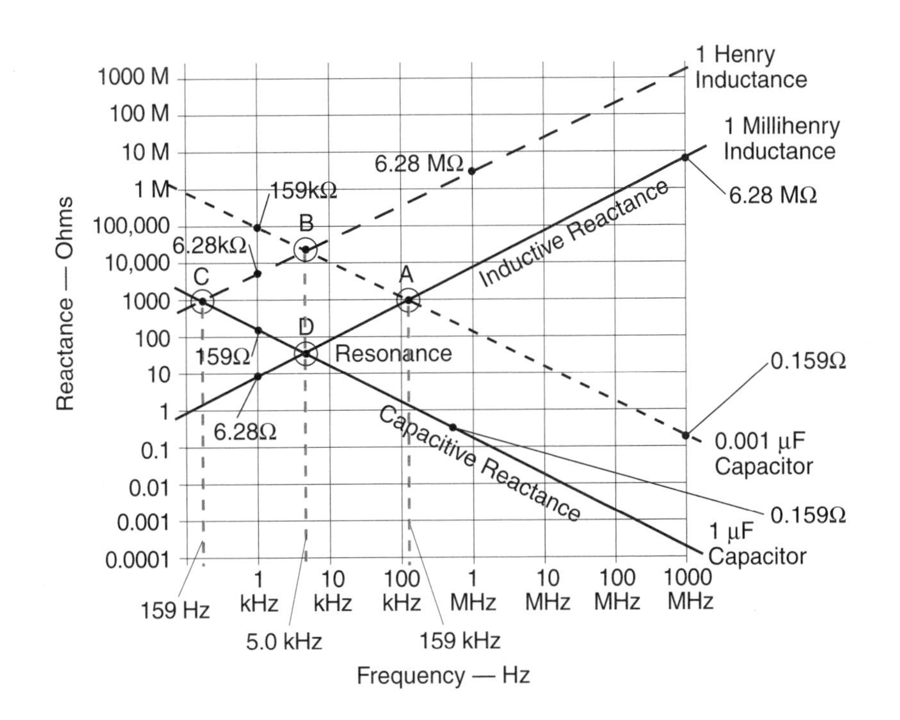


*Figure 4-6. Variation of inductance and capacitive reactance with frequency (illustration not to exact log-log scale).*

This point of resonance results in a special kind of circuit, called a tuned circuit, that is very useful in high-frequency circuits and in communications systems. A tuned circuit is shown in *Figure 4-7a*. It contains an inductance of 100 microhenries in parallel with a 0.01 microfarad capacitance driven by a voltage $V_{in}$ through a resistor R. If the amplitude of $V_{in}$ is kept constant as the frequency of $V_{in}$ is varied from below 100 kHz to above 200 kHz, the frequency response will be as shown in the computer plot in *Figure 4-7b*. The resulting frequency response is similar to the band-pass filter discussed in *Figure 2-18*, but with much sharper sides. One can see that the sharper the sides away from the resonant frequency point the greater the ability to select the resonant frequency out of other signals at other frequencies. This is the big advantage of tuned circuits — their frequency selection capabilities and the fact that they can be designed to operate in any frequency range, particularly the high-frequency bands.

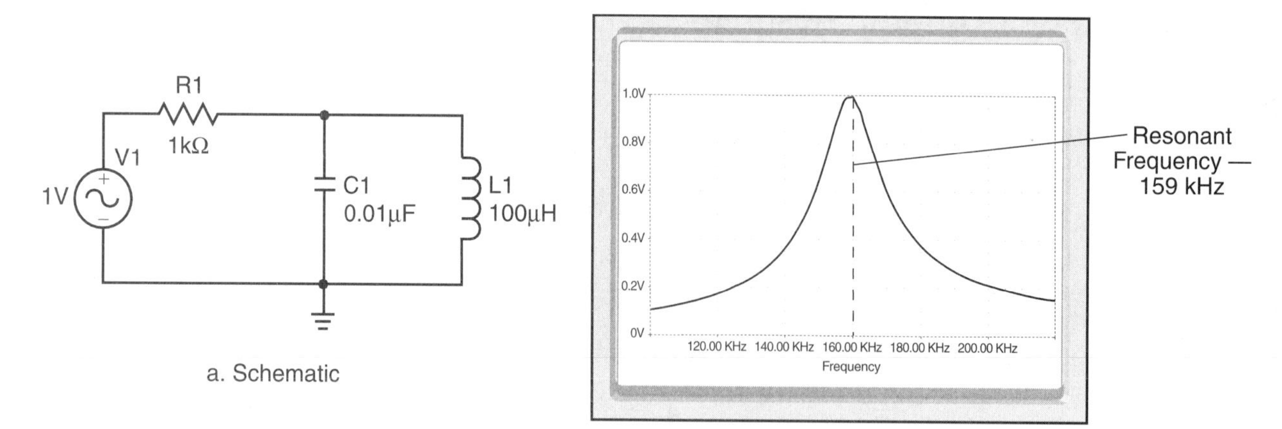


a. Schematic

b. Plot of Frequency Response

For Resonance $X_L = X_C$

$$X_L = 2\pi fL \qquad X_C = \frac{1}{2\pi fC}$$

At Resonance:

$$2\pi f_r L = \frac{1}{2\pi f_r C}$$

Solving for $f_r$:

$$f_r^2 = \frac{1}{(2\pi)^2 LC} \qquad fr = \frac{1}{2\pi\sqrt{LC}}$$

L = in **henrys**
C = in **farads**
f = in **Hz**

$f_r$ for circuit where L = 100 µH and C = 0.01 µF

$$f_r = \frac{1}{6.28\sqrt{100 \times 10^{-6} \times 0.01 \times 10^{-6}}}$$

$$f_r = \frac{1}{6.28\sqrt{1 \times 10^{-12}}} = \frac{1}{6.28 \times 1 \times 10^{-6}}$$

$f_r = 0.159 \times 10^6 = 159$ kHz

c. Calculations

*Figure 4-7. Characteristics of an LC Tuned Circuit.*

## Quality Factor, Q

There are both series resonant circuits as well as parallel resonant circuits, as shown in *Figure 4-8*. The same conditions apply for each and, as shown in *Figure 4-8*, there is a quality factor, Q, for a tuned circuit. The value of Q is the reactance at resonance divided by the resistance, $R_S$, for series resonant circuits ($X / R_S$), or the parallel resistance, $R_P$, divided by the reactance for parallel resonant circuits ($R_P / X$). Since the $X_L$ and $X_C$ are equal and cancel each other, the remaining parallel impedance at the resonant frequency is a large resistance, $R_P$, equal to Q $X_L$ or Q $X_C$. The Q of the tuned circuit, as shown, indicates the sharpness of the frequency response curve around the resonant frequency. The higher the Q the sharper the frequency selection. The sharpness of the response curve also is indicated by the bandwidth (BW), as shown in *Figure 4-8*.

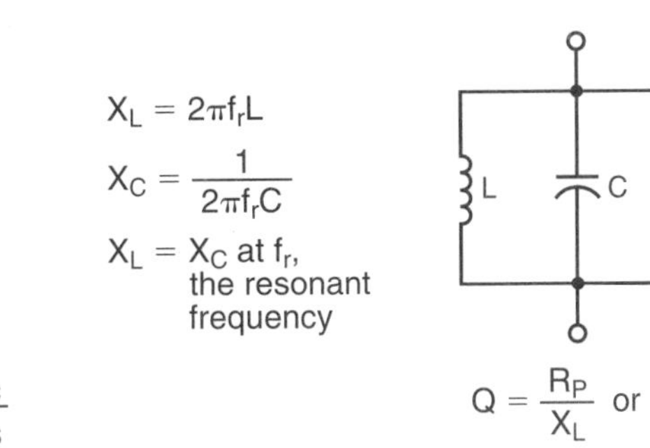


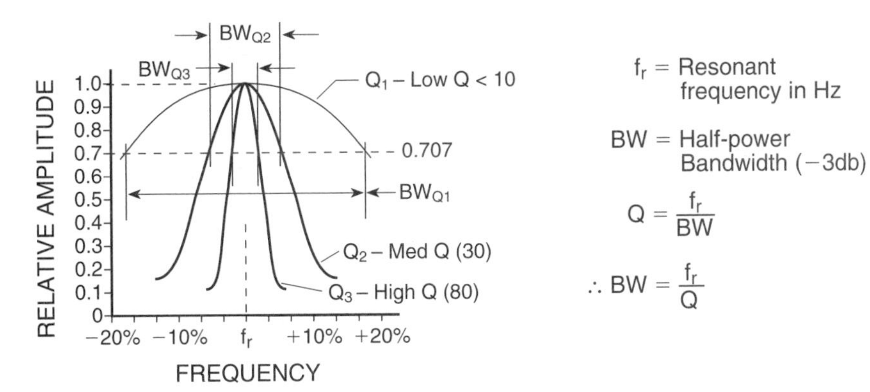


*Figure 4-8. Q is the quality factor for a tuned circuit.*
*Source: Advanced Class, 2ED, G. West, © 1992, 1995, Master Publishing Inc., Lincolnwood, IL.*

### *Example 4: Calculating Resonant Frequencies*

Calculate the resonant frequency of the following tuned circuits from *Figure 4-6*.

Use the equation for resonant frequency $f_r = \frac{1}{2\pi\sqrt{LC}}$

| | | |
|---|---|---|
| A. L = 1 mH | C = 0.001 μF | |
| B. L = 1 H | C = 0.001 μF | |
| C. L = 1 H | C = 1 μF | Remember: |
| D. L = 1 mH | C = 1 μF | $m = 10^{-3}$ |

And these two additional combinations:

| | | |
|---|---|---|
| E. L = 10 μH | C = 1 μμF | $\mu = 10^{-6}$ |
| F. L = 1 μH | C = 1 μμF | $\mu\mu = 10^{-12}$ |

***Solution:***

$$\text{A. } f_r = \frac{1}{6.28 \times \sqrt{1 \times 10^{-3} \times 1 \times 10^{-9}}} = \frac{1}{6.28 \times 1 \times 10^{-6}} = 0.159 \times 10^{+6} = 159 \text{ kHz}$$

$$\text{B. } f_r = \frac{1}{6.28 \times \sqrt{1 \times 1 \times 10^{-9}}} = \frac{1}{6.28 \times \sqrt{10 \times 10^{-10}}} = \frac{1}{6.28 \times (3.163 \times 10^{-5})} =$$

$$\frac{1}{19.86 \times 10^{-5}} = 0.0503 \times 10^{+5} = 5.03 \text{ kHz}$$

$$\text{C. } f = \frac{1}{6.28 \times \sqrt{1 \times 1 \times 10^{-6}}} = \frac{1}{6.28 \times 1 \times 10^{-3}} = 0.159 \times 10^{+3} = 159 \text{ Hz}$$

$$\text{D. } f_r = \frac{1}{6.28 \times \sqrt{1 \times 10^{-3} \times 1 \times 10^{-6}}} = \frac{1}{6.28 \times \sqrt{10 \times 10^{-10}}} = \frac{1}{6.28 \times (3.163 \times 10^{-5})}$$

$$\frac{1}{19.86 \times 10^{-5}} = 0.0503 \times 10^{+5} = 5.03 \text{ kHz}$$

$$\text{E. } f_r = \frac{1}{6.28 \times \sqrt{10 \times 10^{-6} \times 1 \times 10^{-12}}} = \frac{1}{6.28 \times \sqrt{10 \times 10^{-18}}} =$$

$$\frac{1}{6.28 \times (3.163 \times 10^{-9})} = \frac{1}{19.86 \times 10^{-9}} = 0.0503 \times 10^{+9} = 50.3 \text{ MHz}$$

$$\text{F. } f_r = \frac{1}{6.28 \times \sqrt{1 \times 10^{-6} \times 1 \times 10^{-12}}} = \frac{1}{6.28 \times \sqrt{1 \times 10^{-18}}} = \frac{1}{6.28 \times 10^{-9}} =$$

$$0.159 \times 10^{+9} = 159 \text{ MHz}$$

The resonant frequency points for A, B, C, and D are shown in *Figure 4-6*.

## An FET Tuned-Circuit Amplifier

*Figure 4-9* shows an LC-tuned amplifier using a junction FET as the active device. The circuit is made up of an inductance, capacitors, a transistor, and resistors. The tuned circuit of L1 and C1 is the load for the FET, which supplies the signal and, at resonance, the load is $R_P = QX_L$ in parallel with R3. This circuit is designed to act as a band-pass filter, amplifying signals within the passband and attenuating signals outside the passband.

The tuned-circuit amplifier offers a way to achieve gain and a particular response at high frequencies. The circuit frequency response will have sloping vertical sides similar to *Figure 4-7*. The amplifier circuit passband can be controlled by the design of the resonant circuit as long as the transistor has the necessary gain. Tuned-circuit amplifiers can be designed over almost any portion of the radio frequency spectrum, even up to the gigahertz (GHz) region, which is one thousand million cycles. However, the tuned circuit's performance will be severely limited if the transistor cannot operate at the design frequency.

The RF tuned amplifier circuit shown in *Figure 4-9* has a resonant frequency of 10 MHz. The active device is an mpf102 N-channel junction FET that operates in the depletion mode. The FET is chosen because of the ease of design. Supply power and biasing are straight-forward. Input impedance is high, making it easy to drive. Since this is a depletion-mode device, there is current from drain to source with zero volts between gate and source. The voltage across R1 provides self-bias so that the drain to source current is adjusted below its maximum value and in a linear operating range on the load line provided by the tuned circuit. At resonance, this load is R3 (10 kΩ) in parallel with $QX_L$. Input signal variations change the current from source to drain and the amplification of the output voltage is according to the frequency response of the tuned circuit.

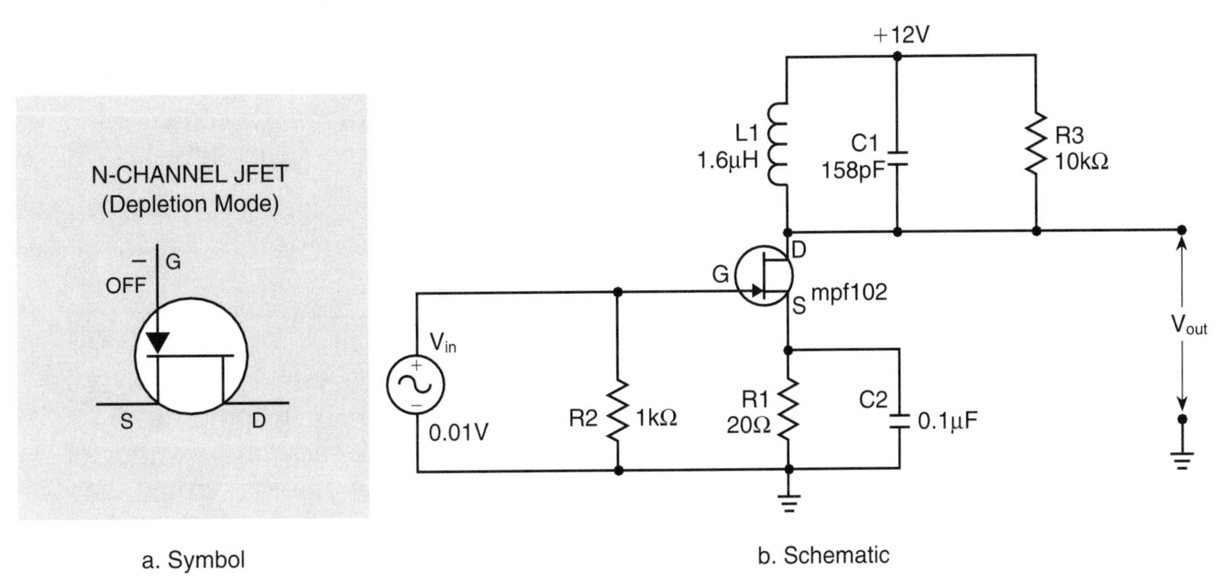

*Figure 4-9. RF tuned amplifier circuit.*

## Gain Equals –gm x $Z_L$

*Figure 4-10* is a computer-generated frequency response curve of the circuit in *Figure 4-9*. This plot shows that the tuned amplifier has a voltage gain of 30.0 dB, a bandwidth of 100 kHz, and the Q of the circuit is 100. The gain of this type of stage can be approximated using the equation $A_V = -gm \times Z_{load}$. The quantity gm, the transconductance (sometimes called mutual transconductance), is specified in the manufacturer's data sheets for FETs. As discussed previously, gm has values of a few thousand microsiemens. Also, as discussed previously, the Siemen (symbol S) is a unit of conductance and is the inverse of resistance. $Z_{load}$ is the net impedance seen by the drain circuit. At 10 MHz, $X_L$ = 100.5 ohms, and $R_P = QX_L$ or 10,000 ohms. Therefore, $Z_L$ is the parallel combination of 10,000 ohms and 10,000 ohms, or 5,000 ohms.

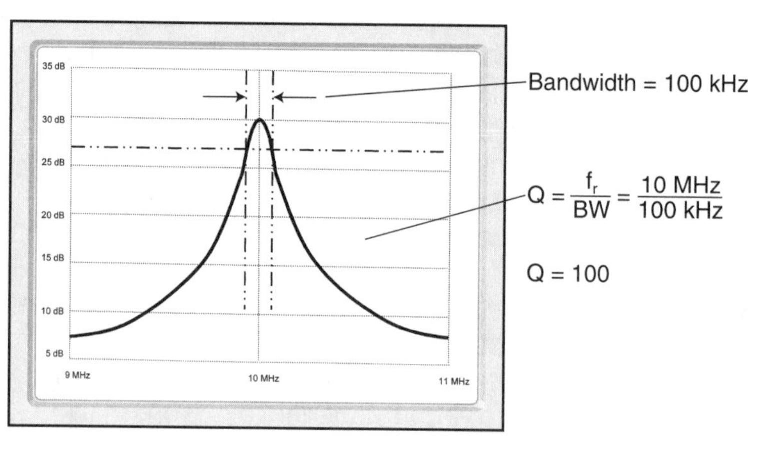


*Figure 4-10. Computer-generated frequency response curve for JFET tuned amplifier.*

## BJT Tuned-Amplifiers

While bipolar junction transistors are used extensively in tuned circuits, such circuits are a bit more difficult to design due to smaller impedance levels within the circuits. Since the BJT requires more impurity diffusions into the silicon substrate to form the necessary diode junctions, the device uses more silicon area. Thus, the frequency response of the transistor itself may limit its performance compared to the much smaller geometry FET, especially at higher frequencies. However, the high current gain of a BJT compared to a lower gm for an FET may win designs for the BJT, especially when driving lower impedances.

## General Purpose Analog Amplifiers

In communications systems there are a number of general-purpose amplifiers used for a variety of applications. One such device is the operational amplifier — commonly called the "op-amp." The op-amp is a dc-coupled amplifier that receives its name from its use in analog computers to provide scaling, computation, and isolation in computer networks used to solve real-time control problems. *Figure 4-11a* shows the symbol for and operating characteristics of a general-purpose op-amp. The op-amp typically has two inputs into the flat side of the triangle commonly called the inverting and non-inverting inputs, respectively. The output is at the point of the triangle. A positive-going input signal at the inverting input will produce a negative-going output. The output will be out-of-phase (180° phase shift) from the input signal. Conversely, a positive-going input into the non-inverting input will produce a positive-going output. The output is in-phase (0° phase shift) with the input. In the case of the inputs, there is no in-between phase shift — either the input produces an output that is in phase or an output that is 180° out of phase.

## Ideal Op-Amp vs. IC Op-Amps

*Figure 4-11b* lists the characteristics of an ideal op-amp. It has an input impedance, $Z_{IN}$, of infinity. It has a gain of infinity. It's output impedance is zero. It has an infinite bandwidth and the circuit is perfectly balanced so that there is no DC offset between $V_{IN}$ and $V_O$. IC op-amps do not meet these ideal characteristics completely, but come very close in practical applications. For example, voltage gains for some devices is in the millions, and a few hundred thousand is very common. As a result, only a very very small voltage will drive the output to its limit (called saturation). The op-amp circuit normally operates with negative feedback so the circuit uses only a portion of the available gain.

If one were to measure the impedance across the input terminals of an IC op-amp, it would be a high resistance. This high impedance is a very desirable characteristic because a circuit with high input impedance does not load the circuit that drives it. The IC op-amp output, on the other hand, will exhibit low output impedance (low output resistance), which is desirable because the circuit can deliver current into a load with little loss in output. The actual input and output impedances depend upon the circuit design for a particular application. While the ideal op-amp has an infinite frequency response (infinite bandwidth), practical op-amps are much more limited in this characteristic. Many of them are limited to low-frequency operation because of their IC construction. However, latest small-geometry techniques for ICs have increased the frequency response dramatically. No offset voltage between $V_O$ and $V_{IN}$ also is an ideal op-amp characteristic. Since op-amps are dc amplifiers, whatever dc level is across the inputs will produce a signal at the output. If $V_{IN} = 0$ and there is a $V_O$ output, this is called an offset. In IC op-amps, terminals are usually provided to inject an input current to counteract and adjust any offset so that $V_O = 0$ when $V_{IN} = 0$.

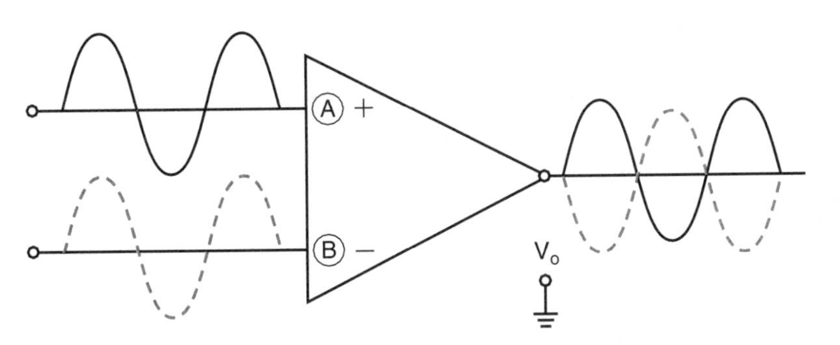


(A) Non-Inverting Input.
A positive voltage on this input will produce a voltage at the output that is in-phase (is more positive).

(B) Inverting Input.
A positive voltage on this input will produce a voltage at the output that is out of phase (is less positive).

a. General-Purpose Operational Amplifier

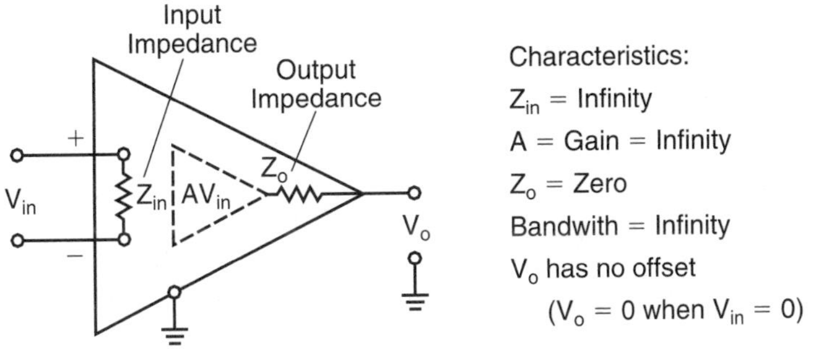


b. Ideal Operational Amplifier

*Figure 4-11. Operational amplifiers (op-amps).*

## Typical Audio Op-Amp

A typical op-amp is shown in *Figure 4-12*. It has a wide range of uses in analog electronic applications, mostly as an amplifier in the audio range where signals are fairly large, or in low-frequency oscillator circuits. *Figure 4-12a* shows the terminal connections schematically, while *Figure 4-12b* shows the IC package. The MC1741 op-amp operates from a plus and minus power supply (+/- 12 volts), as do most op-amps, and *Figure 4-12c* shows a common circuit for providing the power. The middle of the power supply is the reference ground. The op-amp circuit receives its ground reference through an input resistor returned to ground. Note the two extra leads, $OS_1$ and $OS_2$. They are used to balance out the offset. Not all circuits require that the offset be balanced.

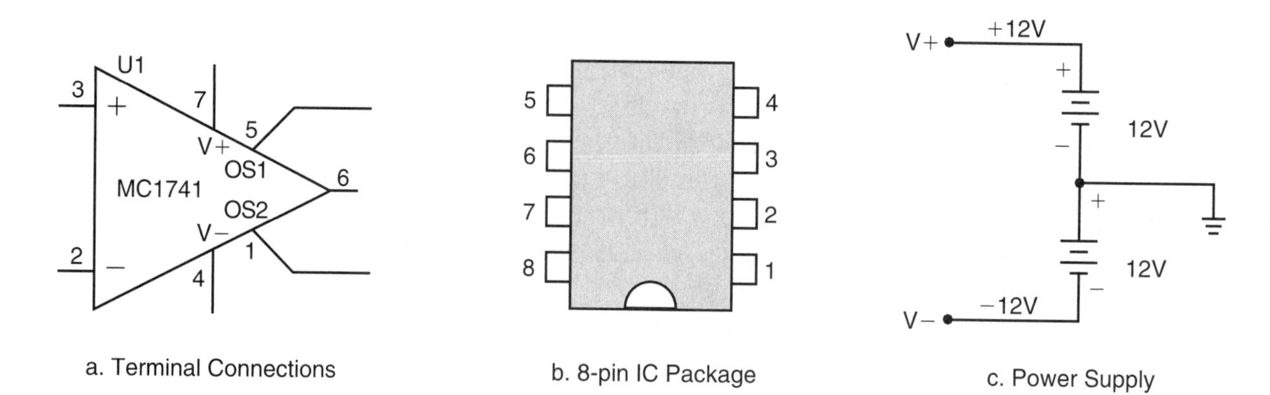


*Figure 4-12. A MC1741 Op-Amp.* *(Courtesy of Motorola, Incorporated.)*

## Op-Amp Circuits

The standard configurations of the op-amp, when used as a voltage amplifier, will result in a linear (analog) amplifier. Whatever the input signal waveform, the output waveform will be a larger carbon copy. An important design consideration is that the amplifier not introduce a large amount of distortion when used in a small-signal application.

The two most common op-amp circuits for voltage amplifiers are shown in *Figure 4-13. Figure 4-13a* is an inverting amplifier and *Figure 4-13b* is a non-inverting amplifier. Note that these amplifiers have a resistor, $R_f$, from the output to the inverting input that supplies negative feedback. Negative feedback is produced when some of the output signal is reintroduced back to the input out of phase with the normal input. This feedback signal counteracts the input signal to reduce and set the overall voltage gain of the amplifier. Negative feedback increases the stability and the frequency response (bandwidth) of the circuit and improves (reduces) distortion.

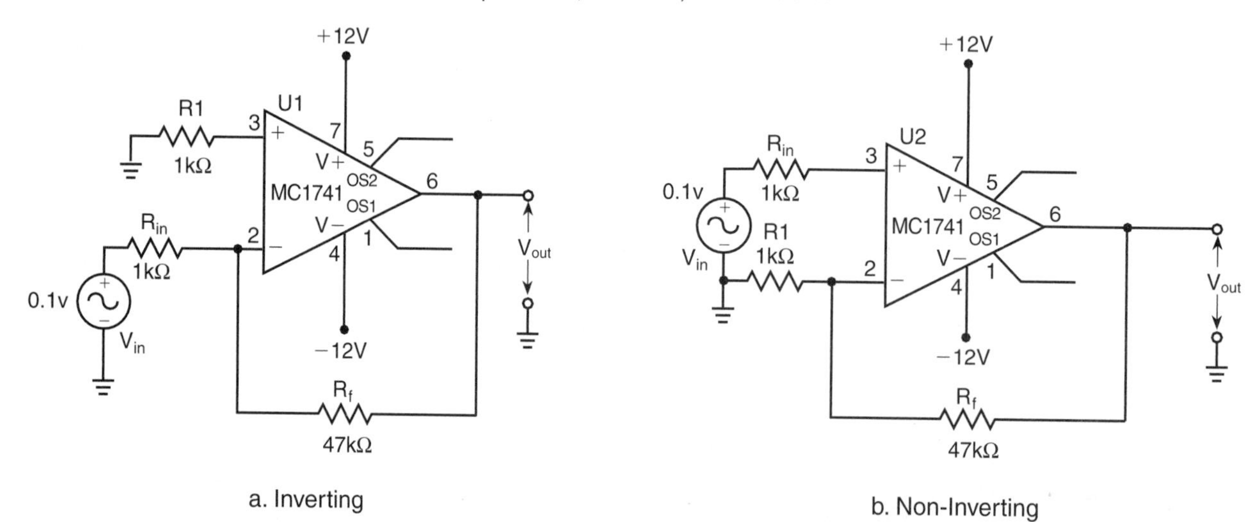

*Figure 4-13. Common Op-Amp voltage amplifiers.*

With negative feedback, and the fact that the op-amp has such a high gain without feedback, the circuit gain of the amplifier can be set very accurately by selecting the ratio of two external resistors. Here are the equations used to calculate the gain, $A_V$, for the circuits shown in *Figure 4-13:*

**Inverting**

$$A_V = -\frac{R_f}{R_{in}}$$

**Non-Inverting**

$$A_V = 1 + \frac{R_f}{R_{in}}$$

The minus sign indicates a 180° phase shift.

### *Example 5: Calculating Op-Amp Gain*

Using the circuit schematics of *Figure 4-13,* calculate the gain of the two amplifiers.

***Solution:***

**Inverting**

$$A_V = -\frac{R_f}{R_{in}} = \frac{47000}{1000} = -47$$

**Non-Inverting**

$$A_V = 1 + \frac{R_f}{R_{in}} = 1 + \frac{47000}{1000} = 48$$

The voltage gain equations are only as accurate as the resistor ratio is controlled. If 1% precision resistors are used, the gain is more accurate than when 20% resistors are used. The limitation on these equations is that the closed loop voltage gain must be much smaller than the "open loop" gain — the gain of the op-amp without feedback. Common voltage gains with feedback are a few hundred or less, depending upon the design. The circuits in *Figure 4-13* can be easily built to verify the gain. Using negative feedback, here are general results you can expect as the amount of feedback is increased. The greater the amount of feedback:

a) the lower the gain,
b) the lower the distortion, and
c) the greater the bandwidth.

*Figure 4-14* is a computer-generated plot of the frequency response of the MC1741 op-amp circuit of *Figure 4-13a*. Since this is an inverting amplifier, the gain is really -47, where the minus sign means a 180° phase shift. The plot is the customary way of presenting amplifier frequency response — frequency along the Y axis and gain in decibels along the X axis. Voltage gain in decibels is calculated using the equation $A_V = 20 \log_{10} (V_{OUT} / V_{IN})$. $V_{IN}$ is kept constant as frequency is varied. The point marked 3 dB in *Figure 4-14* is the special point that defines the bandwidth. It is the point where the voltage gain is reduced to 0.707 of its mid-band value (say at 3 kHz). This amplifier has a bandwidth of 20 kHz.

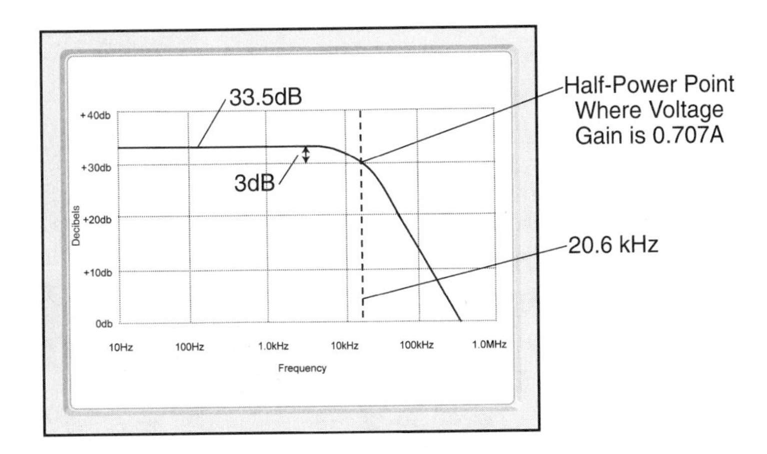


*Figure 4-14. Frequency response plot of the inverting MC1741 op-amp of Figure 4-13.*

Op-amps come in numerous types and performance ranges, and electronic designers will select the particular op-amp that best fits their application. Variations of op-amp circuits are used extensively for signal-processing in the low-frequency range, such as filtering and non-linear gain functions. Others are used for active filter design, where the op-amp is the active device in the circuit. One of the most interesting op-amp design variations is the LM3080, which provides a variable gain block. A pin on the device allows the gain to be adjusted electronically with a voltage level rather than with a component value change. This device will be used in Chapter 5 to demonstrate AM modulation.

## Power Amplifiers

Communications systems, as well as other analog systems, use large numbers of power amplifiers. Unlike the small-signal amplifier, the power amplifier must be able to handle large currents and voltages simultaneously. Both BJT and FET devices are used in power amplifier circuits, and specially-designed FETs that provide power gain have found wide acceptance. However, at power output levels above a few kilowatts, vacuum tubes are still the devices of choice for new designs because cooling semiconductor devices to handle heat dissipation adds too much cost to a system. Power amplifiers are grouped into four basic classes according to their bias point on the load line that we originally discussed in *Figure 3-7* — classes A, AB, B, and C, and repeated again pictorially in *Figure 4-15*. Class AB has an operating point between class A and class B.

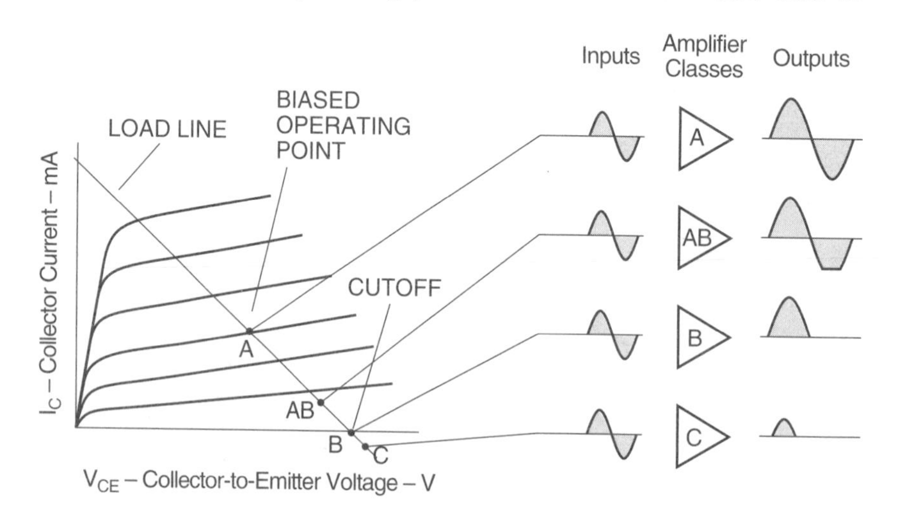


*Figure 4-15. Bias points of various classes of transistorized power amplifiers.*
*Source: Advanced Class, 2nd Edition, G. West, © Copyright 1992, 1995 Master Publishing, Inc.*

## Class-A Amplifiers

The small-signal amplifiers that were discussed are class-A amplifiers. They operate at point A and are completely linear. If the current and voltage levels in a small-signal amplifier are increased, it becomes a power amplifier, but it is very inefficient (usually less than 25%), so this approach is rarely taken. As discussed in *Figure 3-7*, and shown again in *Figure 4-15*, the input signal continuously varies the input current over the complete 360° of the sine-wave signal.

## Class-B Amplifiers

The class-B amplifier also is a linear amplifier, but with the biasing at point B, as shown, it operates only over one-half an input signal cycle (only for 180°). As a result, circuits using class-B amplifiers almost always use two devices. *Figure 4-16* shows two types of class-B amplifier circuits. The one in *Figure 4-16a* is called a complementary-symmetry amplifier; the one in *Figure 4-16b* is called a push-pull amplifier.

### *Complementary-Symmetry Amplifiers*

The complementary-symmetry amplifier uses similar NPN and PNP bipolar transistors. Each transistor operates for half the input cycle, meaning there is input current only half the time in each class-B device. With no input signal, the points X and Y are at the same voltage. The base of Q1 is a diode junction voltage 0.7V more positive than point X, and the base of Q2 is a diode junction voltage 0.7V more negative than point X. Both transistors are right at the verge of conducting. When an input signal drives point X more positive, the increase in voltage is transferred through the forward-biased diode, D1, to the base of Q1 causing Q1 to conduct and raise the output voltage more positive. Since X is driven positive, the base-emitter voltage of Q2 becomes too small for conduction and is turned off. Therefore, Q1 is the only device contributing to the output voltage during the positive-going half cycle of the input signal.

The conditions are reversed when point X is driven negative by the input signal. Q2 conducts and Q1 is turned off. Q2 is the only device contributing to the output voltage. The current through R1, D1, D2 and R2 from $+V_{CC}$ to $-V_{CC}$ keeps the diodes forward-biased. The circuit needs plus and minus power supplies. While the overall circuit is still linear, each device still operates as class B. This type of amplifier finds most of its uses in applications such as audio amplifiers and low-frequency drive circuits for industrial systems. A complementary-symmetry amplifier only has a voltage gain of 1 since each transistor is essentially operating as an emitter-follower, but it does provide a large current gain and thus a power gain.

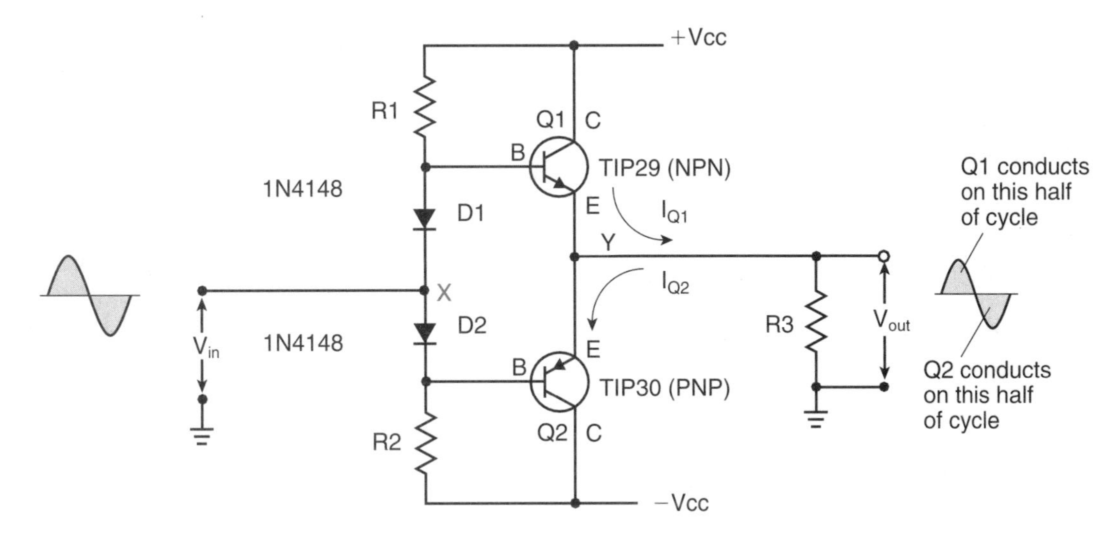

a. Complementary-Symmetry Power Amplifier

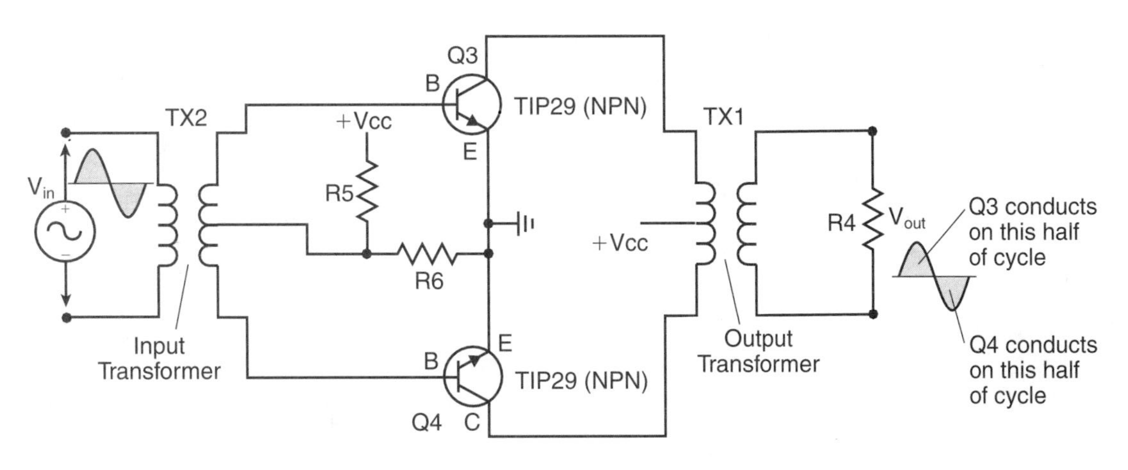

b. Push-Pull Amplifier

*Figure 4-16. Class-B power amplifiers.*

## *Push-Pull Amplifiers*

The push-pull amplifier shown in *Figure 4-16b* uses transformer coupling at both its input and its output. This type of amplifier will have a voltage gain greater than 1 and also exhibit current gain. As with the complementary-symmetry amplifier, each transistor operates for only one-half of the input signal cycle. When the input signal is positive, transformer action produces a positive voltage on the base of Q3 and causes it to conduct. The changing current in the collector is transferred through the output transformer to the load R4. A negative input signal causes Q4 to conduct and its changing current in the output transformer produces an output voltage across R4. The bias provided from $+V_{CC}$ through R5 and R6 to the bases of Q3 and Q4 is such that Q3 and Q4 are just at the verge of conduction with no signal. This amplifier uses two of the same type transistors and only a positive voltage power supply.

## Class-C Amplifiers

Referring again to *Figure 4-15*, class-C amplifiers use a transistor that is biased below cutoff (point C). The transistor does not conduct current until the input signal drives it above the cutoff point (point B). Since the transistor used in a class-C amplifier is biased below cutoff, there will only be collector current during the time that the input drives the transistor into conduction. The input current and the collector current will flow for less than one-half an input signal cycle (<180°). The class-C amplifier is not a linear amplifier since the collector current output will look like "spurts" of current even if the input voltage waveform is a sine wave.

The amplifier normally has a tuned resonant circuit to couple to the load. One of the characteristics of a tuned circuit is that, if it is driven by spurts of current, it will have a slowly-decreasing sine wave at the frequency of the input signal. It delivers power to the load when the transistor is not conducting. Each spurt of current will replace the lost energy and the output sine wave will appear to be constant due to the characteristics of the tuned circuit. A class-C amplifier usually is designed to operate at high frequencies and over a small band of frequencies. A class-C amplifier is not able to follow the amplitude changes of the input waveform; therefore, it is not good as a linear amplifier, but it does find many applications as an FM amplifier. It also is used in AM modulation circuits, but will not amplify an AM modulated carrier once the modulation is produced. Class-C amplifiers typically have high power gain and are very efficient.

## *A Bipolar Class-C Amplifier*

*Figure 4-17* is a class-C amplifier circuit using a bipolar transistor — a 2N2219. The circuit illustrates the circuit techniques, but does not produce high power due to the limitations of the transistor used to demonstrate the operation. Choosing a proper device for an application will produce the power output required. The LC (inductance-capacitance) tuned circuit acts as a transformer to convert the actual load to a lower impedance so the amplifier can deliver more power. The tuned circuit resonates and stores energy at the frequency of the collector current spurts, in this case, 10 MHz. As explained, it supplies power to the load when the transistor is not in conduction. The maximum power available from an amplifier of this type is: $P_O = (V_{CC})^2 / (2 \times R_L)$. If the collector did not have the inductor going to the power supply as part of the load, the power would be limited to: $P_O = (V_{CC})^2 / (4 \times R_L)$. Because of the characteristic of inductors that generates a counter voltage within it as the current through it changes, an inductive load produces twice the voltage swing in the collector circuit as a resistor. $V_{CC}$ is the dc power supply voltage and $R_L$ is the actual transformed load on the transistor collector. It differs from the actual load resistor because of the transformer action of the LC tuned circuit. The calculated power using the equation is the maximum available. In practice, amplifiers do not achieve quite enough voltage swing to have the maximum indicated power out.

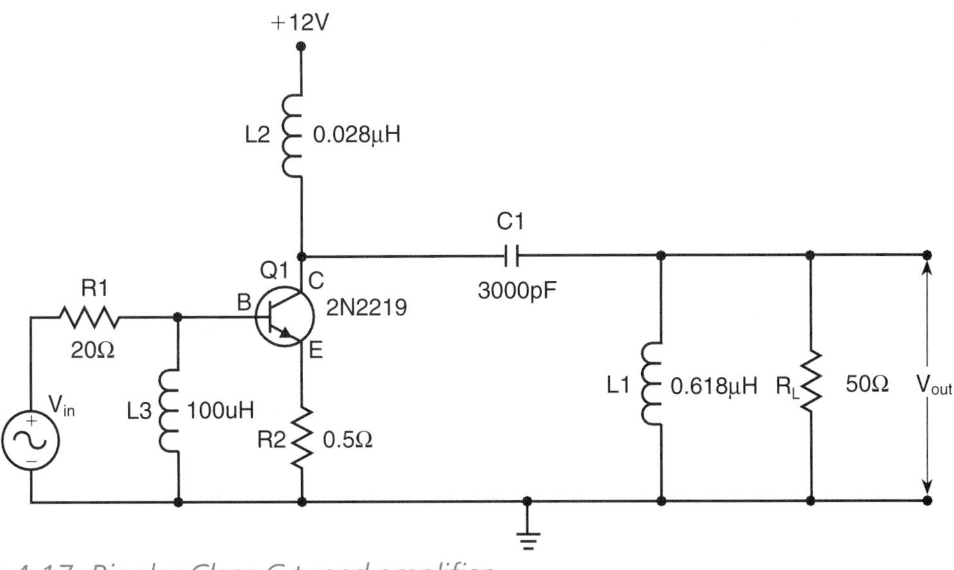


*Figure 4-17. Bipolar Class-C tuned amplifier.*

### *Example 6: Calculate the Power Output of a Class-C Power Amplifier*

Use the circuit shown in *Figure 4-17* for values and calculate the maximum power output. Assume a net load on the collector of 25 Ω.

***Solution:***

$$\text{Since } V_{CC} = +12\text{V and } R_L = 25\ \Omega,\ P_{MAX} = \frac{12^2}{(2 \times 25)} = 2.88 \text{ watts}$$

Some of the waveforms from the amplifier are shown in *Figure 4-18*. This is a typical example of a class-C amplifier. Note that the collector current flows in "pulses" while the output across the load is almost a pure sine wave. The active device is only being turned on during the pulse time. The small glitches in the output waveform amount to distortion and are the point at which the transistor switches on and off. The magnitude of the voltage across the load resistor is almost 70 volts peak-to-peak, indicating the transformer action of the resonant circuit. The same type of performance is obtained using power FETs as well as vacuum tubes. Since the vacuum tube is a high-voltage device, the tuned circuits will differ slightly but the principle is the same. This type of class-C amplifier finds wide use in industrial power applications and in communication systems transmitters where large amounts of RF power are required. Using vacuum tubes, power amplifiers are built that have RF power outputs as large as one megawatt.

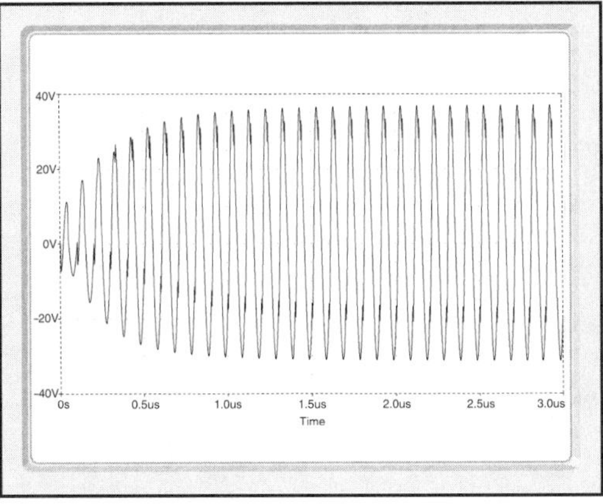


a. Output Voltage (10 MHz)

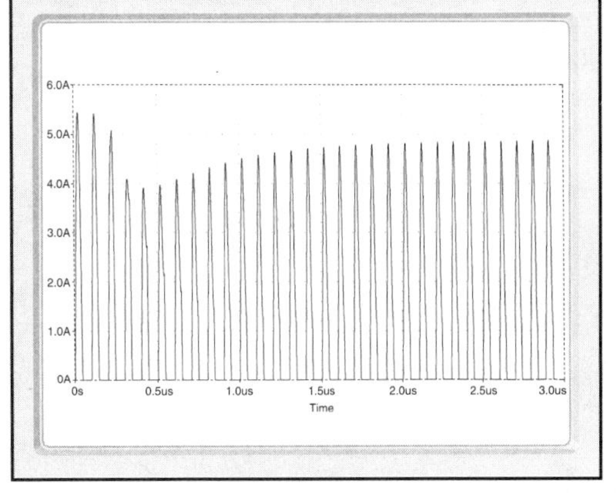


b. Collector Current

*Figure 4-18. Waveforms from the Class-C amplifier.*

## Linear RF Amplifiers

In many RF applications, a linear amplifier is required that will pass the input signal to the output as a near-carbon copy but at a higher power level. This design is either a class-B amplifier using similar circuits to those in *Figure 4-16*, except designed for RF frequencies, or a class-AB tuned amplifier. Most common are the class-AB amplifiers, which look very similar to a class-C amplifier, except that the active device is biased so that it always conducts a small amount of current (point AB on *Figure 4-15*). Class-AB amplifiers are found in SSB (single-sideband) systems, which require linear amplifiers since the SSB modulation is produced at a very low power level.

# The Amplifier's Close Friend — The Oscillator

As discussed in Chapter 2, the oscillator and amplifier are closely related electronically. The purpose of the oscillator is to create an ac waveform of constant or varying frequency by converting energy from a dc power supply. The waveform often is a sine wave in communications work, but other periodic waveforms may be generated. Recall that an oscillator is really an amplifier with some of the amplifier's output fed back to the amplifier's input, but in a different manner than that described for the op-amp. In the op-amp, *negative* feedback is out of phase with the input in order to stabilize the amplifier. In the oscillator, *positive* feedback is in phase so that the output reinforces the input, which causes the circuit to oscillate at a given frequency. The oscillation occurs at the frequency where the total circuit phase shift is 360° and the system has enough gain (at least a net gain just greater than 1) to overcome any losses. The basic statement that an oscillator will oscillate when the total phase shift from input to output back to input is 360° and the system gain is equal to or greater than 1 is call the Barkhausen Criterion, after Heinrich Barkhausen, a German physicist.

## Phase-shift Oscillator

*Figure 4-19a* is an example of a phase-shift oscillator circuit using an LF411 FET op-amp as the active device. The three-section resistor-capacitor network contributes 180° phase

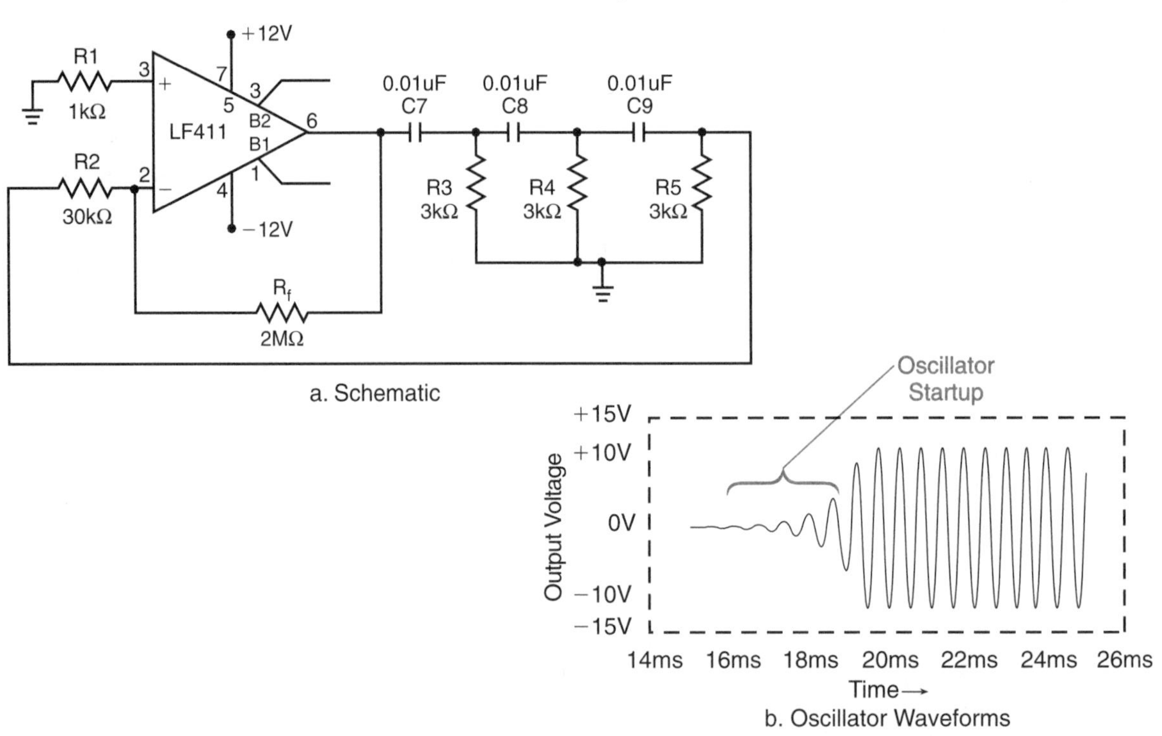


*Figure 4-19. A phase-shift oscillator circuit using an LF411 FET Op-Amp.*

shift and the amplifier contributes 180°. The network can actually contribute more than 180°, but the circuit will oscillate at the point where the total phase shift is 360° and the gain is greater than one. Because the phase shift through the RC network is 180°, the frequency of oscillation can be approximated using the equation f = 1 / 15.2 RC. R and C are the component values in the phase-shift network, often called a β (Beta) network. The approximation is close enough for most applications, and the oscillator will provide a good sine wave output with careful component selection. *Figure 4-19b* is a computer-generated plot of the phase-shift oscillator just as it begins to oscillate.

### Example 7: Calculating Phase-Shift Oscillator Frequency

Using the component values of *Figure 4-19a*, calculate the oscillator's frequency.

**Solution:**

$$f = \frac{1}{15.2\ RC} = \frac{1}{15.2 \times (3 \times 10^{3}) \times (0.01 \times 10^{-6})} = \frac{1}{15.2 \times (3 \times 10^{-5})} =$$

$$\frac{1}{45.6 \times 10^{-5}} = 0.02193 \times 10^{+5} = 2193\ \text{Hz}$$

## IC Oscillators

There are many integrated circuit oscillators available, and *Figure 4-20* shows three different types. One of the most used circuits is the 555 timer circuit when fairly low-frequency timing waveforms are required. It is called a one-shot multivibrator and its output is not a sine wave but a pulse that has a variable time depending on the values in an external RC coupling network. The 8038 function generator is popular for use as a

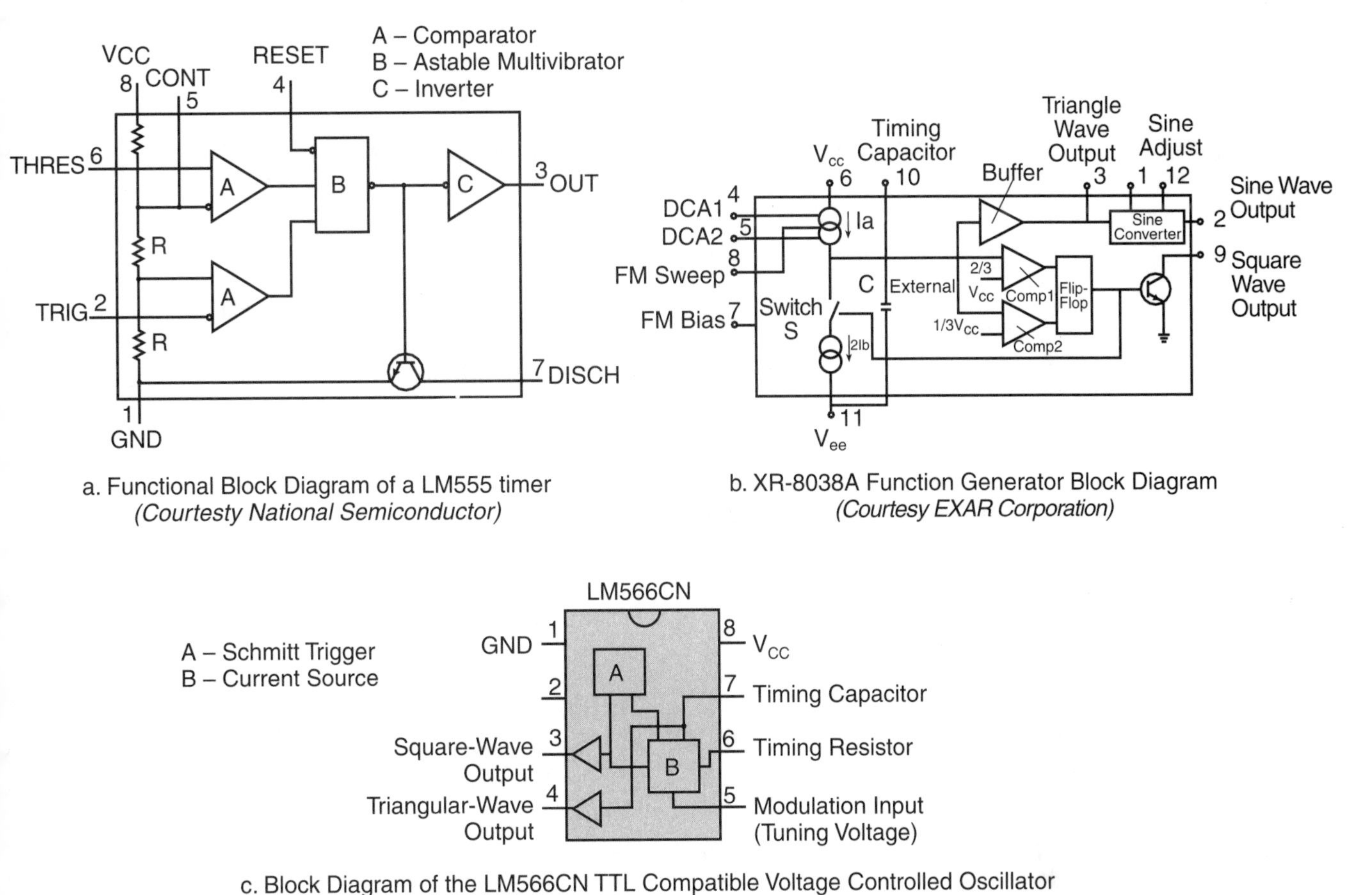


a. Functional Block Diagram of a LM555 timer *(Courtesty National Semiconductor)*

b. XR-8038A Function Generator Block Diagram *(Courtesy EXAR Corporation)*

c. Block Diagram of the LM566CN TTL Compatible Voltage Controlled Oscillator *(Courtesty National Semiconductor)*

*Figure 4-20. Different types of IC oscillators.*

sine-wave oscillator where the frequency can be varied easily by varying an external resistor. The 566 voltage-controlled oscillator (VCO) also is widely used. It allows the frequency of the oscillator to be changed electronically with a tuning voltage. VCO's also are built with discrete parts when the frequency gets high enough so that IC's can no longer operate.

## Discrete Component Oscillators

At higher frequencies, discrete active devices and components often are used to construct oscillators. The two RF oscillator circuits shown in *Figure 4-21a* and *b* are fundamental types referred to as Colpitts and Hartley oscillators, respectively. They are usually built using transistors but, in some cases, may use special ICs. These oscillators use a tuned resonant circuit to set the frequency. Their stability is limited by the variability of the components used to construct the circuit. The Colpitts circuit uses two capacitors that are split to provide the feedback, and the Hartley uses two inductors. The oscillators can be tuned to frequency by varying the inductances or capacitances. The Colpitts tends to be a favored circuit because variable capacitors are easier to use for tuning than inductors. When the frequency can be changed, these circuits are referred to as variable frequency oscillators (VFO's). Both the Colpitts and the Hartley will oscillate at approximately the calculated resonant frequency of the tuned circuit: $f = 1/2\pi\sqrt{LC}$.

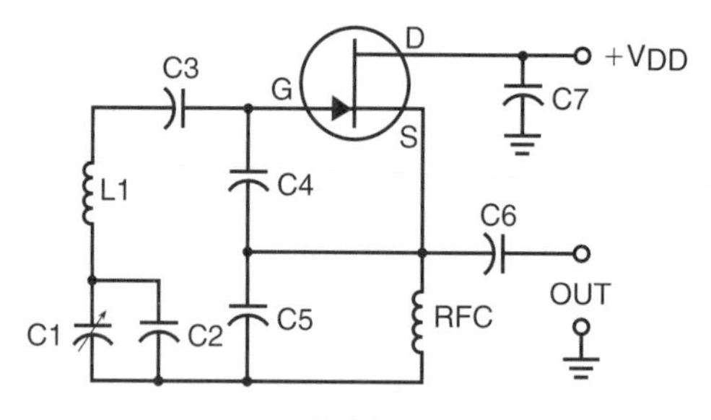


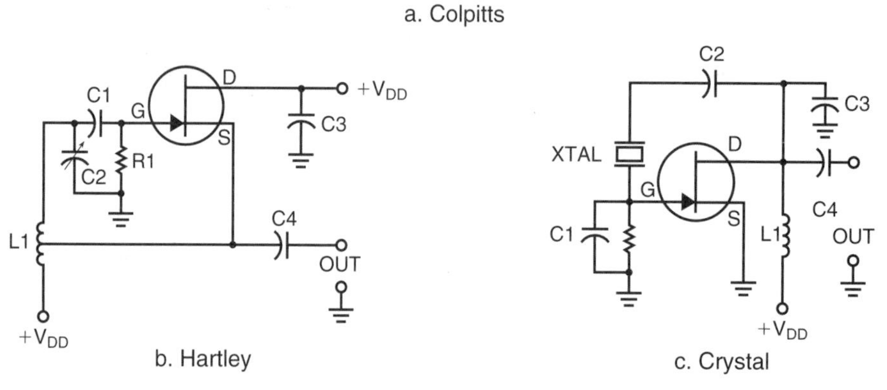


*Figure 4-21. Discrete component oscillators.*
*(Source: GROL Plus, F. Maia, G. West ©1996, Master Publishing, Inc., Lincolnwood, IL.)*

## Crystal Oscillators

In communications circuits, it is often very important that the frequency of the oscillator be very stable. In such cases, the accuracy of an oscillator will be measured in Hz, even for oscillators operating in the MHz range. If the application does not require the oscillator to be tunable and the ultimate in frequency stability is required, the tuned circuit can be replaced with a piezoelectric device called a crystal. The most common material used is quartz and a properly prepared and mounted slice of quartz will behave very similarly to a tuned circuit. An example of a crystal oscillator circuit is shown in *Figure 4-21c*. It oscillates at the resonant frequency of the crystal. For extreme stability, crystal oscillators are sometimes mounted in a small, constant-temperature enclosure to keep the oscillator from drifting in frequency as temperature changes. These devices are usually referred to as temperature-compensated crystal oscillators (TCXO).

## Frequency Synthesizer

Few circuits have made such a significant difference in communications systems as the frequency synthesizer. With the invention of the synthesizer, system designers were able to depend on oscillators that were both stable and tunable because the frequency synthesizer allows oscillators to be stable like a crystal oscillator and tunable like a VFO. Modern IC synthesizers have become quiet popular for tuning radios and TVs, and in systems where variable frequencies are required. This is the case despite the fact that extra circuitry is needed.

Synthesizers use a special circuit called a phase-lock loop (PLL) to cause a variable oscillator to behave as though it were locked in frequency to a crystal reference oscillator. One of the less desirable features of this type of circuit in some applications is the fact that it tunes in steps rather than continuously. If the system is channelized, the steps are not noticeable; on the other hand, if the application requires continuous tuning, the frequency steps must be very fine. *Figure 4-22* is a block diagram of a basic synthesizer using a PLL.

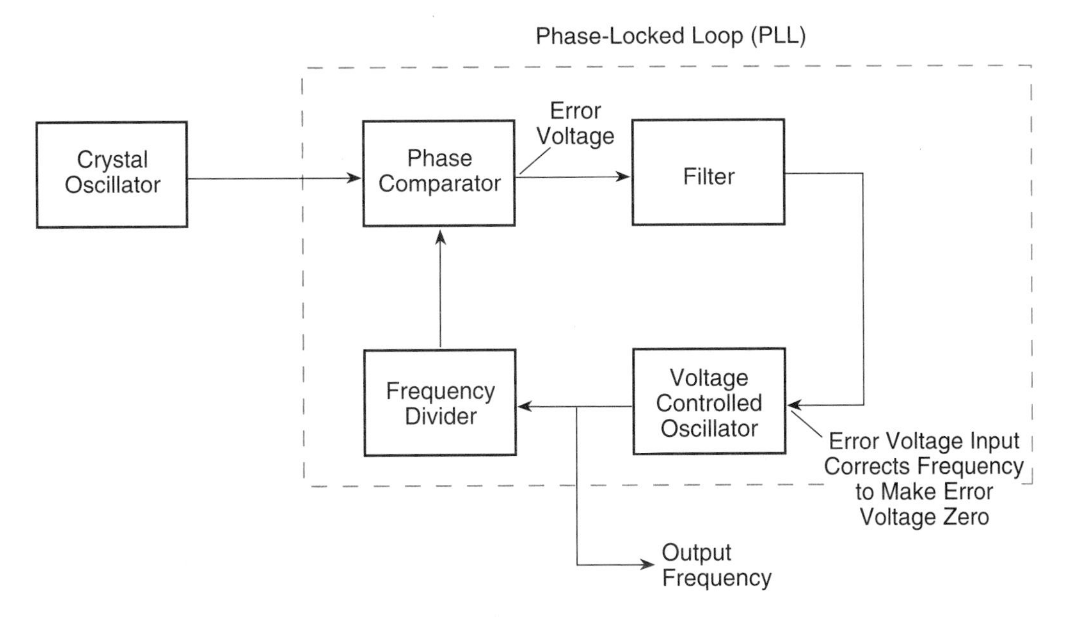


*Figure 4-22.Block diagram of a basic synthesizer using a PLL.*

This circuit functions by comparing the crystal oscillator frequency to the VCO frequency and generating an error voltage if they are different. The error voltage is of such a polarity that it forces the VCO to change its frequency so the error voltage is zero. This locks the VCO frequency to the crystal frequency. However, in *Figure 4-22*, the VCO does not connect directly to the phase comparator; it is divided down by some integer number before comparison. Changing the amount of frequency division changes the frequency at which the VCO "locks," making the oscillator tunable while also locked to the crystal oscillator. The crystal oscillator is very stable, the divider network is digital so it is stable, making for a very stable circuit for many applications. Many IC manufacturers are producing frequency synthesizer ICs, making them economical and readily available.

## Summary

In this Chapter, we have learned about amplifier circuits, both small-signal and power. We have identified discrete component amplifier circuits and general purpose amplifiers in IC form called op-amps. We have learned that power amplifiers are Class A, B, AB, or C, and that an amplifier's cousin is the oscillator. In the next Chapter, circuits that perform the modulation function will be explored.

# Quiz for Chapter 4

1. Small-signal amplifiers
   a) Have large power gain.
   b) Increase voltage or current amplitude.
   c) Operate at large currents.
   d) Act at high signal levels.
2. Small-signal amplifier circuits use:
   a) NPN and PNP bipolar transistors.
   b) N-channel and P-channel JFETS.
   c) N-channel and P-channel MOSFETS.
   d) All of the above.
3. Configurations of amplifier circuits are:
   a) Common-emitter, common-collector, common-base.
   b) Common anode, common cathode, common mode.
   c) Common primary, common secondary, common center tap.
   d) None of the above.
4. Biasing a transistor for circuit operation is
   a) Adjusting the reverse-bias on the anode.
   b) Setting the dc operating point in the middle of its operating range.
   c) Forward-biasing the gate of a MOSFET.
   d) Setting the reverse breakdown of a Zener diode.
5. Negative feedback in a common-emitter amplifier circuit
   a) Helps stabilize the dc operating point of the circuit.
   b) Produces oscillation.
   c) Is caused by a transistor collector resistor.
   d) Should never be used.
6. Amplifier frequency response has
   a) Amplifier bias on the X-axis and frequency on the Y-axis.
   b) Amplifier gain in dB on the X-axis and frequency on the Y-axis.
   c) Amplifier gain in dB on the Y-axis and frequency on the X-axis.
   d) Amplifier collector voltage on the Y-axis and frequency on the X-axis.
7. Amplifier circuits using FETs
   a) Have high input impedance.
   b) Have gain equal to $A_V = -gm\ R_L$.
   c) Amplify input voltage changes into output current changes.
   d) All of the above.
8. Tuned circuit amplifiers
   a) Depend on a resonant circuit for operation.
   b) Are general purpose linear amplifiers.
   c) Are used as high-frequency small-signal amplifiers.
   d) a and c above.
9. The quality factor, Q, of a tuned circuit
   a) Equals the frequency where $X_C = X_L$.
   b) Determines the cost of the circuit.
   c) Determines the bandwidth of a tuned amplifier.
   d) Is the collector voltage divided by the collector current.
10. The voltage gain of an FET tuned-circuit amplifier equals
    a) Collector voltage times collector current.
    b) $-gm \times Z_L$.
    c) Gate voltage times gate current.
    d) Collector voltage divided by gate voltage.
11. A general-purpose, low-frequency amplifier used many times in a communications system is called
    a) An operational amplifier, or "op-amp."
    b) A tuned-circuit amplifier.
    c) An over-driven amplifier.
    d) A non-linear amplifier.
12. An integrated-circuit op-amp has
    a) Infinite gain at very high frequencies
    b) Characteristics very close to those of an ideal op-amp.
    c) Ability to provide amperes of output current.
    d) None of the above.

# Quiz for Chapter 4

13. The four classes of operation of power amplifier actives devices are
    a) Class A, Class B, Class AB, and Class C.
    b) Class ABC, Class ABD, Class AB, and Class C.
    c) Class W, Class X, Class Y, and Class Z.
    d) All of the above.
14. Transistors used in complementary symmetry power amplifiers are operating in
    a) Class A.
    b) Class C.
    c) Class AB.
    d) Class B.
15. A power amplifier active device that is biased to operate to conduct current in bursts is operating in
    a) Class A.
    b) Class AB.
    c) Class C.
    d) Class B.
16. An oscillator is an amplifier with
    a) Positive feedback from output to input.
    b) Negative feedback from output to input.
    c) Positive feedback from input to output.
    d) Negative feedback from input to output.
17. An oscillator circuit has
    a) A phase shift of 180° and a gain of less than one.
    b) A phase shift of 360° and a gain at least equal to one.
    c) A phase shift of 270° and a gain at least equal to one.
    d) A phase shift of 360° and a gain less than one.
18. An oscillator circuit that has the stability of a crystal oscillator but also is tunable is a
    a) Tuned-circuit oscillator.
    b) Voltage-controlled oscillator.
    c) Phase comparator.
    d) Frequency synthesizer.

**Answers:**
1b, 2d, 3a, 4b, 5a, 6c, 7d, 8d, 9c, 10b, 11a, 12b, 13a, 14d, 15c, 16a, 17b, 18d

# Questions & Problems for Chapter 4

1. Using the approximate equation for common-emitter stage biasing, calculate $V_B$, $V_E$, $I_E$ and $V_C$ when R1 = 4700Ω, R2 = 1200Ω, R3 = 4400Ω, R4 = 22,000Ω, and $V_{CC}$ = 10V.
2. What is $h_{FE}$ and what is ß?
3. If the mid-band gain of an amplifier frequency response is 36dB, what will the amplifier gain be at $f_L$ and $f_H$?
4. What is another name for a common-collector amplifier?
5. How do you get a very large voltage gain using small-signal amplifiers?
6. What are the disadvantages of using FETs in small-signal amplifiers?
7. If an FET small-signal amplifier uses a device with a gm = 8,000 µmhos (siemens), what is the voltage gain when $R_L$ = 10,000 ohms?
8. What is the $R_P$ of a parallel resonant circuit that is resonant at 20 MHz when L = 20 µH and the Q of the circuit is 20?
9. What capacitor is required to be in parallel with L = 20 µH to make the circuit resonant at 20 MHz?
10. If a tuned circuit has a Q of 50 at 200 MHz, what is the circuit bandwidth?
11. What is the voltage gain of an inverting op-amp when $R_{in}$ = 2 kΩ and $R_f$ = 200 kΩ?
12. Push-pull amplifiers operate with the active devices in what class?
13. Tuned-circuit amplifiers operate with the active device in what class?
14. If an amplifier normally has a phase shift from input to output of 180°, how much phase shift is required in the negative feedback loop from output to input to make the circuit an oscillator?
15. What input to a VCO corrects the frequency to make the error voltage zero?

*(Answers on page 210.)*

# CHAPTER 5
# Modulation



The objective of modulation is to take a baseband signal that contains information (such as a voice, music, or video signal) and translate the signal to a new frequency so that it can be transmitted easily and accurately over a communications channel. We will learn that the modulation function and the mixing function are similar. They both need a device in their circuitry with a non-linear characteristic in order to produce sum and difference frequencies from the signals being processed. Mixing usually will use only one of the resulting frequencies, while modulation may or may not use both resulting frequencies.

An early form of modulation still occasionally used today is the On-and-Off keying used to transmit Morse code. This keying action turns on or off a continuous sine wave (CW) signal. Since the advent of CW in the early-1900s, we have seen the development of amplitude modulation (AM), single-sideband modulation (SSB), frequency modulation (FM) — either by direct frequency changes or by phase angle changes (PM) — and digital modulation systems. Single-sideband is closely related to AM modulation.

When one uses an oscilloscope to view a composite AM signal, it can be understood that the information is carried by the amplitude of the composite signal. With FM systems, the information is carried by the frequency of the composite signal. The composite signal for both AM and FM is the carrier plus signals located on either side of the carrier called sidebands — one is called the upper sideband (USB) and the other is called the lower sideband (LSB). We will study AM first — including SSB, which is the transmission of only one sideband — and then look at FM.

## Amplitude Modulation

The earliest analog modulation system to be developed was AM (amplitude modulation). AM gained wide acceptance as the first radio communications system used to transmit voice and music because it is extremely easy to detect and the circuitry required for detection is considerably simpler than for other types of modulation. AM has an output in which a carrier signal has sideband energy, generated by the modulation process, added to or subtracted from the carrier. The net composite signal has an envelope, or magnitude, controlled by the modulating signal. The carrier is a steady, unchanging signal and represents the "middle" of the required communications channel. There is no information in the carrier, it is all in the sideband signals. In commercial AM broadcast radio systems, the carrier frequency is the one you "tune in" to receive the station.

The mathematics used to describe the composite signal — the modulated carrier with its sidebands — is rather complicated. Using sine-wave notation, the general equation is as follows:

$$E = \underbrace{E_{cx} \sin (2\pi f_{cx} t)}_{\text{Carrier}} + \underbrace{mE_{cx} [\sin (2\pi f_{cx} t)] [\sin (2\pi f_m t)]}_{\text{Carrier Times Modulation}}$$

where E = composite signal in **volts**
$E_{cx}$ = peak carrier amplitude in **volts**
$f_{cx}$ = frequency of carrier in **Hz**
$f_m$ = frequency of modulating signal in **Hz**
m = is a ratio to specify the degree of modulation called the **modulation index**
t = time in **seconds**

Here we see that the composite signal after modulation is the carrier frequency signal plus a term that includes the multiplication of the carrier signal times the modulating signal. When this equation is expanded using trigonometry, the equation of an AM wave with a single frequency modulation is:

$$E = \underbrace{E_{cx} \sin (2\pi f_{cx} t)}_{\text{Carrier}} + \underbrace{\frac{mE_{cx}}{2} \cos 2\pi (f_{cx} + f_m)t}_{\text{Upper Sideband}} - \underbrace{\frac{mE_{cx}}{2} \cos 2\pi (f_{cx} - f_m)t}_{\text{Lower Sideband}}$$

As noted, the first term is the carrier, while the second and third terms are the upper and lower sidebands, respectively. Note that the upper sideband is a result of $(f_{cx} + f_m)$ and the lower sideband is the result of $(f_{cx} - f_m)$, which are the sum and difference frequencies generated by modulation. The spectral plot of an AM signal is shown in *Figure 5-1*. This is similar to *Figure 2-5d*. Since the AM signal has two sidebands that are copies of the original signal, the occupied bandwidth will be twice that of the original signal. For example, a baseband voice signal of about 3 kHz requires a channel bandwidth of about 6 kHz. This is because the AM modulation process produces two copies of the baseband signal.

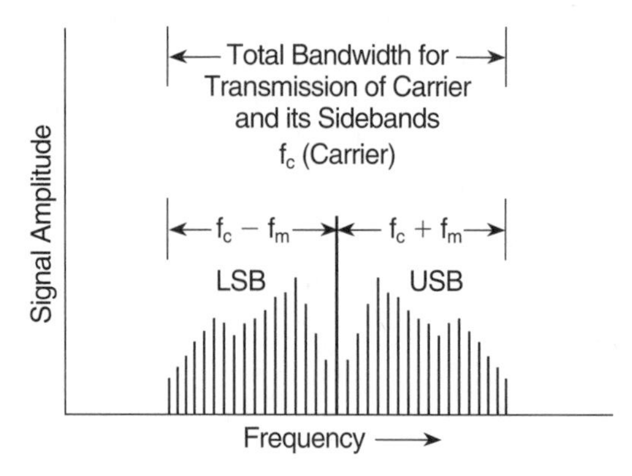


*Figure 5-1. Spectral plot of an AM signal.*

## AM Basics

*Figure 5-2* is a block diagram of an AM system and *Figure 5-3* shows its composite AM output plotted against time. The computer-generated plot in *Figure 5-3* shows the characteristic shape of an AM signal viewed with time as the variable on the horizontal axis. The baseband signal forms an "envelope" around the carrier. If there is no modulation signal, the AM signal will be a single sine wave at the carrier frequency; that is, the carrier signal will be there alone with no sidebands. The typical shape of the output of the AM signal is due to the sidebands being added to and subtracted from the carrier. The envelope frequency is the same as the modulating frequency.

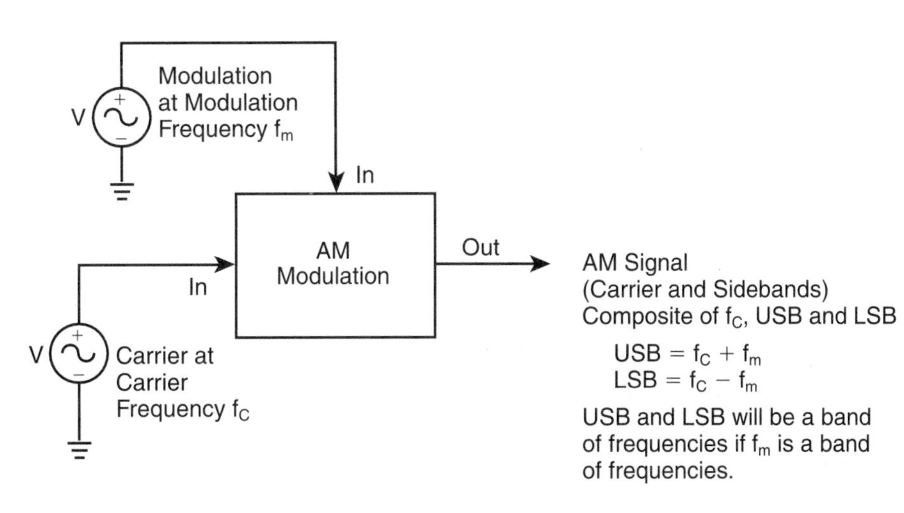


*Figure 5-2. Block diagram of an AM modulator.*

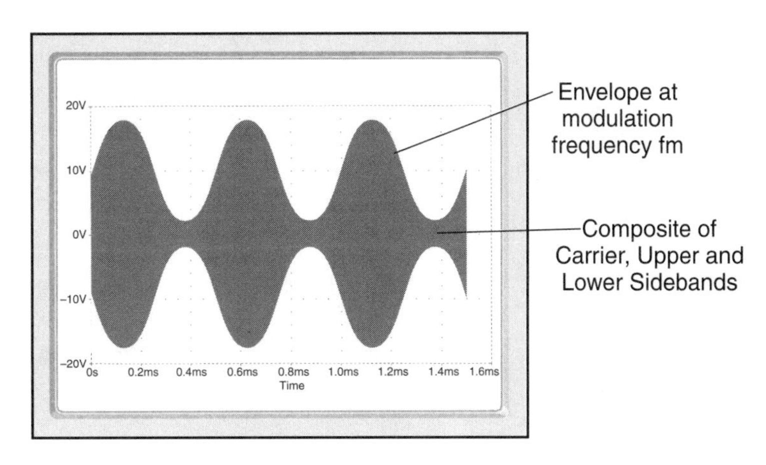


*Figure 5-3. Computer plot of an AM signal, amplitude vs. time.*

In *Figure 5-1*, since the modulating signal is a band of frequencies in the baseband, the sidebands are "blocks" of frequencies resulting from the sum and difference between the carrier and all the signal frequencies in the baseband spectrum. If the modulating signal had only been a single frequency, the upper and lower sidebands would only be single frequencies on either side of the carrier, as shown in the spectral plot of *Figure 5-4*.

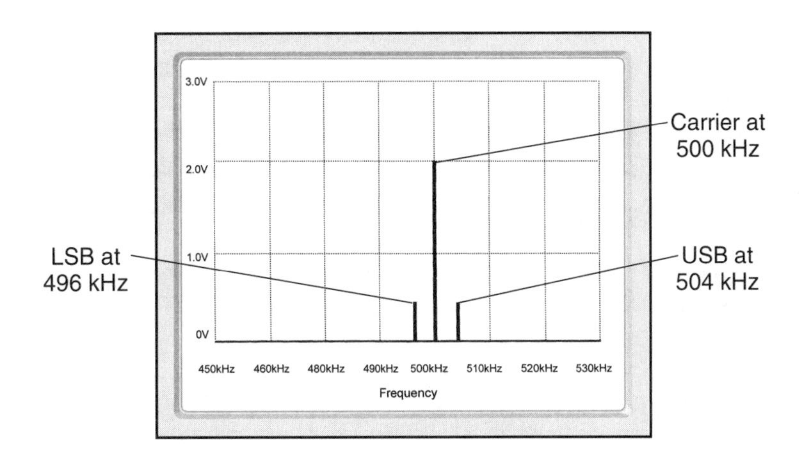


*Figure 5-4. Spectral plot of a 500-kHz carrier modulated with a single-frequency 4-kHz signal.*

## Power in Carrier and Sidebands

In *Figure 5-4*, a carrier of 500 kHz is modulated with a single frequency of 4 kHz. The spectral plot indicates that the magnitude of the sidebands are not as large as the magnitude of the carrier. In fact, the voltage amplitude of each sideband is about one-half that of the carrier at 100% modulation. The power in any signal is proportional to voltage squared; therefore, the power in each sideband would be one-quarter that in the carrier. The maximum power permissible in each sideband is one-quarter of the power in the carrier; therefore, since there are two sidebands, the total maximum power in the sidebands is one-half the power in the carrier when m = 1 (100% modulation). Attempts to increase sideband power above this level will result in over-modulation and cause severe distortion of the signal. Thus, over-modulation should always be avoided in order to preserve the integrity of the information. The power relationship in an AM signal is as follows:

$$P_{out} = P_c\left(1 + \frac{m^2}{2}\right)$$ where:

$P_{out}$ = total composite signal power in **watts**
$P_c$ = carrier power in **watts**
m = modulation index or percent modulation (100% modulation is m = 1)

When m = 1, the *total composite output*, if the *carrier power is given*, is:

$$P_{out} = P_c\left(1 + \frac{1}{2}\right) = P_c + \frac{P_c}{2}$$

$$P_{out} = \frac{3}{2} P_c$$

or, if the *total composite output* is given, the *carrier power* is

$$P_c = \frac{2}{3} P_{out}$$

and since sideband power is $\frac{P_c}{2}$, then

$$\text{sideband power} = \frac{P_c}{2} = \frac{1}{2} \times \frac{2}{3} P_{out} = \frac{1}{3} P_{out}$$

When m=1, one-third of the total power will be in the two sidebands and two thirds will be in the carrier. For instance, if the carrier power is 1000 watts then there will be 500 watts in the sidebands — with 250 watts in each sideband — with the composite signal having a total of 1500 watts. If the modulation index varies, the carrier power will always be 1 kW but the power in the sidebands will vary with the percent modulation, with the peak never exceeding 500 watts total sideband power.

### *Example 1: Calculating Carrier and Sideband Power*

Calculate the power in the carrier and sidebands of AM modulated signals at a total composite power output of 1500 watts, 12,000 watts, and 30,000 watts. Assume 100% modulation.

***Solution:***

| $P_{out}$ | $P_c = 2/3\ P_{out}$ | Sideband Power = $1/3\ P_{out}$ |
|---|---|---|
| 1500 | 1000 | 500 (250 in each sideband) |
| 12,000 | 8000 | 4,000 (2000 in each sideband) |
| 30,000 | 20,000 | 10,000 (5000 in each sideband) |

## Low-Level AM

There are two basic ways to produce AM modulation. One is referred to as low-level modulation and the other is high-level modulation. Low-level AM modulation is used where overall signal power is small and there is no need to operate at high power levels. Typically, low-level AM is used for control applications where transmission is over short distances or in communications systems where low-power amplifiers are used.

*Figure 5-5b* shows a typical low-level modulation system using a 3080 linear IC as the active device. It comes in an 8-pin dual-in-line plastic package, and the circuit schematic is shown in *Figure 5-5a.* The 3080 is a gain-controlled amplifier whose gain can be changed according to the bias current that flows in pin 5. A bias current of about 10 μa to 1000 μa will cause the gain to change by a factor of 1000 to 1 (over three decades) by changing the transconductance of the amplifier. When the signal on pin 5 is the modulating signal, the gain control feature is used to produce AM modulation. The effect is to multiply the carrier by 1 plus the input signal, which produces an output that is a characteristic AM waveform. The signal is produced by the LM3080 circuit at very low power levels and then amplified by additional linear amplifiers to preserve the overall characteristic shape of the AM signal as the power level is increased.

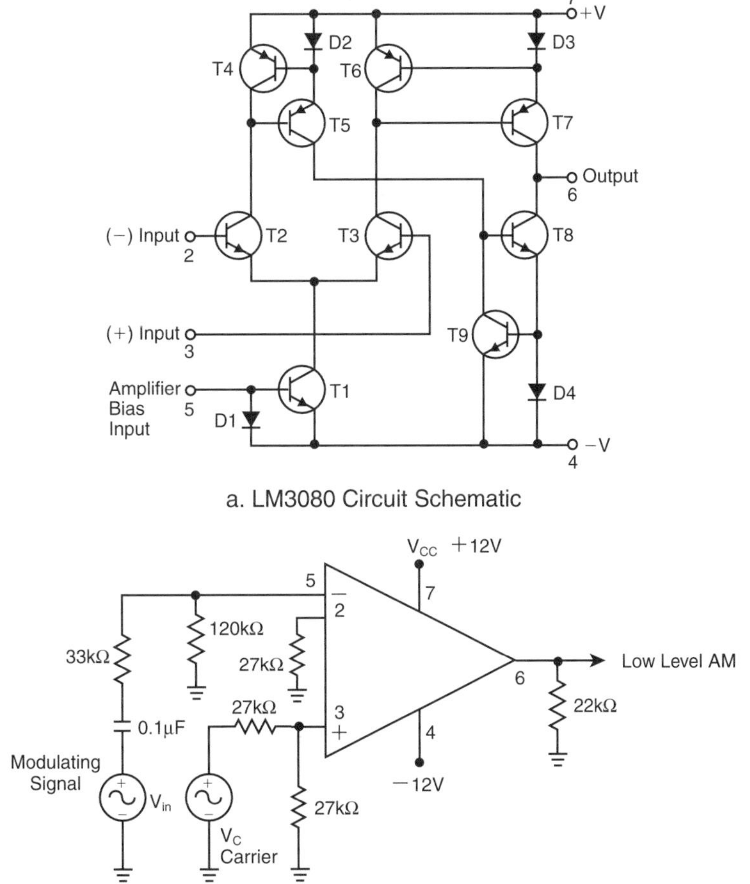


*Figure 5-5. Low-level AM modulator. (Courtesy of National Semiconductor)*

## High-Level AM

It is more common to produce AM at high power levels using the process shown in *Figure 5-6*. The circuit of *Figure 5-6a* is the same tuned class-C amplifier used in *Figure 4-17* and operates in the same way except the transformer, TX1, is added in the collector circuit of the bipolar transistor, Q1. This particular tuned circuit operates at 10 MHz, but can be built to operate at different carrier frequencies. The transformer is designed to handle a high-power-level modulating signal. With the transformer in the collector circuit after the tuned circuit and before the power supply, the modulating signal acts as if it is in series with the power supply. As a result, the modulating signal voltage is combined with the power supply voltage fed to the class-C amplifier adding and subtracting voltage to and from the power supply, and providing the amplifier with a power supply voltage that varies as the modulation signal varies. Since the output voltage of the amplifier depends upon the combined voltage of the power supply, the carrier is "modulated" by the changing modulation voltage. The transformer is driven by amplifiers that increase the modulating signal to the required power level. The effect is to multiply the carrier signal by 1 + the modulating signal, which is the multiplication needed for modulation. To achieve 100% modulation when the carrier power is 1 kW, the modulation amplifiers, which produce all of the sideband power, would have to produce 500 peak watts. *Figure 5-6b* shows the 10-MHz output carrier frequency waveform modulated with a 500-kHz signal. Compare this to the output voltage, $V_O$, of *Figure 4-18* in Chapter 4 to see how the carrier has been modulated.

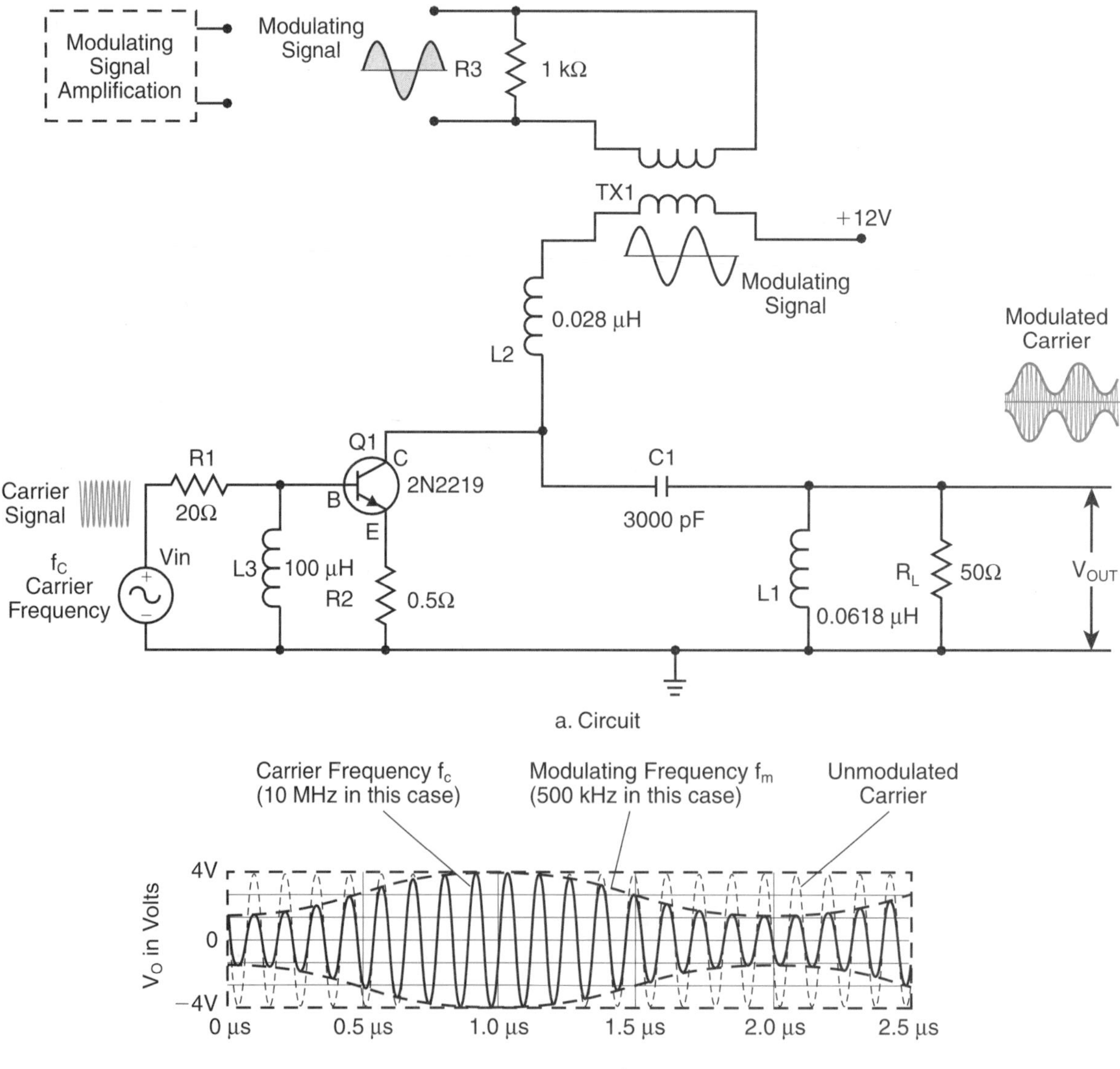


*Figure 5-6. High-level AM modulator.*

The circuit of *Figure 5-6a* is the type most commonly used by commercial AM broadcast radio stations, or anytime that high power and high efficiency is important. High power operation (over 1-kW) requires the use of a vacuum tube rather than a transistor. The class-C amplifiers can operate with efficiencies as high as 80% to 90%.

## Single Sideband — SSB

The occupied bandwidth of an AM signal is twice that of the baseband signal, and the upper and lower sidebands are identical, except for their frequency orientation. The lower sideband is "flipped over" or inverted. All of the baseband information is contained in one sideband signal. So, why do you need both sidebands and, for that matter, why do you need the carrier? SSB is a transmission technique whereby only one sideband is transmitted. The other sideband and the carrier are removed before transmission.

When you remove one of the sidebands and the carrier, the resulting signal is called Single Sideband Suppressed Carrier (SSBSC), or just SSB. The spectral plot of an SSB signal will look like *Figure 5-7* with only the upper sideband being transmitted (it could have been the lower sideband). The sideband that is not used is filtered out leaving only one sideband and no carrier. Note that the lower sideband signal appears inverted in frequency from the upper sideband in *Figure 5-7.* If the lower sideband is used, the difference frequencies of the modulating signal are in different relative positions on the frequency spectrum from the upper sideband. Increasing the modulation frequency causes a corresponding frequency *increase* in the output signal for the upper sideband, while it causes a *decrease* in frequency of the output signal for the lower sideband.

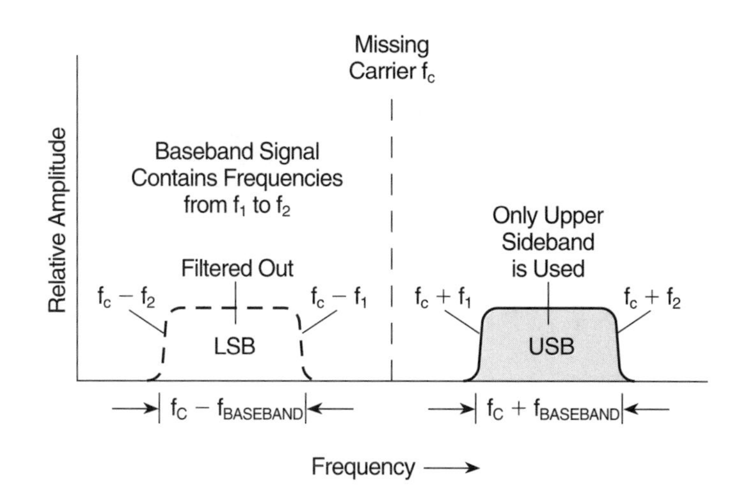


*Figure 5-7.Spectral plot of a SSB signal.*

Unlike high-level AM, where modulation occurs just before the signal is transmitted, note that SSB signals are always generated at low power levels, and an SSB system requires the use of a linear amplifier to increase the power level of the output signal for transmission over a communications channel.

### Advantages & Disadvantages

There are three principle advantages to an SSB system when compared to AM. First, the signal only needs one-sixth the power to be at the same detected level as an AM signal. Remember from *Example 1* that, in an AM signal, only one-third of the total composite signal power can be in both sidebands, which means that just one-sixth of the total power is in one sideband, giving SSB a possible six times advantage. But the net power advantage of SSB is only three times because the linear amplifiers needed to boost the power level of the SSB signal are only about 50% to 60% efficient. Even so, because of the power advantage, the overall efficiency of the transmitter is increased. Because no carrier power is transmitted, an SSB transmitter with the same power output as an AM transmitter can use a power supply with a lower power capability.

The second advantage of SSB is that the signal will occupy only the bandwidth needed for the baseband signal, which is one-half the bandwidth required by an AM system, to transmit the same information. A third advantage occurs at the receiver. At high frequencies, SSB exhibits less selective fading during transmission compared to AM.

The disadvantage of SSB is that it is more difficult to detect than an AM signal because it lacks the carrier. The carrier must be inserted at the receiver in order to detect the original information. There are typically two methods to do this. In the first, an oscillator at the receiver — called a *beat-frequency oscillator* — inserts the carrier. In the second, a pilot signal is transmitted along with the SSB information signal to reconstruct the carrier at the receiver. Since the carrier may not be inserted at the correct frequency, SSB is not always as accurate as AM in reproducing baseband signal frequencies unless the pilot carrier is used. However, for difficult environments, such as high-frequency radiotelephone systems, marine radio, and amateur radio communications, SSB is the preferred voice transmission technique because of its power and bandwidth advantages.

## SSB Balanced Modulator

In Single-Sideband (SSB) transmission the carrier is not required, and a balanced modulator is used to suppress the carrier signal while producing a modulated double-sideband signal for further processing. The carrier signal is coupled into the balanced modulator circuit in such a way that only the sideband signals appear at the output. In a balanced modulator, if there is no input modulation signal there will be no output signal. As a modulation signal is applied, the balance of the circuit is upset, providing modulation, and the resulting sum and difference frequencies appear in the sideband outputs. Both sideband signals are present, but no carrier. To help understand how a balanced modulator works, let's look at *Figure 5-8*.

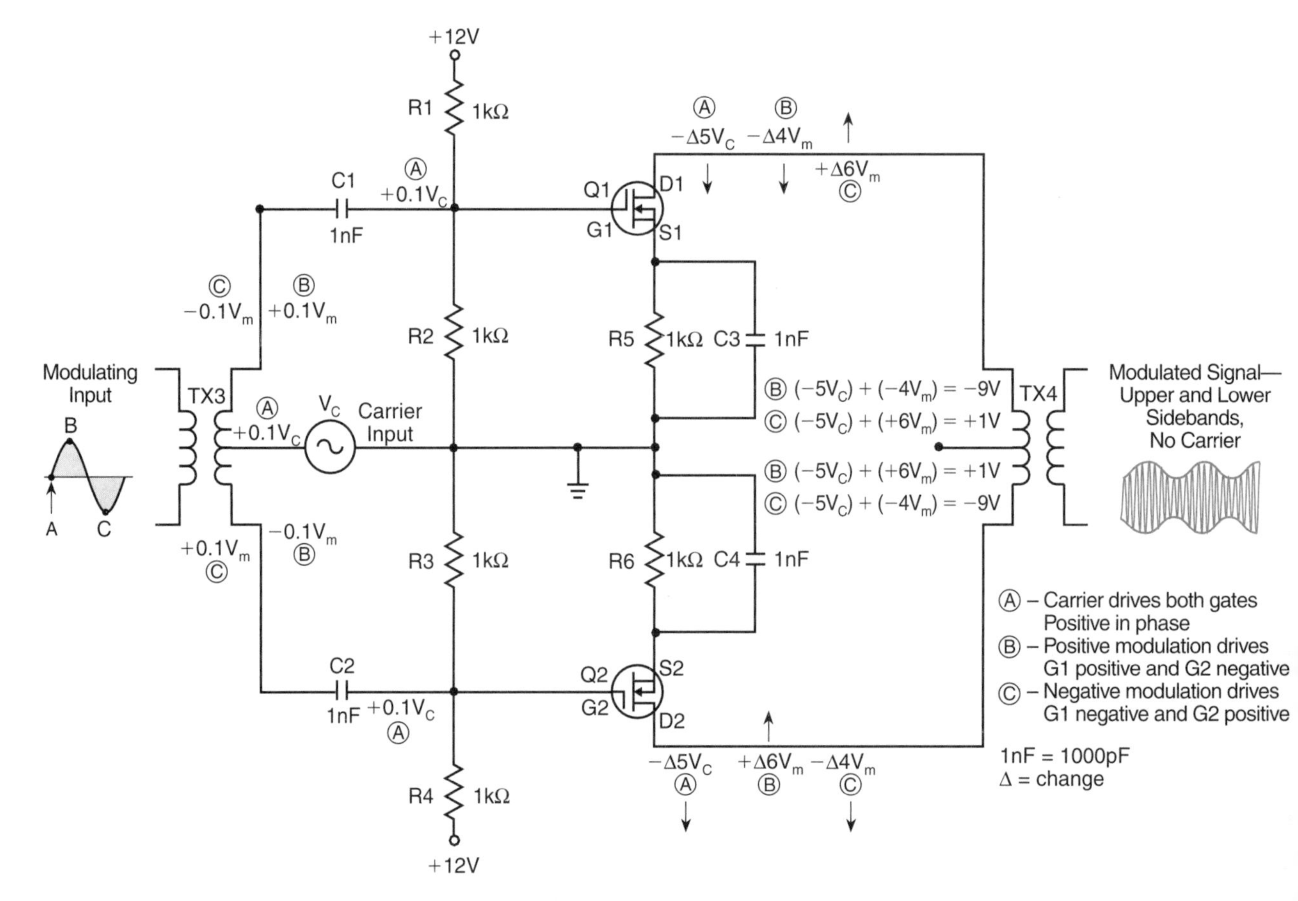


*Figure 5-8. An SSB balanced modulator.*

Assume there is no modulation signal (point A on the modulating input.) The carrier frequency is coupled to the center tap of the secondary of the input transformer; therefore, if it is going positive by 0.1V as at Ⓐ, it drives both gates of the FETs positive at the same time (in phase). The drains of the FETs change -5V due to the gain in the circuit. However, since both ends of the primary of the output transformer move in the same direction and the same amount, there is no net voltage change across the primary of TX4 and no signal is transferred. This would occur for the full cycle of the carrier if there is no modulating signal. The voltages on the drains are continually equal, and thus, there is no carrier signal at the output — the carrier is suppressed, and no energy is transferred.

Now apply a modulating signal that goes positive 0.1V as the carrier goes positive 0.1V (point B). The modulating signal is coupled through the center-tapped secondary such that the gate of Q1 is 0.1V positive and the gate of Q2 is 0.1V negative (as at Ⓑ), in other words, out of phase. The additional 0.1V added to the carrier positive 0.1V on the gate of Q1 by the modulation causes an additional -4V change (non-linear region) on Q1's drain. At the same time, the negative 0.1V on the gate of Q2 drives its drain in the opposite direction (+6V) from the carrier causing a cancellation of the -5V carrier change (but 1V more because of the nonlinearity). The net result is that Q1's drain is changed to -9V and Q2's drain is changed to +1V, transferring energy into the output secondary.

Applying a negative modulation signal as the carrier goes positive 0.1V (point C) causes the reverse effect on the drains of Q1 and Q2 (shown as Ⓒ). The net result is that Q1's drain is at a change of +1V and Q2's drain is at a change of -9V, again transferring energy, but in the opposite phase, to the output secondary.

Because of the non-linearity and the unbalancing of the circuit by the modulating signal, the required multiplication occurs, the sum and difference frequencies are produced, and the output is *the upper and lower sidebands with the carrier suppressed.*

## SSB Generation

*Figure 5-9a* is a block diagram of a system for generating a single-sideband signal. It consists of the modulating signal source, the carrier signal source, a balanced modulator,

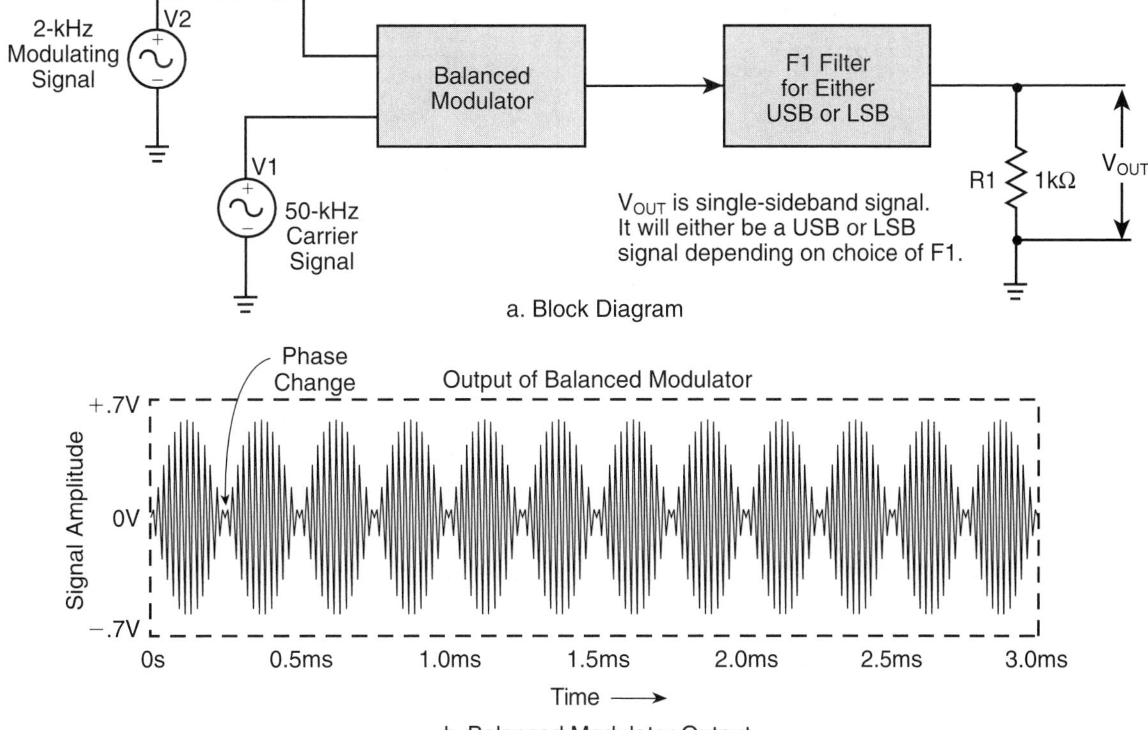


*Figure 5-9. Generating an SSB signal.*

and a filter. The balanced modulator outputs the USB and LSB signals with the carrier suppressed. The output of the balanced modulator is then fed to a high-quality, sharp-sided, band-pass filter that will remove one of the sidebands. This is called the filter method of generating an SSB signal. The band-pass filter usually is fabricated from a piezoelectric device (a crystal) that produces a filter with very sharp cutoff characteristics to remove the unwanted sideband. *Figure 5-9b* shows the balanced modulator output. Note the phase change in the signal at the point where the modulation input crosses its 0 axis. This is characteristic of a balanced modulator.

*Figure 5-10a* is a spectral plot of the balanced modulator output and *Figure 5-10b* is the filter output when filtering out the upper sideband. The 50-kHz carrier is modulated with a 2-kHz signal. With the carrier balanced out, and the upper sideband filtered out, the result is the lower sideband output at 48 kHz with a bandwidth of just 2 kHz.

The signal output from the filter is at a low level. It must be amplified by linear power amplifiers. Even though Class-C amplifiers work in AM, they will not work in SSB since they will not faithfully reproduce the input waveform. This is a disadvantage for SSB because the linear amplifiers are only about 50 to 60% efficient.

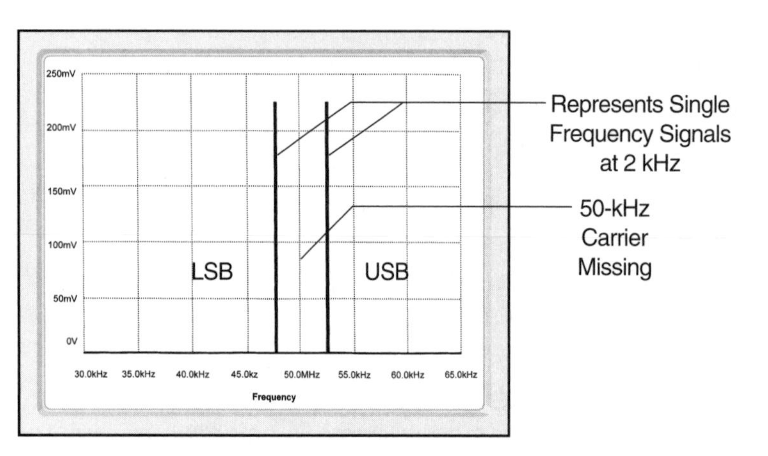


a. Output of Balanced Modulator

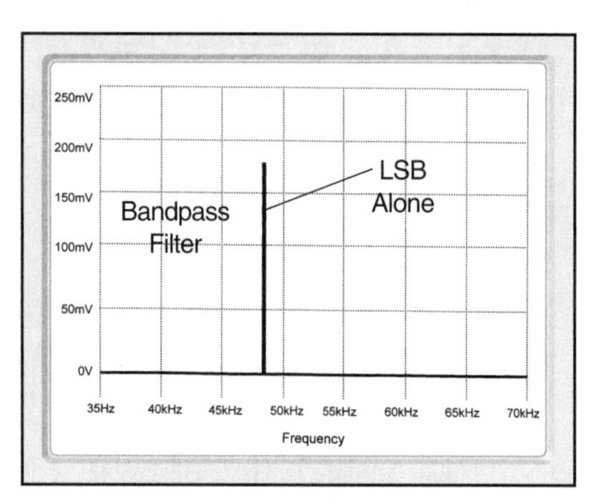


b. Output when USB is Filtered Out

*Figure 5-10. Spectral plot of SSB signals using USB.*

## Diode Balanced Modulator

*Figure 5-11* is an example of a balanced modulator circuit using a diode bridge. This circuit often is referred to as a ring modulator and operates by unbalancing the ring to disrupt the balancing out of the carrier. If there is no modulation input, the carrier is suppressed. When a modulation signal is applied at the center tap of the output primary, D1 and D2 conduct when the cycle is positive and D3 and D4 conduct when the cycle is negative. This conduction through the non-linear diodes disrupts the balance for the carrier and produces the multiplication required and the resulting double-sideband output.

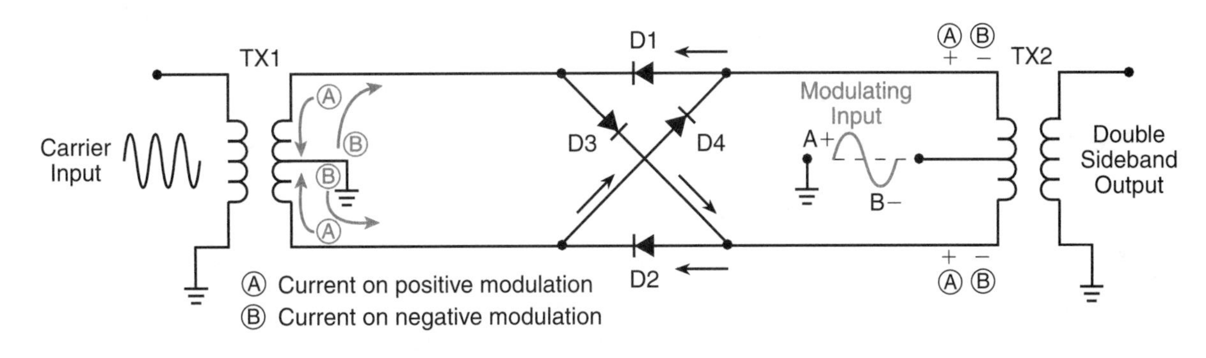


*Figure 5-11. Diode balanced modulator.*

## Types of Balanced Modulators

There are many types of balanced modulator circuits, from those using diodes to those using bipolar or field-effect transistors to those made as integrated circuits. We will show an integrated circuit in the next chapter on mixers (MC1496) that also can be used as a balanced modulator. All of the circuits are striving to get the largest carrier suppression along with the cleanest double-sideband output. Some use many more components than others to achieve this, but all circuits must use very-carefully-matched components, whether they are diodes, transistors or transformers. Here is where integrated circuits have the advantage — because the components are made at the same time on the silicon wafer they will have matched characteristics.

## AM and SSB Signal-to-Noise

AM and SSB do not perform well in a noisy environment. Lightening, motor noise, static, engine noise, and the like will amplitude modulate the carrier in the same manner as the baseband information signal. Such noise will affect both systems in a similar manner. A transmitter power increase helps both systems equally because it results in the same increase in the output S/N ratio.

SSB does have an advantage when there is an unstable communications channel, such as high-frequency skip propagation. SSB exhibits much less selective fading. Selective fading occurs when signals fade, *but only one of the sidebands fades out due to propagation effects.* This causes a severe loss of signal in AM but has less effect on SSB.

# Angle Modulation

Angle modulation is a general term for frequency modulation (FM) and phase modulation (PM). These two types of modulation techniques are very closely related. In analog communications, FM is the most commonly-used modulation technique for voice and entertainment systems. PM finds wide use in digital systems.

In angle modulation, the frequency or the phase angle of a sine wave is changed. So, in FM, the frequency of a "carrier" is changed and the information is carried in the frequency change. In PM, the phase of a sine wave is changed from its unmodulated state and the information is "carried" in the phase angle change, the result of which is an instantaneous frequency change. Studying *Example 2* will help in identifying the difference.

## Example 2: Frequency and Phase Changes

1. With time as the horizontal axis, plot the sine waves generated by two vectors rotating at the same frequency (angular velocity = $2\pi f_t$) with vector B leading vector A by a phase angle of $\theta = 60°$. Assume that f = 1 MHz and the same amplitude for the two vectors.

**Solution:**

Waveforms A and B will be as shown. One cycle in 1 µsec. Remember $\frac{1}{f}$ = period.

2. Now plot vector A with f = 2 MHz.

**Solution:**

Waveform A now has one cycle in 0.5 µsec.
One can see the direct effect of the frequency change. Two cycles now appear in the time of one cycle before. The change in frequency is direct when f is changed by modulation.

3. Now plot vector B again, this time with $\theta = 120°$.

**Solution:**

Waveform B now starts ahead of A by 120°. One cycle still occurs in 1 µsec.
Changing θ has changed the time relationship between vector A and vector B. Phase (angle) modulation (PM) changes θ according to the input information. It has the indirect effect of causing instantaneous frequency changes. That is why FM and PM are linked together.

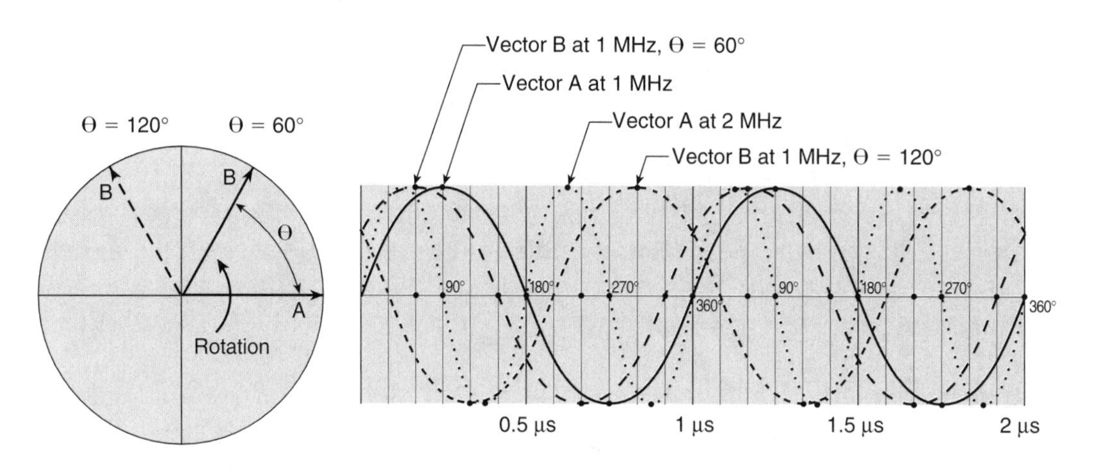


*Figure 5-12* is an exaggerated view of the two modulations showing where a large net frequency change and a large net phase change (180° in this case) occurs. The frequency and phase changes are abrupt to illustrate the basic concept for the two modulation techniques.

Frequency modulation is proportional only to the *amplitude* of the modulating wave, while phase modulation may be proportional to both the *amplitude and frequency* of the modulating wave.

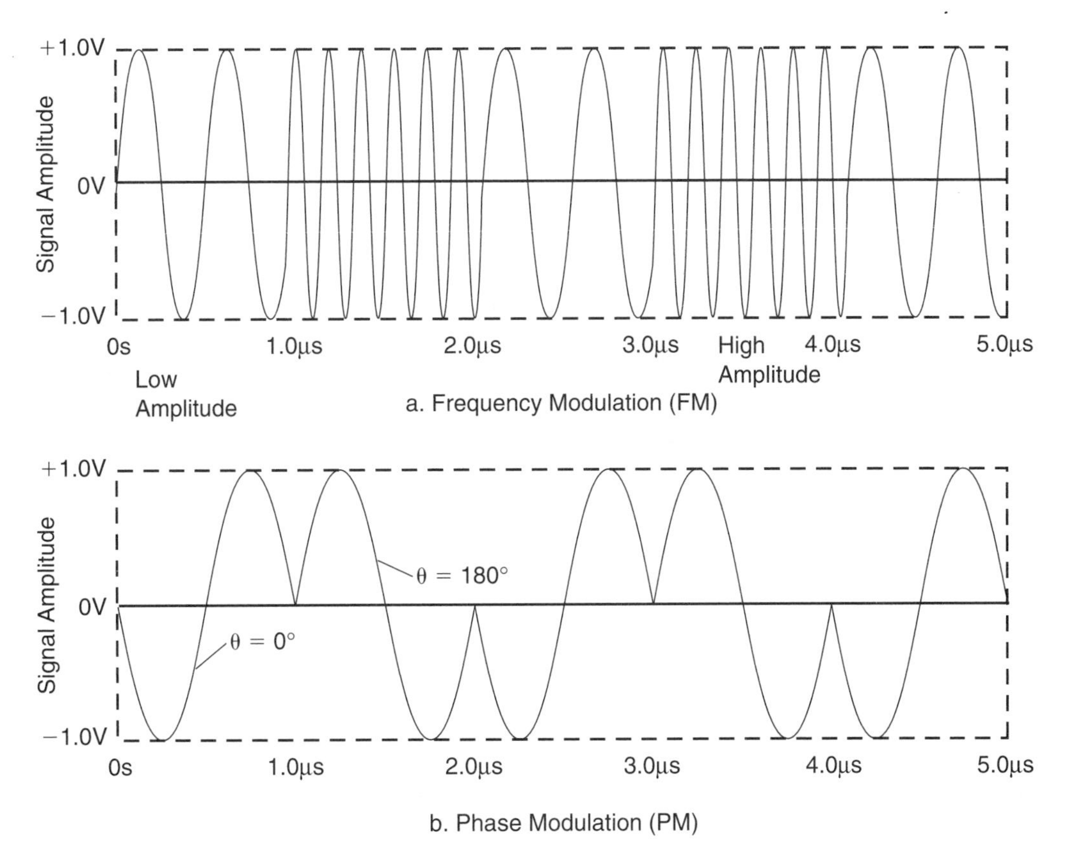


*Figure 5-12. FM and PM modulation.*

## FM Modulators

### *VCO Modulators*

FM is produced when the frequency of the carrier is changed instantaneously according to the modulating signal. One way to produce FM, shown in the block diagram of *Figure 5-13a* , uses a voltage controlled oscillator (VCO). The VCO will have an output frequency that is proportional to the voltage amplitude of a signal at a control pin. When the signal at the control pin is the modulating voltage, the output frequency of the oscillator changes as the amplitude of the modulating voltage changes, producing FM. *Figure 5-13b* shows the variation of output frequency plotted against input voltage. The VCO will be at the carrier frequency, $f_c$, when there is no modulation input and the control input is at the required initial bias voltage. The output of the VCO is shown in *Figure 5-13c*. Notice how the period of the carrier changes indicating a net change in frequency. At any given instant of time, the carrier frequency is referred to as having an "instantaneous frequency," which will be proportional to the voltage level of the input modulating baseband signal. As shown in *Figure 5-13b*, if the amplitude of the modulating signal is increased, the carrier frequency will go higher; if the amplitude of the modulating signal is reduced, then the carrier frequency will go lower. This frequency change is generally referred to as *deviation*, and the maximum change in the carrier frequency is the carrier's *maximum deviation*. The amount of deviation of the carrier depends only upon the amplitude of the modulating signal. The *deviation ratio* is defined for FM as the *maximum change in the carrier frequency* divided by the *maximum frequency in the modulating signal*, expressed as follows:

$$M_{FM} = \frac{f_{c\,max}}{f_{m\,max}}$$

where: $M_{FM}$ = deviation ratio

$f_{c\,max}$ = the maximum carrier frequency change (deviation) in **Hz**

$f_{m\,max}$ = the maximum modulating frequency in **Hz**.

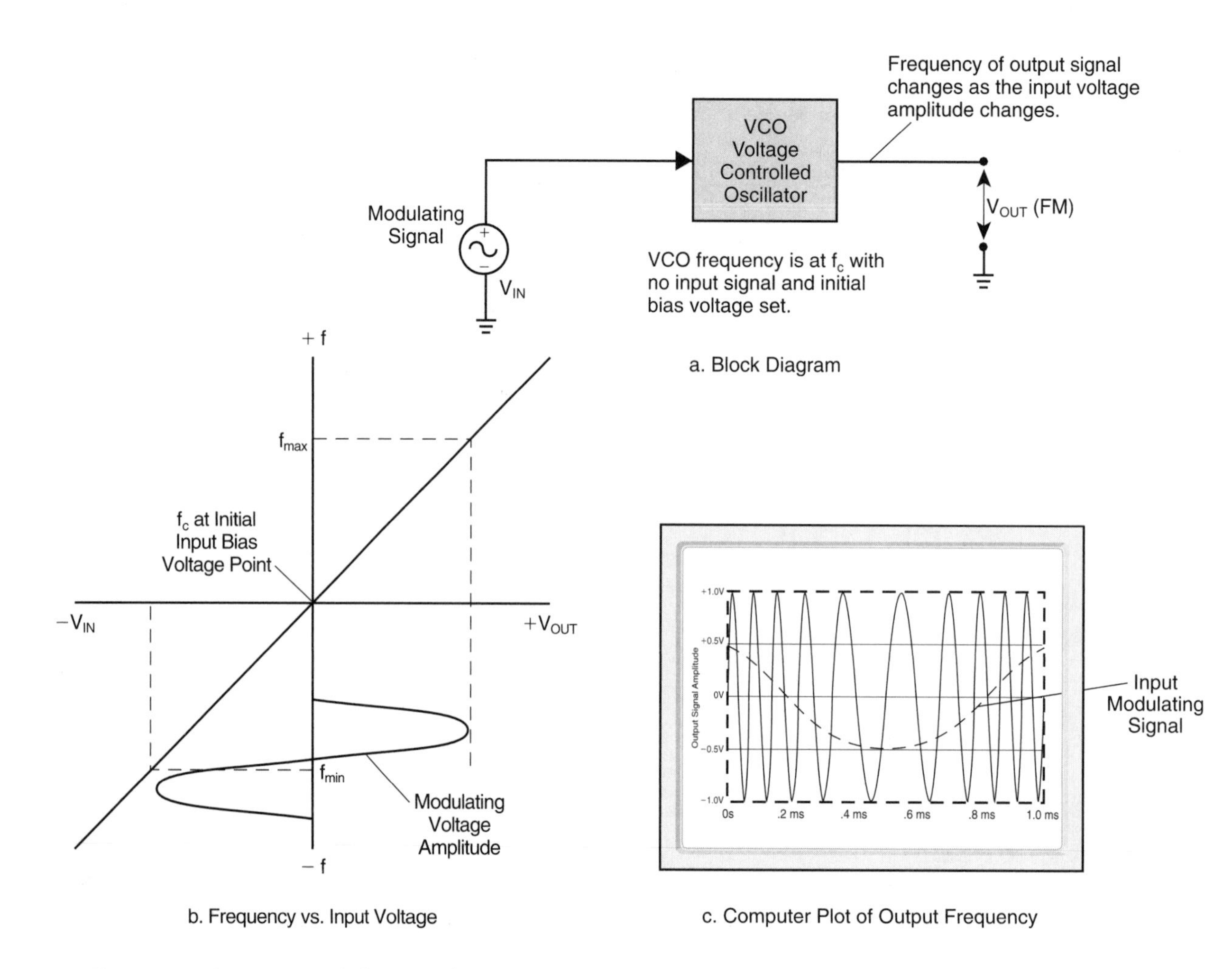


*Figure 5-13. FM modulator using VCO.*

## Example 3: Calculating Deviation Ratio

Use $M_{FM} = \frac{f_{c\,max}}{f_{m\,max}}$ to calculate the deviation ratio when $f_{c\,max}$ and $f_{m\,max}$ have the values shown.

**Solution:**

| $f_{c\,max}$ | $f_{m\,max}$ | $M_{FM}$ |
|---|---|---|
| 400 kHz | 20 kHz | 20 |
| 100 kHz | 20 kHz | 5 |
| 75 kHz | 15 kHz | 5 |
| 30 kHz | 10 kHz | 3 |
| 5 kHz | 3 kHz | 1.67 |

Another type of voltage-controlled oscillator uses an LC oscillator with a voltage variable capacitor (varactor diode) in place of one of the regular capacitors that form the tuned circuit. *Figure 5-14* illustrates the 2N2222 bipolar circuit. D1 is the varactor diode whose capacitance changes according to the voltage across it. It is a Colpitts oscillator similar to *Figure 4-21a*. The modulating voltage is applied to the diode and changes the capacitance of the diode according to the modulating signal, causing the oscillator frequency to change, producing FM.

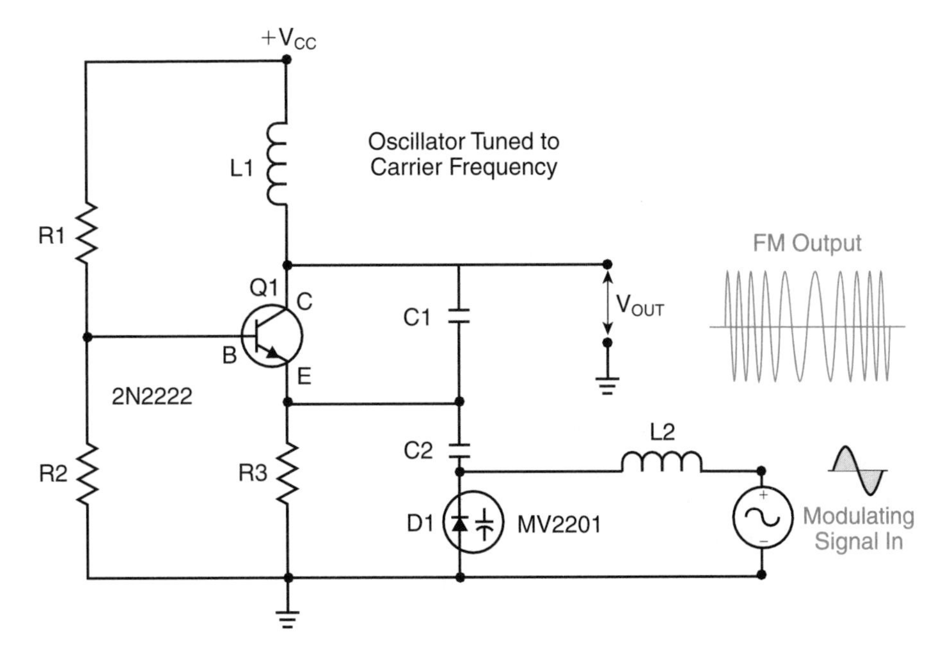


*Figure 5-14. FM modulator using varactor diode.*

## Phase Modulation

Phase modulation can be produced by a reactance modulator, which depends on changing the phase of the carrier by varying the circuit reactance of a coupling circuit or circuit in parallel with a tuned circuit. *Figure 5-15b* shows an equivalent circuit of the circuit in *Figure 5-15a*. As the JFET conduction (the value of R) is varied by the modulating signal, the relative phase of the carrier is changed because the relative values of R compared to C change the reactance of the output coupling circuit thus producing the desired PM carrier.

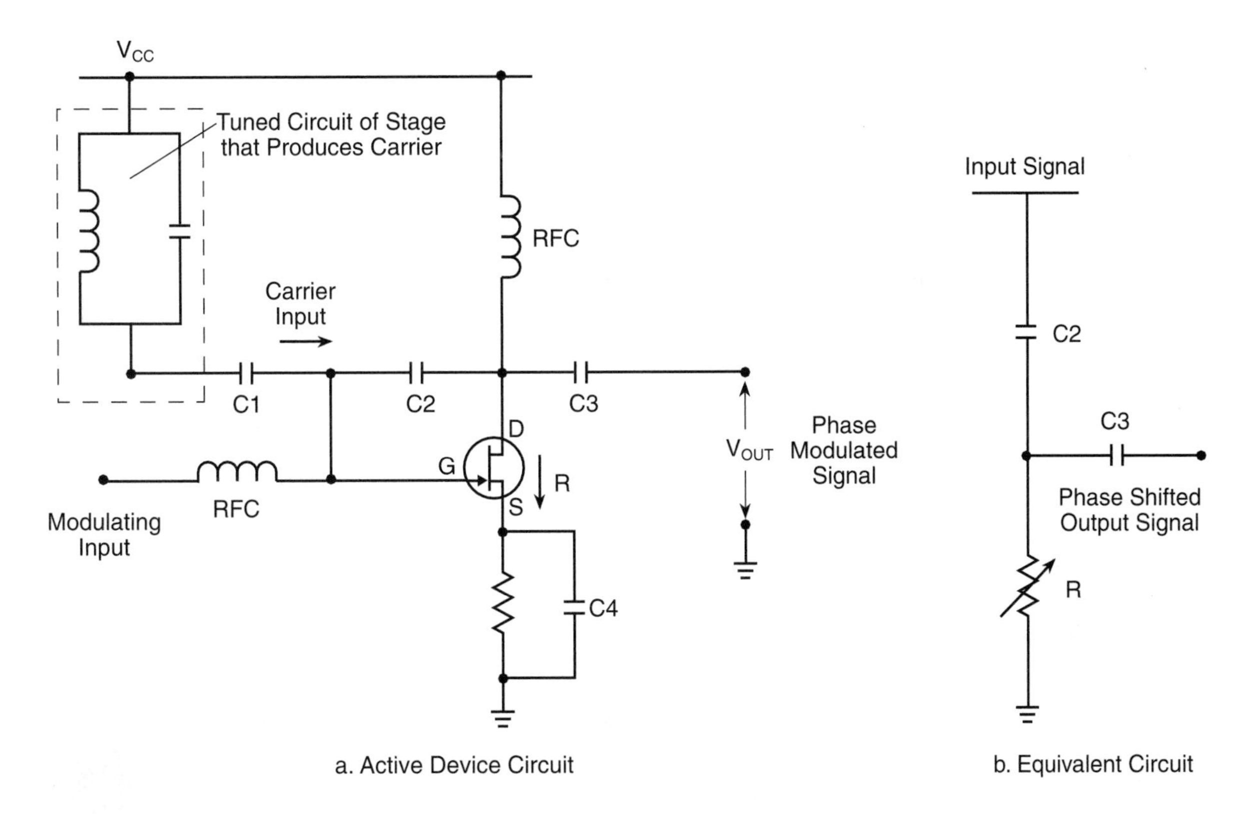


*Figure 5-15. Reactance modulator.*

# FM versus AM

## Greater Number of Sidebands

As in AM, the baseband information signal is translated up to the carrier frequency as a consequence of FM modulation. The effect of the frequency being deviated is essentially different than the AM process. In AM, when the AM modulation is a single frequency, the sum and difference frequencies result. In FM, one sine wave is used to modify another sine wave and multiple sums and differences result, depending on the deviation ratio. This is illustrated in *Figure 5-16a*. Sidebands appear on either side of $f_c$ at $f_m$, 2 $_{fm}$, 3 $f_m$, etc., depending upon the deviation ratio. *Figure 5-16b* shows that there are five significant sidebands for a deviation ratio of 3, and *Figure 5-16c* shows three significant sidebands for a deviation ratio of 1.67. The multiple sidebands are formed in both the upper and lower sideband positions, separated by the modulating frequency. What also is of note is that the carrier amplitude, $f_c$ , *varies* rather than being a constant, as in AM. Note that the *composite* FM signal is at a *constant* amplitude. Of course, if multiple frequencies are contained in the modulating signal, complete bands of frequencies for the sidebands result over the entire bandwidth.

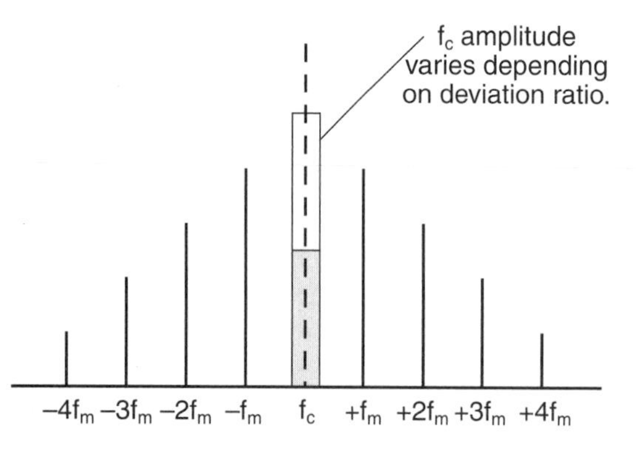


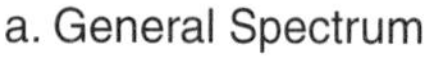

a. General Spectrum

b. Deviation Ratio of 3

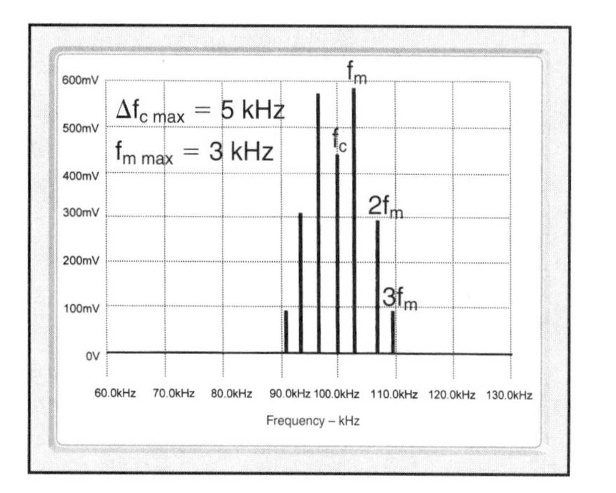


c. Deviation Ratio of 1.67

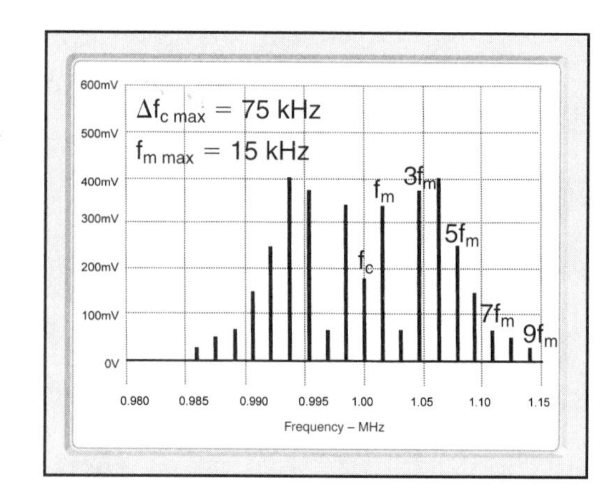


d. Deviation Ratio of 5

*Figure 5-16. Frequency spectrum of FM signals.*

## Greater Required FM Bandwidth

The multiple sidebands cause the FM signal to occupy a greater bandwidth than AM or SSB signals. An approximation of the required bandwidth for an FM signal can be calculated using the following equation (referred to as Carson's Rule after the developer of the relationship).

$BW = 2\ (f_{c\ max} + f_{m\ max})$ where: $BW$ = bandwidth in **Hz**
$f_{c\ max}$ = maximum carrier frequency deviation in **Hz**
$f_{m\ max}$ = maximum modulation frequency in **Hz**

### *Example 4: Calculating Bandwidth*

Calculate the required bandwidth of a 20-kHz maximum carrier frequency deviation modulated by a baseband audio signal of 100 Hz to 3 kHz:

***Solution:***

Use Carson's Rule

$BW = 2\ (20\ kHz + 3\ kHz) = 46 kHz$

## Total Power Output Constant for FM

Another important feature of FM is the fact that the total amount of power in an FM signal is constant, regardless of modulation. Unlike AM systems, there is no fixed relationship between the power in the carrier and the sidebands. Effectively, the power in the carrier is transferred to the sidebands as modulation occurs. This can be seen in *Figure 5-16* by observing that the carrier is not a constant amplitude as the deviation ratio changes. In a few special cases, the carrier may disappear completely as all of the energy goes into the sidebands. Unlike SSB AM, FM signals can be amplified by Class C non-linear power amplifiers because the information is in the change of frequency, not in the amplitude of the signal.

# Typical FM Systems

## Narrow Band

An example of a commonly-used FM system is the narrow-band FM system used by police, fire, marine VHF, and other services for voice communications. Such systems have a maximum carrier deviation of about 5 kHz. In the past, these channels were allocated 25- to 30-kHz spacing, but today are often used with only 15-kHz spacing. Up to three sidebands may form on each side of the carrier, as shown in *Figure 5-16c*, with the first two being the most significant.

### *Example 5: Applying Carson's Rule*

Using Carson's Rule, calculate the bandwidth of the narrow-band system described above. Assume a maximum voice frequency of 3 kHz and a deviation ratio of 1.67

***Solution:***

Use $M_{FM} = \frac{f_{c\ max}}{f_{m\ max}}$ to solve for $f_{c\ max}$, then Carson's Rule.

$1.67 = \frac{f_{c\ max}}{3kHz}$

$f_{c\ max} = 1.67 \times 3 = 5.01\ kHz$

$BW = 2\ (5.01 + 3) = 2\ (8.01) = 16\ kHz$

### Commercial FM

Commercial FM entertainment systems use a much wider deviation of 75 kHz and the baseband, hi-fidelity audio frequency may be up to 15 kHz. This gives a deviation ratio of 5. Carson's rule predicts an occupied bandwidth of 180 kHz. The commercial FM channel is 200 kHz wide — adequate to handle the predicted bandwidth. *Figure 5-16d* shows the spectral distribution for a deviation ratio of 5. As the deviation ratio increases, note the large number of multiple sidebands and also the variation in the amplitude of the carrier.

## FM SYSTEM CHARACTERISTICS

### FM Signal-to-Noise (S/N)

Since FM systems occupy more bandwidth than AM and SSB signals, there must be some reward for using this extra bandwidth. A principal reward is improved S/N ratios. The FM signal is much less susceptible to interference from noise, such as thunderstorms and man-made noise, which affect the amplitude of an RF signal rather than the frequency. The amplitude noise will be removed in the FM receiving system, so the FM system has protection from amplitude interference.

### Processing Gain

As a non-linear system, FM has a special property called processing gain. This effect will cause the S/N ratio *after* demodulation to be greater, but to produce better output S/N ratios it is "paid for" by using greater bandwidth. Consequently, for systems where a large bandwidth is available, improved signal-to-noise ratios may be achieved because of processing gain. FM systems exhibit a threshold where the signal-to-noise curve changes, and tend to work best above this threshold point. Systems with high deviation ratios generally do not perform as well as systems with low deviation ratios at low S/N ratios.

### Quieting

Another interesting property of FM is called "quieting." When an FM carrier is being transmitted with no modulation, the noise in the system output appears to diminish. This effect is due to detector circuits which, when no carrier is present, see large frequency differences with noise, but when the carrier is present, the frequency changes caused by the noise have less effect on the detector, which makes the noise appear to be less.

### Capture

"Capture" is another important aspect of FM systems. When an FM station is fairly strong on a channel it tends to capture the channel — that is, if another station that is weaker attempts to use the same channel the interfering station will not be heard. This can be a benefit because it cuts down on interference from sometimes unwanted stations. This effect is not always desirable, however. For example, aircraft VHF communications use AM modulation, not FM, because air traffic controllers want to hear when multiple transmissions are occurring and don't necessarily want the strongest signal to completely dominate the channel.

## Summary

In this Chapter we have learned that modulation is used to translate a baseband signal containing information to a higher frequency so that it can be transmitted efficiently. There are two basic forms of modulation, amplitude modulation (AM), and frequency modulation (which includes direct frequency modulation, FM, and phase modulation, PM). Single Sideband (SSB) is a form of AM modulation.

AM is normally produced at high power levels and contains a carrier and two sideband signals. SSB also uses AM modulation, but the carrier and one sideband are removed so that only the remaining sideband is transmitted. SSB uses less power and less bandwidth than AM, but SSB signals are more difficult to detect.

FM is produced at low power levels and has a carrier and multiple sidebands. The AM carrier is constant power and the FM carrier power varies with modulation. FM signals generally occupy more bandwidth than an AM signal. FM has improved S/N (signal-to-noise) ratios when compared to AM and SSB, and a strong FM signal will "capture" a channel, which eliminates interference from weaker signals on the same channel.

In the next Chapter, we will discuss a cousin of modulation — mixing.

# Quiz for Chapter 5

1. What is the purpose of modulation?
   a) To improve the tonal quality of the signal at the receiver.
   b) To limit the frequency deviation for improved signal transmission.
   c) To move the information signal to a new position in the frequency spectrum so it can be transmitted easily and accurately.
   d) To permit engineers to view a signal on an oscilloscope.
2. Which type of modulation requires more bandwidth, SSB or AM?
   a) AM
   b) SSB
   c) They require the same amount of bandwidth.
   d) Bandwidth does not apply to these types of modulation.
3. Why is AM used for entertainment radio broadcasting in the 550-kHz to 1600-kHz frequency range?
   a) Because AM stations broadcast to customers that are within line-of-sight range.
   b) Early radio engineers didn't know how to generate signals above 1600 kHz.
   c) This frequency band was mistakenly assigned to amateur radio.
   d) AM is very easy to detect and the required detection circuitry is simple.
4. Generally does FM require more or less bandwidth than AM modulation?
   a) FM and AM require equal amounts of bandwidth.
   b) FM requires greater bandwidth.
   c) AM requires greater bandwidth.
   d) Bandwidth does not apply to these types of modulation.
5. Why is SSB modulation good for voice transmission in the high-frequency region?
   a) It requires more bandwidth so it can accommodate more channels.
   b) It is less susceptible to "selective fading."
   c) SSB doesn't work at all at high frequencies.
   d) Thunderstorms and man-made noise do not affect SSB.
6. How many sidebands are present in SSB, AM, and FM?
   a) SSB has one; AM has two; and FM has two or more.
   b) They all have the same number of sidebands.
   c) SSB and AM each have two; FM has four.
   d) SSB and FM have multiple sidebands; AM has one.
7. What does 100% modulation mean for AM?
   a) The signal level is greatest and reception is best at 100% modulation.
   b) The power in the sidebands is equal to the power in the carrier.
   c) It is the maximum modulation possible without distortion.
   d) It is the minimum modulation necessary to successfully transmit a signal.
8. What is meant by deviation with regard to an FM signal?
   a) It is the "drift" in frequency that requires one to retune their radio from time to time.
   b) It is the difference between the carrier frequency and the baseband frequency.
   c) It is the method used for producing FM stereo.
   d) It is the maximum carrier frequency change caused by the modulating frequency.
9. Why is FM used in entertainment radio broadcasting in the 88MHz to 108MHz range?
   a) FM uses greater bandwidth, which is better accommodated at higher frequencies.
   b) Because FM stations can broadcast for very long distances.
   c) Radio manufacturers in the early 1950's agreed to this standard.
   d) The amount of power to the FM antenna requires higher frequencies.
10. What are the beneficial characteristics of FM modulation?
    a) There are none. AM signals are better than FM signals.
    b) Improved S/N; processing gain; quieting; and capture.
    c) FM has better tonal quality that SSB, and requires less bandwidth.
    d) FM has more sidebands and is easier to produce than AM.

**Answers:**
1 c, 2 a, 3 d, 4 b, 5 b, 6 a, 7 c, 8 d, 9 a, 10 b.

## Questions & Problems for Chapter 5

1. In AM modulation, all of the __________ is in the sidebands.
2. If a 4-kHz baseband signal is used to AM modulate a carrier at 1-MHz, what are the sideband frequencies?
3. A 1000 watt carrier is 50% AM modulated. How much power is in both sidebands? In each sideband?
4. A single 1-kHz signal is used to AM modulate a 1-MHz carrier. What is the total bandwidth of the output signal?
5. In a single sideband signal, the __________ is suppressed and one sideband is filtered out.
6. A __________ modulator is used to generate SSB.
7. In __________ modulation, the phase of a signal is changed. In __________ modulation the frequency of the signal is changed. Both forms of modulation are termed __________ modulation.
8. An FM carrier of 1000 watts is deviated by 3-kHz. How much power is in the modulated signal?
9. Calculate the deviation ratios for the following FM signals:

| Maximum carrier frequencies ($f_{c\ max}$): | Maximum modulation frequencies ($f_{m\ max}$) |
|---|---|
| 120 MHz | 8 kHz |
| 200 MHz | 20 kHz |
| 70 MHz | 10 kHz |
| 60 MHz | 12 kHz |

10. FM systems have a quality called _______ where the stronger station signal will completely dominate the channel and eliminate interference from stations broadcasting on the same frequency with a weaker signal.

*(Answers on page 211.)*

# CHAPTER 6

# Mixing & Heterodyning

Recall that in Chapter 2 we showed that mixing is a function that translates the information signal to a different position in the frequency spectrum, ideally without affecting the quality of the information signal. Mixers are very common circuits in communications electronics and nearly all radio receiver designs include a mixer, except (possibly) some small, short-range receivers used in control applications. Mixers also are used extensively in transmitters, and anytime an application needs to move a signal from one frequency to another. Mixing, like modulation, uses devices with non-linear characteristics to multiply signals together and produce sum and difference frequencies.

## The Basic Idea...

Let's review the basic process of the mixing function that was explained in Chapter 2 by looking at *Figure 6-1*. Mixing requires at least two frequencies — the first (Signal A in *Figure 6-1*) usually is the modulated signal containing the information, and the second (Signal B) is supplied by an oscillator at a single frequency. Signal B — the oscillator signal — is either higher or lower in frequency but usually larger in amplitude than Signal A. Signal C, the mixer output, has a frequency which is either the sum or difference of the frequencies of Signal A and Signal B. In our example, if Signal A is a modulated signal used to transfer information, Signal C, at the difference frequency, will be modulated the same as Signal A. The result is that the information is translated — moved or converted — from the higher frequency of Signal A to the

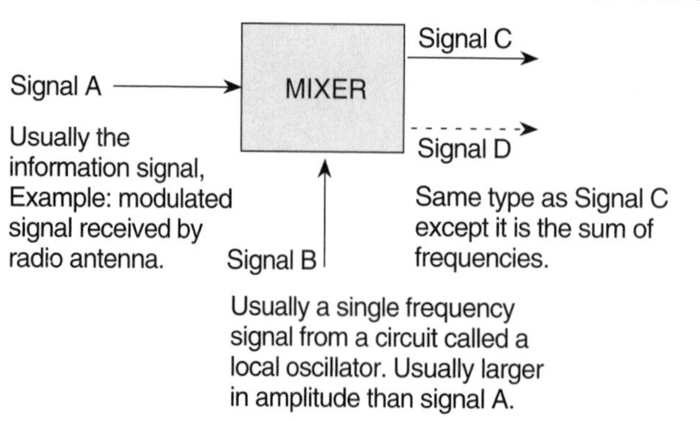


*Figure 6-1. Block diagram of a mixer circuit.*

lower (difference) frequency of Signal C. This is a very important function in communications systems because it allows signals to be *translated* to lower (or higher) frequencies where they can be processed in different ways.

The circuit devices used to accomplish the mixing function have non-linear transfer characteristics. As a result, Signal D, the signal that is the sum of the frequencies of Signal A plus Signal B, also is created. Mixers may translate signals either up or down in frequency, but only one of the translated signals will be used, and the other rejected, or filtered-out. We will call the circuit that does the frequency conversion a mixer. At times, people may refer to the mixer as a "converter." Our definition of a *converter* will be the *mixer combined with the local oscillator*. Mixers are used in both transmitters and receivers. Certain receivers, called dual-conversion receivers, use two converter circuits and are discussed in Chapter 8.

## Heterodyning

*Heterodyning* is a form of mixing used in receivers. Heterodyning is best illustrated with an example, shown in *Figure 6-2*, of how a mixer is used in an AM broadcast radio receiver. The design of the receiver system requires that Signal C — the output signal of the mixer — always be at 455 kHz so that the signal can be amplified and processed more efficiently and with greater selectivity than if the complete band of frequencies at which Signal A may be received were amplified and processed.

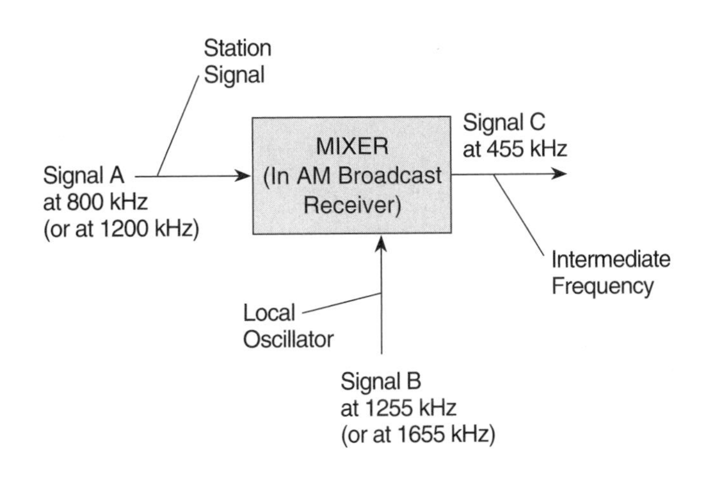


*Figure 6-2. Block diagram of an AM broadcast receiver mixer.*

Let's assume that the receiver is to amplify the information contained in Signal A from a broadcast station at a frequency of 800 kHz (800 on your AM dial). In order for Signal C to be at the difference frequency of 455 kHz it must be mixed with Signal B (from the oscillator) at a frequency of 1255 kHz. Now, let's turn the dial so that Signal A is from a station at a frequency of 1200 kHz (1200 on your AM dial). In order for the output Signal C to be at a difference frequency of 455 kHz, Signal B must be at 1655 kHz. For each station selected on the tuning dial, the local oscillator (Signal B) is adjusted to produce the difference frequency signal of 455 kHz (Signal C). The tuning of the radio to select an individual station in an AM receiver is accomplished by varying the oscillator frequency of Signal B to produce an output frequency for Signal C of 455 kHz. This frequency conversion tuning principle is called *heterodyning*.

Heterodyning in a receiver requires a station signal (Signal A) and a tunable *local oscillator* output (Signal B), and provides a single frequency output (Signal C) called the *intermediate frequency* (IF). The intermediate frequency contains all of the information that modulated the original broadcast signal (Signal A). Heterodyning makes it possible to design the receiver so that it has to amplify only the intermediate frequency of

## *Example 1: Mixing High-Frequency Signals*

Mixing, especially when one signal is much, much greater than another, can be represented by two vectors rotating at their respective frequencies, with one rotating at the tip of the other. A third vector will result (the third side of a triangle) representing the amplitude of the envelope of the mixed signal[1]. The *envelope amplitude* will be at the *difference frequency,* and if signal A varies due to modulation, the envelope will be modulated in the same fashion.

Prove with vector plotting and amplitude measuring that the envelope varies at the difference frequency when $f_2 = 3f_1$, and vector A = 0.5 units and vector B = 3 units. Vector B rotates at frequency $f_1$ and vector A rotates at frequency $f_2$.

### *Solution:*

Since $f_2 = 3f_1$ when B rotates 30°, A will rotate 90° or, conversely, when B rotates 90°, A rotates 270°. Each 30° rotation of B and the corresponding rotation of A are plotted on the vector diagram and the measured amplitude of vector C is plotted on the time diagram.

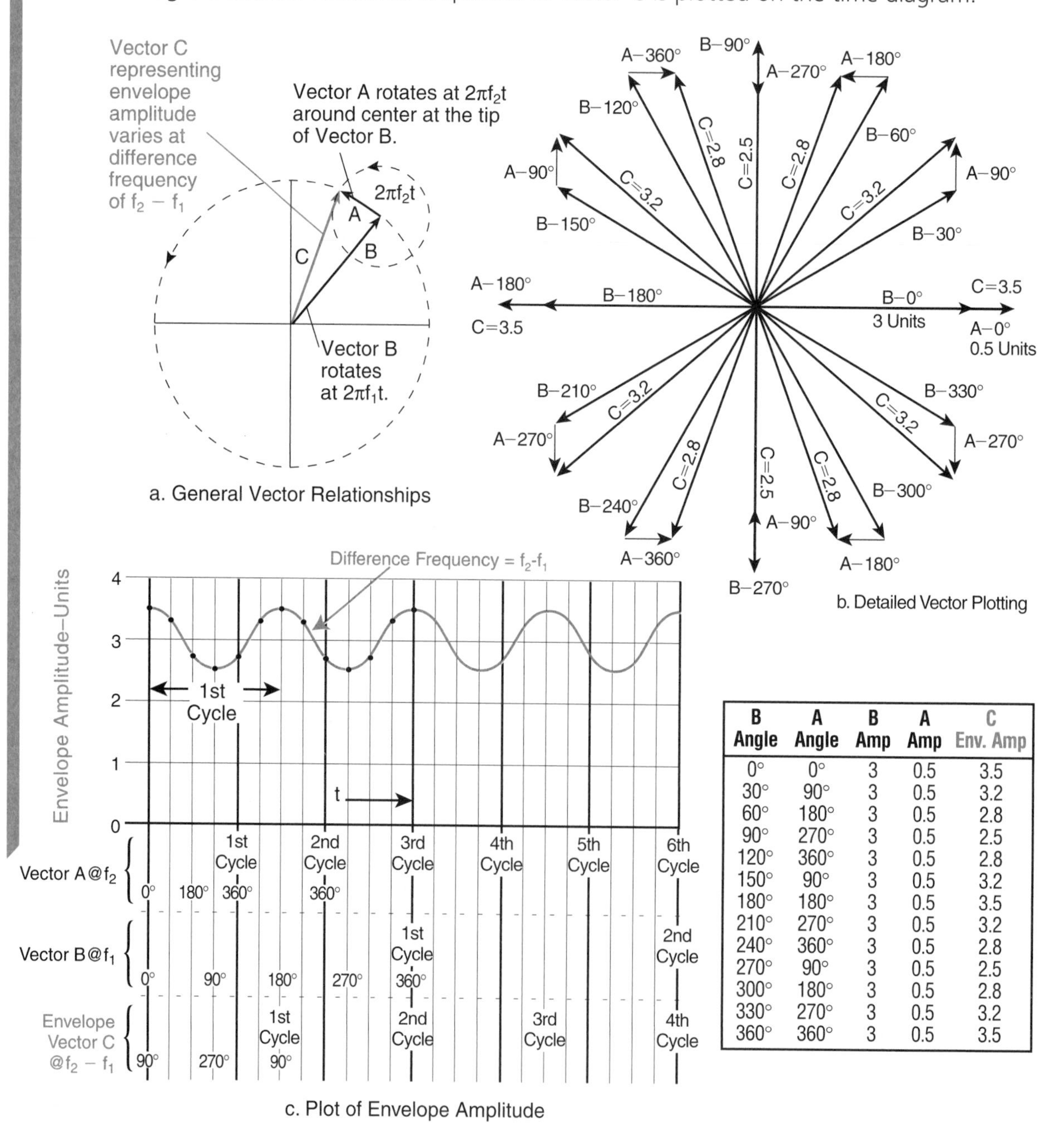


| B Angle | A Angle | B Amp | A Amp | C Env. Amp |
|---|---|---|---|---|
| 0° | 0° | 3 | 0.5 | 3.5 |
| 30° | 90° | 3 | 0.5 | 3.2 |
| 60° | 180° | 3 | 0.5 | 2.8 |
| 90° | 270° | 3 | 0.5 | 2.5 |
| 120° | 360° | 3 | 0.5 | 2.8 |
| 150° | 90° | 3 | 0.5 | 3.2 |
| 180° | 180° | 3 | 0.5 | 3.5 |
| 210° | 270° | 3 | 0.5 | 3.2 |
| 240° | 360° | 3 | 0.5 | 2.8 |
| 270° | 90° | 3 | 0.5 | 2.5 |
| 300° | 180° | 3 | 0.5 | 2.8 |
| 330° | 270° | 3 | 0.5 | 3.2 |
| 360° | 360° | 3 | 0.5 | 3.5 |



[1] *Radio Engineering, F. E. Terman, pp 525-527, Copyright 1932, 1937, 1947, McGraw-Hill Book Company, Inc.*

455 kHz regardless of the frequency of the stations received. This simplifies signal processing and makes it economical to design amplifying circuits and filters that provide excellent sensitivity and selectivity. A very stable, tunable local oscillator is required.

*Figure 6-3* shows the output of a computer simulation that mixes a 5-MHz signal with a 9-MHz signal to arrive at a difference frequency of 4 MHz. Normally, in high-frequency circuits, the separation between the local oscillator frequency and the signal frequency is much greater (more like 10 to 20 MHz), and the amplitude of the oscillator signal is much larger than the station signal. As a result, the variations from a sine wave in the difference frequency signal will be much smaller than those shown in *Figure 6-3* where both signals have equal amplitudes. Only the resultant difference frequency is shown in *Figure 6-3c.* The sum frequency shown in *Figure 6-3d* also is present and must be filtered out if not used.

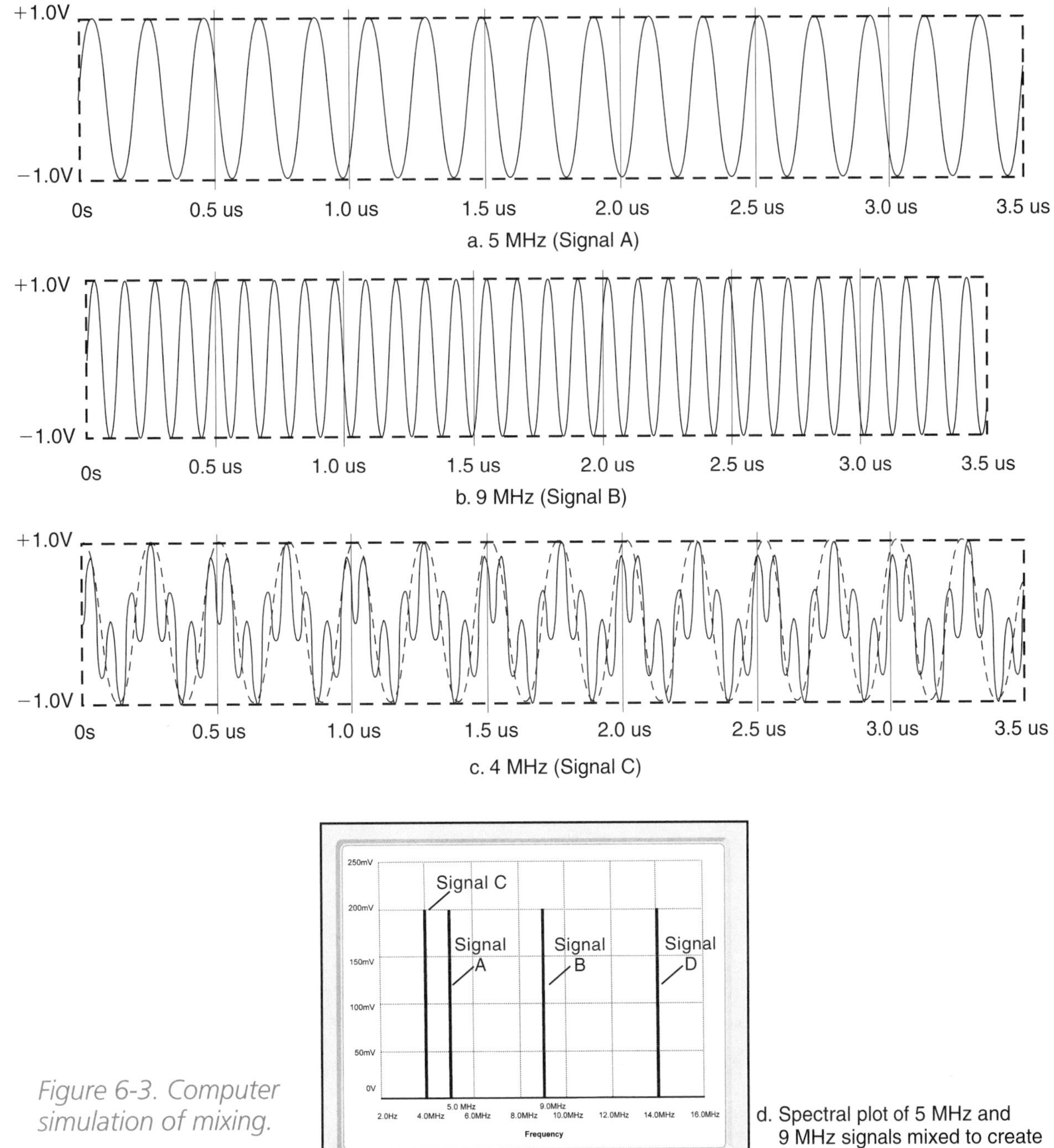


*Figure 6-3. Computer simulation of mixing.*

### *Example 2: Tuning-in AM Stations*

Using *Figure 6-2*, determine the frequency of the local (tuning) oscillator, Signal B, required to tune-in stations at the following AM frequencies: 550 kHz, 720 kHz, 1000 kHz, 1370 kHz, and 1590 kHz.

***Solution:***

Local Oscillator (Signal B) -Station (Signal A) = Intermediate Frequency (Signal C)

| **Local Oscillator (Signal B)** | **- Broadcast Station - (Signal A)** | **= Intermediate Frequency = (Signal C)** |
|---|---|---|
| 1005 kHz | 550 kHz | 455 kHz |
| 1175 kHz | 720 kHz | 455 kHz |
| 1455 kHz | 1000 kHz | 455 kHz |
| 1825 kHz | 1370 kHz | 455 kHz |
| 2045 kHz | 1590 kHz | 455 kHz |

## Single-Stage Mixer Circuits

In most communications systems, mixers are constructed from single-stage semi-conductor circuits. Diodes and both bipolar and FET transistors may be used.

## Diode Mixers

Diodes were the first devices used as mixers because they are simple devices that have the non-linear characteristic required to produce the difference (and sum) of the two frequencies mixed together. *Figure 6-4* shows the non-linear transfer characteristic of a diode. The forward conduction characteristic shows that when the anode voltage is greater than 0.7V with respect to the cathode, small changes in voltage will cause large changes of current. Near the 0.7V region, the ratio of current change to voltage change

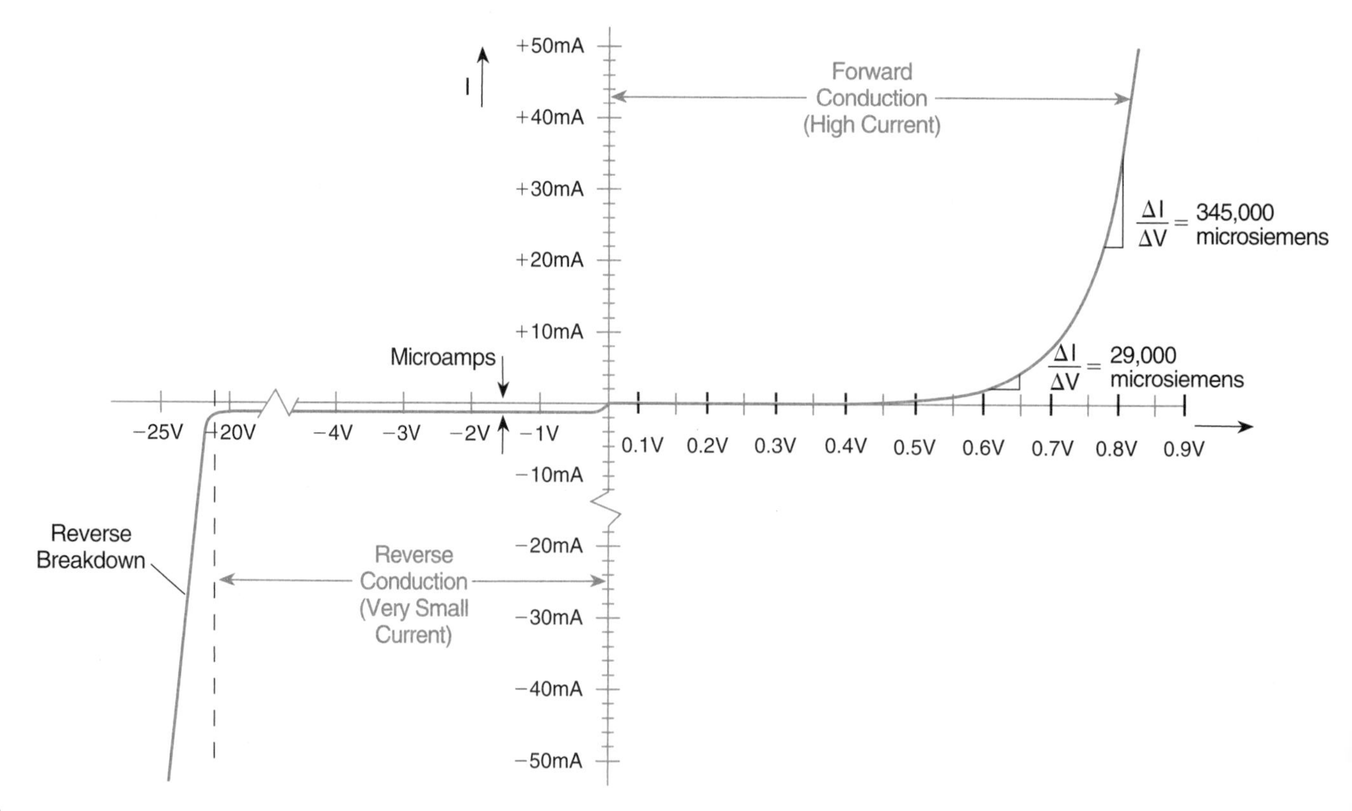


*Figure 6-4. Non-linear transfer characteristics of a diode.*

is 29,000 microsiemens, but for voltages above 0.7V the ratio of current to voltage change is 345,000 microsiemens, illustrating the non-linear characteristic of diodes that makes for good mixers.

*Figure 6-5* shows a diode mixer circuit — in this case a singly-balanced mixer. The diodes contribute the non-linear characteristic and the triple-wound (trifilar) transformers, carefully designed, provide the balancing. The station RF signal (Signal A) is on the primary of the transformer. The local oscillator signal (Signal B) is fed at the center tap of the secondary. The result is that the balancing keeps the station RF signal isolated from the local oscillator input, the local oscillator signal isolated from the RF input, and an output which is a balanced signal at the difference frequency — the intermediate frequency (Signal C). The RF chokes provide high impedance to the respective input signals. Early mixers were built with point-contact diodes with limited frequency performance, but the latest semiconductor technology has produced high-performance Schottky diode mixers up to 100 GHz in frequency.

Diode mixers have large dynamic range, very good broad-band frequency performance, a conversion loss (they don't have gain), and usually require larger amounts of local oscillator power. Their circuits generally are simpler than those that use active devices. There also are doubly-balanced diode mixers designed to provide even greater isolation between RF, local oscillator, and output frequency signals.

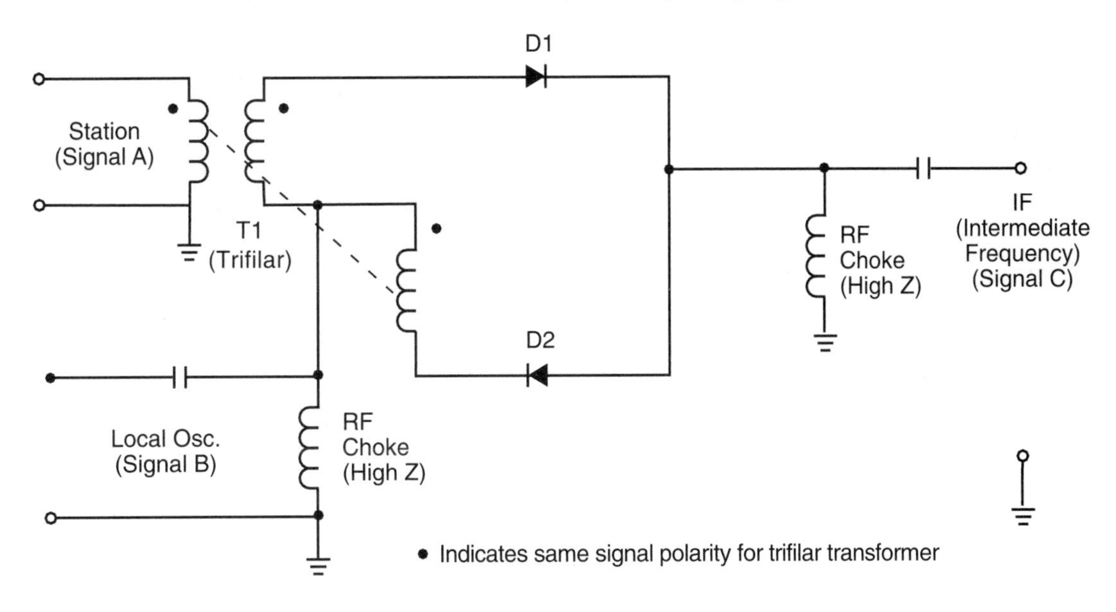


*Figure 6-5. A singly-balanced diode mixer circuit.*

## Active Device Mixer — JFET

*Figure 6-6* is an active device mixer built using a 2N3819 N-channel JFET operating in the depletion mode. The 270-ohm resistor from source to ground provides self-bias, similar to the tuned amplifier in *Figure 4-9* of Chapter 4. The bias point is shown on *Figure 6-7*, which is the transfer characteristics of the gate-to-source voltage plotted against the drain-to-source current of the 2N3819 JFET. The self-bias of approximately 1.2V is developed with approximately 4.4mA through the 270-ohm source-to-ground resistor. Notice that, as the input voltage increases on the gate, the drain current does not decrease in a linear or straight-line fashion. This type of FET has the necessary non-linear characteristic and is sometimes referred to as a "square-law device," since the output tends to be a mathematical square of the input. The LC tuned circuit in the drain of the FET passes only a narrow band of frequencies, so this band-pass filter on the output selects only the difference — or sum — (Signal C) frequency to pass on to the intermediate frequency (IF) amplifiers.

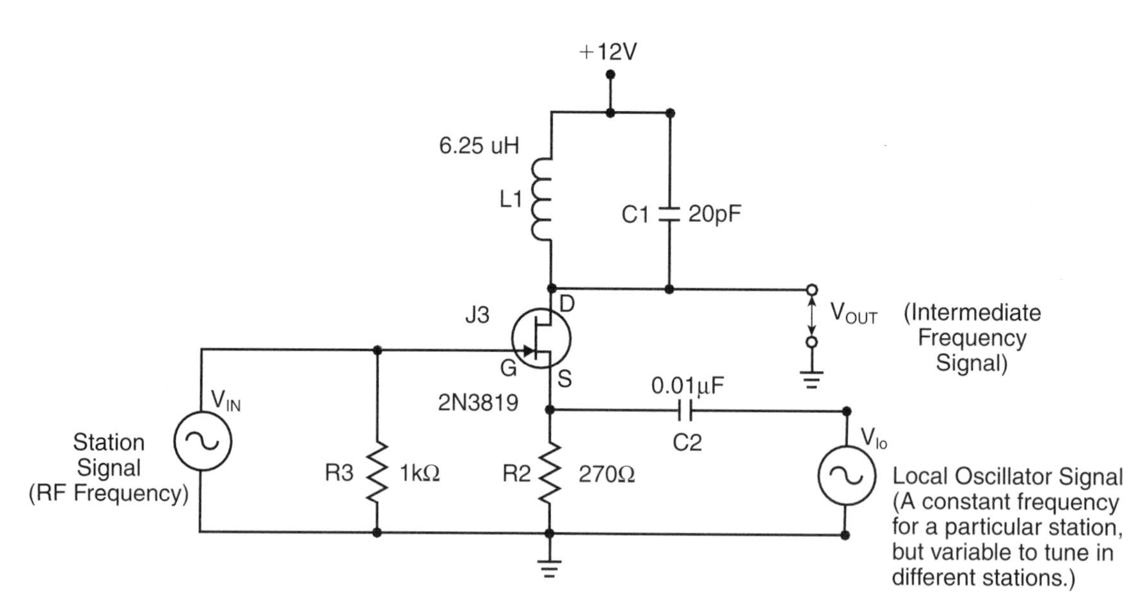


*Figure 6-6. An active-device mixer circuit using a 2N3819 JFET.*

The mixer stage resembles a tuned amplifier where the input levels are adjusted to give the optimum mixing action. The station or RF input signal is coupled to the gate of the FET and the local oscillator is coupled to the source. This coupling technique helps separate the two signals to be mixed, but there is much less isolation between RF and local oscillator signals in this circuit than in the balanced diode circuit, so well-designed filtering is required at the output to eliminate unwanted frequencies before coupling the signal to the IF amplifiers. The mixer "acts" as two types of amplifier circuits at the same time. With regard to the signal on the gate, the circuit is a common-drain amplifier. With regard to the signal coupled to the source, the circuit is configured as a common-gate amplifier. Referring again to *Figure 6-7*, we see that the RF signal on the gate operates over a linear region and the local oscillator signal, which is a much larger amplitude, operates over a non-linear region to produce the mixing action.

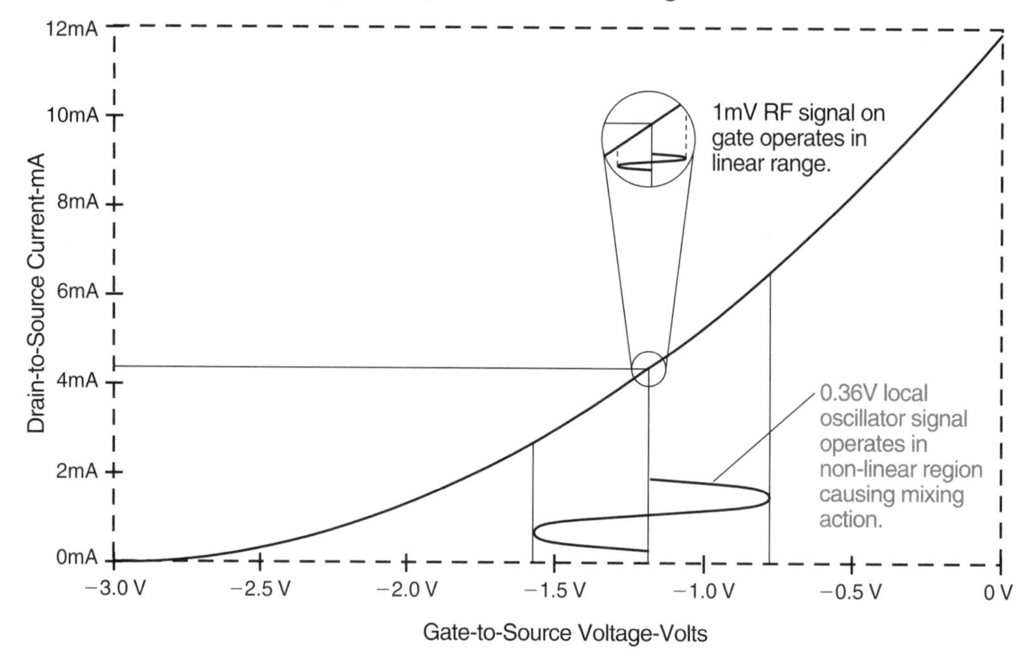

*Figure 6-7. Non-linear transfer characteristics of the 2N3819.*

*Figure 6-8* shows the mixer signals on a frequency spectrum plot when the RF signal (Signal A) is at a frequency of 5 MHz, and the local oscillator signal (Signal B) is at

9-MHz. Note that there are two intermediate frequency signals, the difference frequency at 4 MHz and the sum frequency of 14 MHz. The output may be set to select either the sum or difference frequency, and the circuit in *Figure 6-6* is set to pass the sum frequency of 14 MHz and reject the difference frequency of 4 MHz.

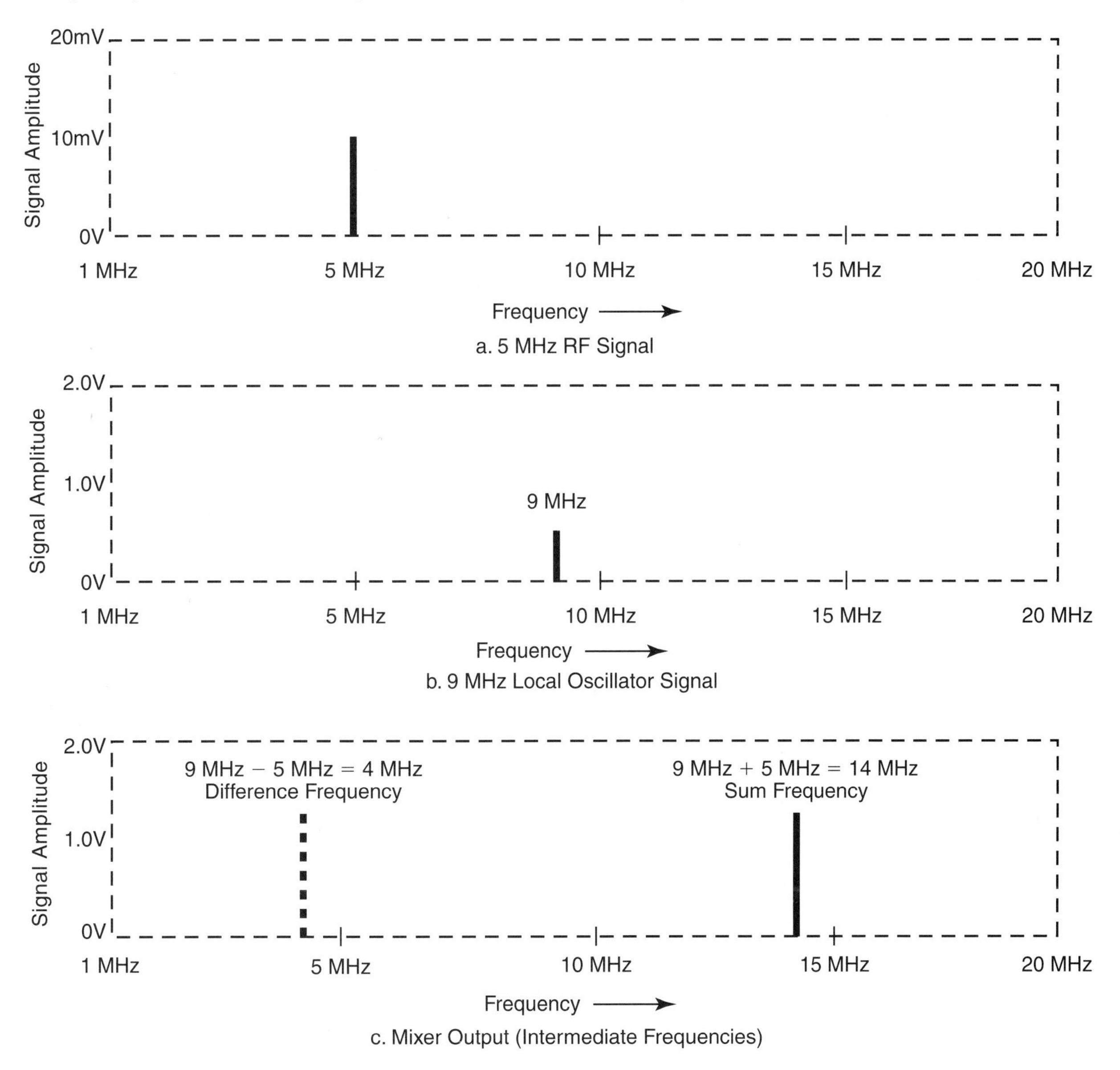


*Figure 6-8. Frequency spectrum plot of mixer signals.*

## Active Device Mixer — Dual Gate MOSFET

A popular mixer design that is very similar to the circuit of *Figure 6-6* is shown in *Figure 6-9*. It uses an N-channel depletion-mode dual-gate MOSFET as the active device. The dual-gate MOSFET has two input gates rather than one, which allows the two signals to be mixed to be applied to individual gates. This provides a big advantage, since it normally is easier to drive the high-input-impedance gate of an MOSFET transistor rather than its source, thus simplifying the design. The output is tuned to either the sum or the difference frequency. The mixer circuits of *Figure 6-6* and *6-9* have gain (rather than conversion loss, as for diodes) and have good sensitivity, dynamic range, and frequency performance. Because there is gain, isolation of signals is a much greater problem and filtering out unwanted signals to pass the required sum or difference frequency requires very careful circuit design and construction. In addition, circuits like the one shown in

*Figure 6-9* usually are biased so that the MOSFET is just at conduction. Therefore, the local oscillator signal causes the transistor to conduct only on one-half of its cycle. Thus, the circuit operates as a class-C tuned amplifier for the mixing.

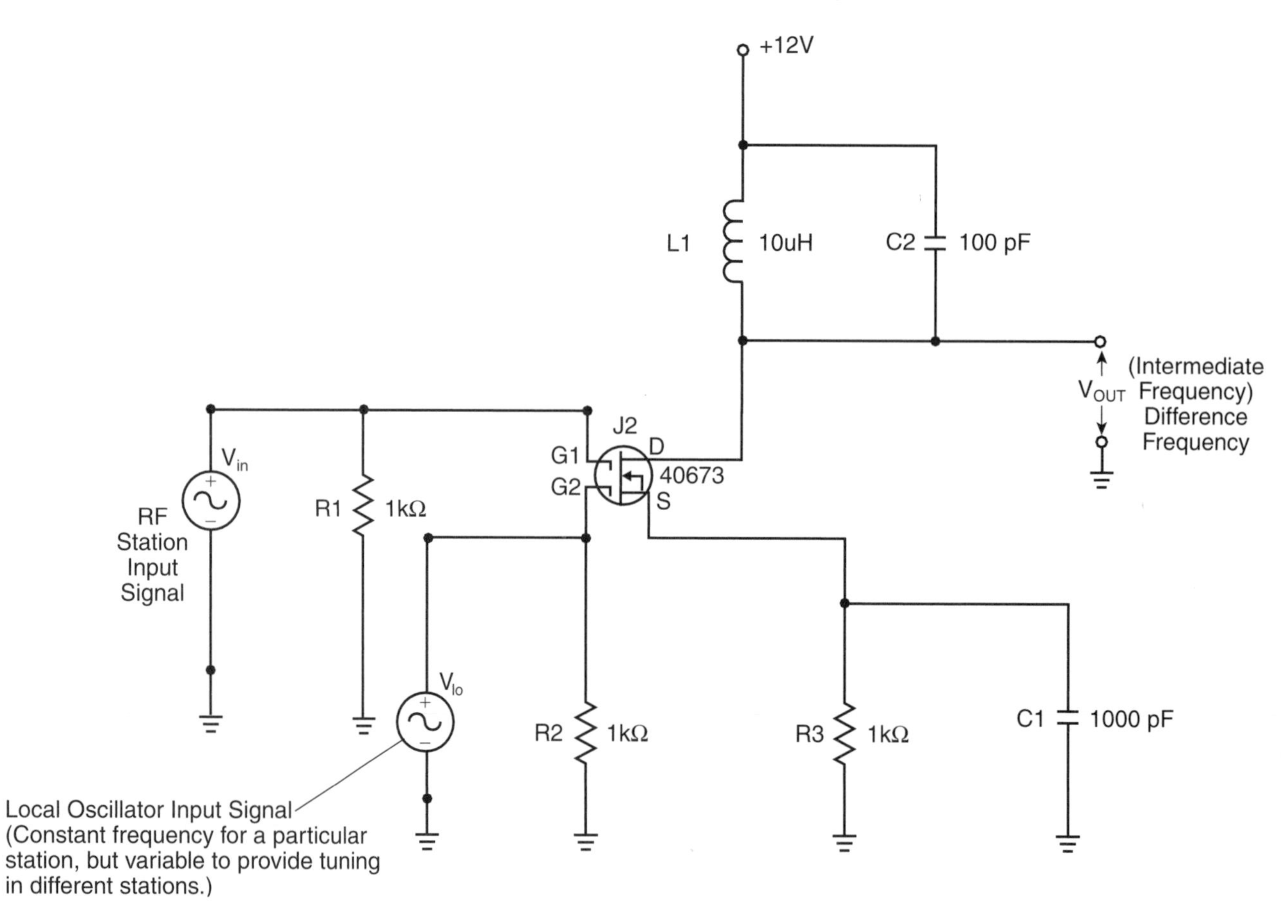


*Figure 6-9. Dual-gate MOSFET mixer circuit.*

## Integrated Circuit Mixers

There are a large number of integrated circuits used as all different kinds of mixers, and for various frequency ranges. One good example is the MC1496 balanced mixer IC, shown in *Figure 6-10*.

The MC1496 is a bipolar silicon IC. It consists of a combination of differential amplifiers. Q1, Q2 and Q3, Q4, are interconnected as a dual differential amplifier. This top dual amplifier is driven by a standard differential amplifier with Q5 and Q6 as the active devices. The differential amplifiers are designed to amplify the difference between the signals on the bases of their respective transistors. Signals at the base of the differential amplifiers drive the base inputs out of phase. If one base goes positive, the other goes negative. Cross coupling the collectors provides full-wave balanced multiplication of the two input voltages to produce the mixing action. The output consists of large amplitudes, compared to the input signals, of the sum and difference frequencies. In this case, the difference frequency is used. The circuit of *Figure 6-10* is tuned to a 9-MHz IF frequency. The signal input and output impedances are 50 ohms. The circuit has gain and good frequency performance.

## Image Frequencies

Because mixing produces sum and difference frequencies of the station signal and the local oscillator signal, there is the opportunity for interfering signals that appear at the RF input to also produce intermediate frequency signals at the mixer output. One of the

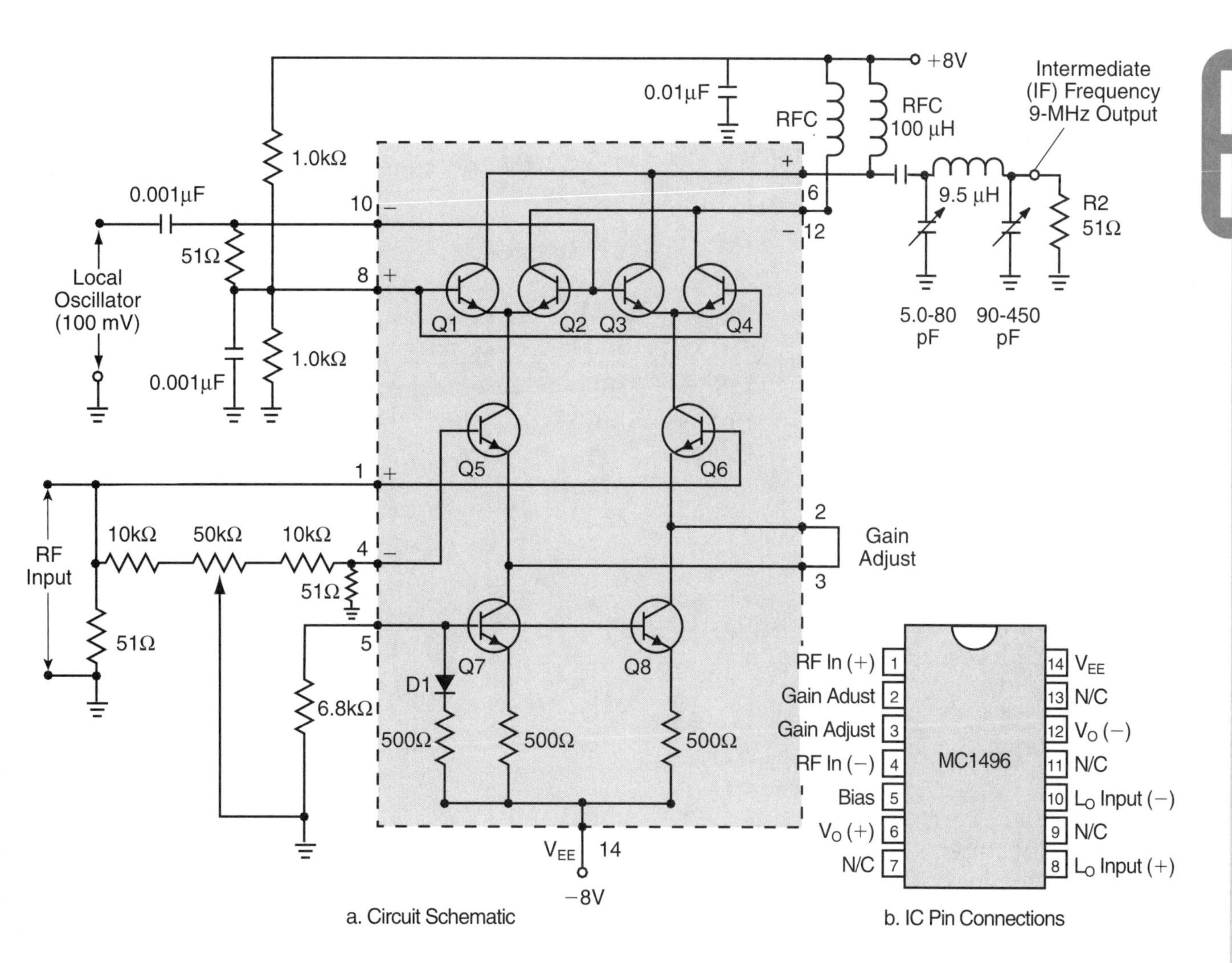


*Figure 6-10. MC1496 bipolar IC balanced mixer.* (Courtesy of Motorola, Inc.)

most prominent signals of this type is one at a frequency that is greater than the station frequency by twice the intermediate frequency. These interfering signals are called *image frequencies*. For example, in AM broadcast receivers, the intermediate frequency is 455 kHz. If a station is at 600 kHz and the local oscillator is changed to 1055 to tune in the station, then a station frequency at 1510 kHz is an *image frequency* because 1510 kHz - 1055 kHz also produces an intermediate frequency of 455 kHz. Image frequencies are produced at frequencies that are *twice the intermediate frequency* away from the station frequency. Image frequency problems are decreased in receivers by making sure the selectivity of circuits ahead of the mixer reduce the amplitude of the image frequency signal to levels where it is not a problem, and by careful selection of the first IF frequency.

## Mixing Applications

Mixing, as we have shown, is very similar to modulation in that it produces the sum and difference frequencies of the two mixed signals through multiplication of the signals with devices that have non-linear characteristics. Whether a mixer is used to translate a signal up in frequency or down in frequency only depends on the particular application.

In addition, we have implied that each of the signals that are mixed together has one frequency. That certainly is not the case. In many applications, the signals to be mixed have multiple frequencies, and the mixing will not be just the multiplication of two signals, but a composite of the multiplication of multiple frequencies.

### *Example 3: Calculating Image Frequencies*

Calculate the image frequencies that could interfere with AM standard broadcast input stations when the IF is 300 kHz, 500 kHz and 700 kHz.

***Solution:***

Use Station Frequency + 2 IF = Image Frequency

| Station Frequency (kHz) | Intermediate Frequencies 300 kHz | 500 kHz | 700kHz | |
|---|---|---|---|---|
| 550 | 1150 | 1550 | 1950 | Image Frequencies |
| 700 | 1300 | 1700 | 2100 | |
| 950 | 1550 | 1950 | 2350 | |
| 1100 | 1700 | 2100 | 2500 | |
| 1350 | 1950 | 2350 | 2750 | |

Check:

| IF | Station | LO | Image Freq | |
|---|---|---|---|---|
| 300 | 550 | 850 | 1150 | 1150 - 850 = 300 |
| 500 | 950 | 1450 | 1950 | 1950 - 1450 = 500 |
| 700 | 1350 | 2050 | 2750 | 2750 - 2050 = 700 |

Also, we may not have stressed enough that the modulation on one of the signals into the mixer is preserved on both the sum and difference frequency signal outputs from the mixer. Thus, the designer can have the freedom to choose the signal that best fits the application.

Mixers are used extensively in many electronic communication systems other than receivers and transmitters. Even though we stated the following at the beginning of this Chapter, it is good to state it again: *"Anytime an application requires that the signal be moved from one frequency to another, a mixer can be used."*

## Summary

In this Chapter, we have learned that processing analog signals together to translate them to higher or lower frequencies is a function called *mixing. Heterodyning* is a radio receiver design technique that uses mixer circuits in conjunction with tunable oscillators to select individual frequencies, translating them to a single intermediate frequency for further signal processing (usually 455 kHz for AM broadcast receivers). While mixing can produce both the sum and difference frequencies between the signals mixed, usually the difference frequency is selected, but the sum frequency is used with equal results. Mixing generally takes place at low signal levels with one signal level larger than the other. Even though mixing is similar to modulation, the frequency difference between signals usually is much greater for mixing than for modulation. Also, mixing occurs at lower power levels than modulation.

In the next Chapter, we will look at transmitters.

# Quiz for Chapter 6

1. How are mixing and modulation similar?
   a) They both translate a signal in frequency
   b) Mixing is easier to do
   c) They both require a linear process
   d) All of the above
2. A converter circuit has what basic components?
   a) A modulator and amplifier
   b) A mixer and amplifier
   c) A mixer and local oscillator
   d) A mixer and a modulator
3. Heterodyning is a special type of mixing that…
   a) Converts radio station signals of different frequencies to the same fixed intermediate frequency
   b) Uses a tunable local oscillator
   c) Is a form of mixing used in receivers
   d) All of the above
4. Diodes are useful as mixers because they…
   a) Handle large amounts of power
   b) Have non-linear characteristics
   c) Have linear characteristics
   d) Require low signal levels to drive
5. Active devices such as FET transistors are good mixers because…
   a) They have very low gain and linear characteristics
   b) They have high gain and linear characteristics
   c) They have high gain and non-linear characteristics
   d) None of the above
6. A modulated signal can be easily translated by a mixer because…
   a) It is larger than the oscillator signal
   b) The mixing process preserves all the frequencies in a translated signal
   c) Signal envelopes will not pass through the mixing process
   d) Mixers only work with sinewaves at a single frequency
7. Image frequencies arise because…
   a) Mixers have single frequency outputs
   b) Mixer circuits may have to much gain
   c) Mixers provide insufficient signal isolation
   d) Mixers produce sum and difference frequency outputs
8. Mixing two signals, each with single frequencies, produces
   a) Signals at twice the sum of the frequencies.
   b) Signals at twice the difference of the frequencies.
   c) Signals with frequencies that are the sum and difference of the original frequencies.
   d) All of the above.
9. The intermediate frequency of heterodyning is used in receiver design
   a) To provide high signal-to-noise ratios.
   b) To provide for sensitivity and selectivity in the receiver.
   c) To lower the receiver gain for lower signal-to-noise ratios.
   d) So that two local oscillators can be used in the mixer circuit.
10. Mixing can have two input signal of the following type:
    a) One a single-frequency input signal and one a saw-tooth wave local oscillator signal.
    b) One a square-wave signal and the other a local oscillator signal.
    c) One a single-frequency oscillator signal and one a modulated carrier signal.
    d) a and c above.
11. Dual-gate MOSFET mixers are an advantage because they:
    a) Provide two outputs that can be used.
    b) Have very-low input impedance.
    c) Have very-high input impedance.
    d) Have no isolation between the inputs.
12. Diode mixers have the following characteristics:
    a) Large gain, no conversion loss, narrow-band frequency range.
    b) Small gain, no conversion loss, narrow-band frequency range.
    c) No gain, conversion loss, broad-band frequency range.
    d) No gain, no conversion loss, broad-band frequency range.

**Answers:**
1 a, 2 c, 3 d, 4 b, 5 c, 6 b, 7 d,
8 c, 9 b, 10 c, 11 c, 12 c

## Questions & Problems for Chapter 6

1. If the input signals to a mixer are 5 MHz and 17 MHz, what will the output frequencies be?
2. If you wish to receive a station transmitting at 1.8 MHz and the receiver IF frequency is 455kHz, what frequency should the local oscillator be tuned to?
3. If the intended received frequency is 11 MHz, what is the image frequency if the IF frequency is 9 MHz and the Local Oscillator is 20 MHz?
4. Is a diode mixer an active device mixer?
5. Why is an FET a candidate to be able to design a good mixer?
6. In order to design a receiver with very good selectivity and sensitivity, a ____________, ____________, and __________ __________ oscillator are required.
7. Semiconductor technology has produced diodes used in diode mixers up to a ____ Hz in frequency.
8. If an FET is used that has the "square law" gate-to-source characteristics like *Figure 6-7,* and the operating point is to be at –1.7V gate-to-source voltage, what size resistor must be used for R2 in the circuit of *Figure 6-6?*
9. A spectrum analyzer is used to detect the frequency output of two mixed signals. One has a frequency of 100 MHz and the other has two frequencies, one at 2 MHz and another at 4 MHz. Where will the signals be detected?
10. In the bipolar balanced mixer of *Figure 6-10,* a __________ __________ __________ of the two input signals produces the mixing action.

*(Answers on page 211.)*

# CHAPTER 7
# Transmitters

Transmitters are a major subsystem in communications systems. A transmitter accepts the input information, prepares that information for transfer on the transmission link, and raises the output power to a satisfactory level for successful transfer of the information. The transmitters we will be discussing in this chapter are for radio communications.

A radio transmitter's purpose is to generate a signal at a specific frequency (the carrier frequency), modulate the carrier with the input baseband signal that contains the information, and increase the power level of the modulated signal for coupling to an antenna for transmission. A transmitter is a complete, functional system and we will show how many of the circuits that provide the functions we have been discussing are integrated together to form the transmitter system.

## System Requirements

There are three very important characteristics for a radio transmitter:

1. Frequency stability
2. Quality of output signal
3. Reliability

Let's discuss each of these separately.

### Frequency Stability

The output frequency of the transmitter needs to remain stable to assure that its signal can be tuned-in at the receiver so that the information can be transferred completely and accurately. The overall frequency stability of a transmitter is determined by the stability of the oscillators used to establish the transmitting frequency. Oscillator quality should be sufficient for the application, and frequency stability is generally easy to achieve with the use of synthesizers and crystal oscillators.

### Quality of Output Signal

The quality of the output signal refers to the level of distortion in the output signal. Distortion is caused when harmonic signals are present in the output signal that were not present in the input signal. Harmonic frequencies that are radiated, along with unwanted energy at other frequencies, are called spurious radiation. These spurious output signals must be suppressed as much as possible — at least below the limits set for a given application. The maximum level of spurious radiation that is allowed in a transmission is set by the FCC.

## Reliability

Reliability is important in the design of a transmitter, just as it is in other electronic systems. Many circuits operate at high power levels and, as a result, generate large amounts of heat. Heat is one of the major contributors to degrading the reliability of most circuit elements, so transmitters must be cooled to maintain their reliability.

# The AM Transmitter

*Figure 7-1* is a block diagram of a general-purpose AM transmitter. This is a simple transmitter and it could be expanded to contain additional capabilities for a different system, but the basic functions will be the same. All of the stages are either transformer-coupled or use capacitors between stages to isolate the dc voltages used for supplying power or for biasing.

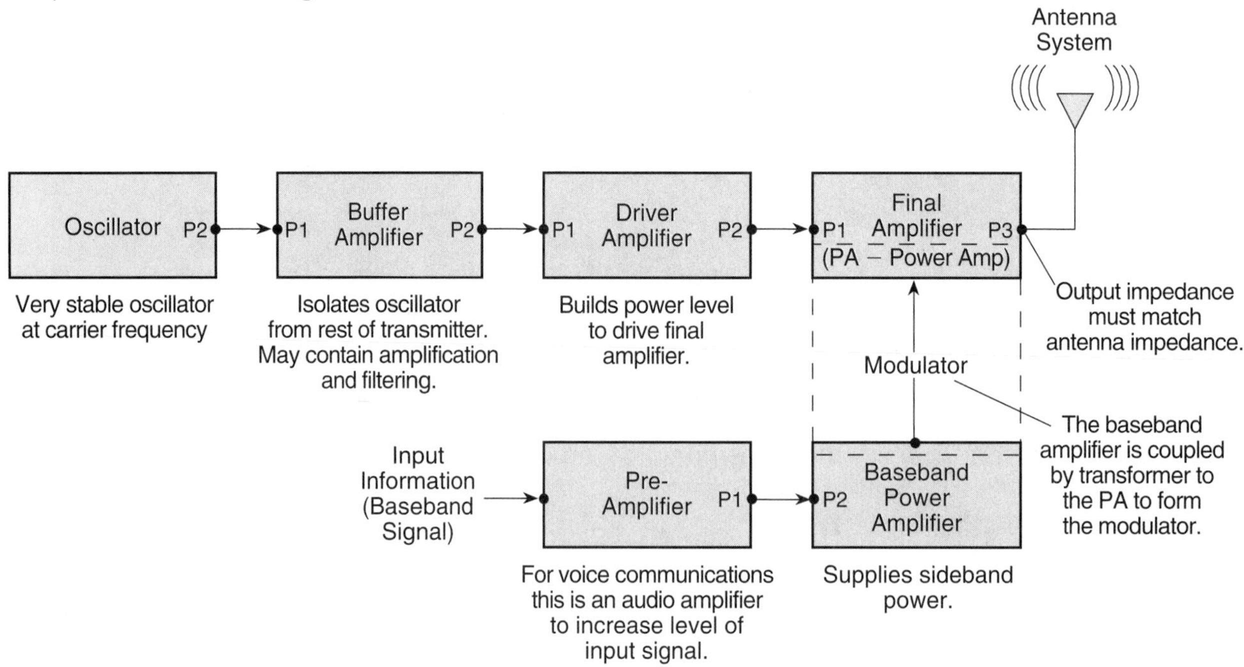


*Figure 7-1. Block diagram of an AM transmitter.*

## Oscillator

The oscillator generates a sine wave of constant amplitude at the carrier frequency. The carrier oscillator is required to be stable in frequency, and it is important that the oscillator be very "clean"— that is, without harmonics or other spurious signal content. The objective of the design is that all of the energy is concentrated at one or a band of frequencies. For some transmitter systems, the oscillator will be fixed in frequency, while for others it will be variable. We will discuss the oscillator in further detail, but for now let's assume that the oscillator is at a fixed frequency.

## Buffer

The oscillator outputs a signal to a buffer stage. Buffering, as the name suggests, is used to isolate or "buffer" the oscillator from the rest of the transmitter so that the oscillator circuit sees a constant load. The buffer prevents any changes in the power stages of the transmitter from being reflected back into the oscillator stage. Because the oscillator operates at a very low power level, the buffer stage will increase the power level of the oscillator signal. If it is an amplifier, it generally is an LC tuned amplifier similar to that shown in *Figure 4-9b.* Besides producing a higher power level, the buffer also may provide some filtering to remove any distortion introduced by the oscillator.

## Driver

After the buffer we find the driver stage. As the name suggests, drivers build the signal power level to be able to "drive" the final power amplifier. There may be one or more driver stages depending upon how much power is to be delivered to the power amplifier. The driver stage usually is a class-C tuned amplifier, like that shown in *Figure 4-9b,* or *Figure 4-17* if still higher power is required. Generally, one can get a power gain of 10 to 50 from these amplifier circuits and the number of stages depends upon the power required to drive the final amplifier to deliver the proper power to the antenna. Like the buffer stage, the driver stage will provide some buffering and filtering since it normally will exhibit the bandpass characteristics associated with a tuned amplifier.

## Final Amplifier

The final amplifier stage is a class-C, high-power amplifier, like that shown in *Figure 4-17.* In many high-power cases, the active device is a vacuum tube. The AM power output will depend upon the system specifications and the type of tube or transistors chosen for the output stage. The tube or transistor circuit must be designed to get rid of heat from the active device so the device will not be damaged or have its reliability reduced. Such devices have a power dissipation rating using particular cooling techniques that are used to guide and dictate the overall design. This final stage is sometimes referred to just as the "PA" or power amplifier.

## Modulator

In AM systems that use high-level modulation, the power amplifier is modulated. The power amplifier will be fed dc power through a transformer that is driven by a baseband amplifier to produce the actual AM modulation. Such a circuit is shown in *Figure 7-2.* The job of the tuned LC network at the output of the PA is to transform the antenna or load to the proper impedance for the amplifying device. The net load that the tube or transistor will "see" is usually a fixed impedance that has been transformed to the correct value for the particular vacuum tube or transistor circuit that is used.

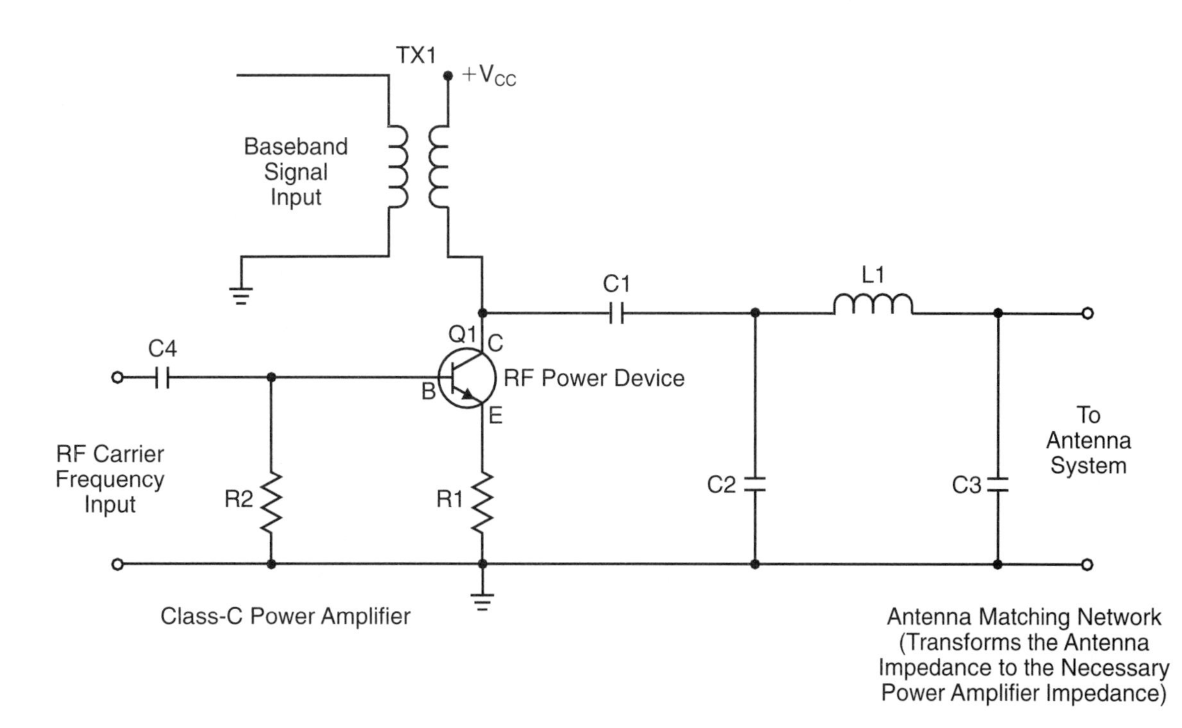


*Figure 7-2. Class-C power amplifier-modulator being modulated with baseband signal.*

## Baseband Amplifiers

The modulation information in this example is a voice baseband audio signal that needs to be amplified after leaving the microphone. The baseband amplifier stage will have one or more amplifiers. Typically, there is a preamplifier, which operates at low levels, and then a power amplifier that will produce enough power to 100% modulate the AM carrier. The amount of baseband power depends upon the system requirements. To achieve 100% modulation, it will be one-half the power required in the carrier to achieve 100% modulation, since all the information power is in the sidebands in an AM system.

# AM Transmitters — More Detail

*Figure 7-3* shows an entire schematic for a AM transmitter system. The oscillator is a fixed-frequency crystal oscillator and the rest of the RF stages are class-C tuned circuit amplifiers.

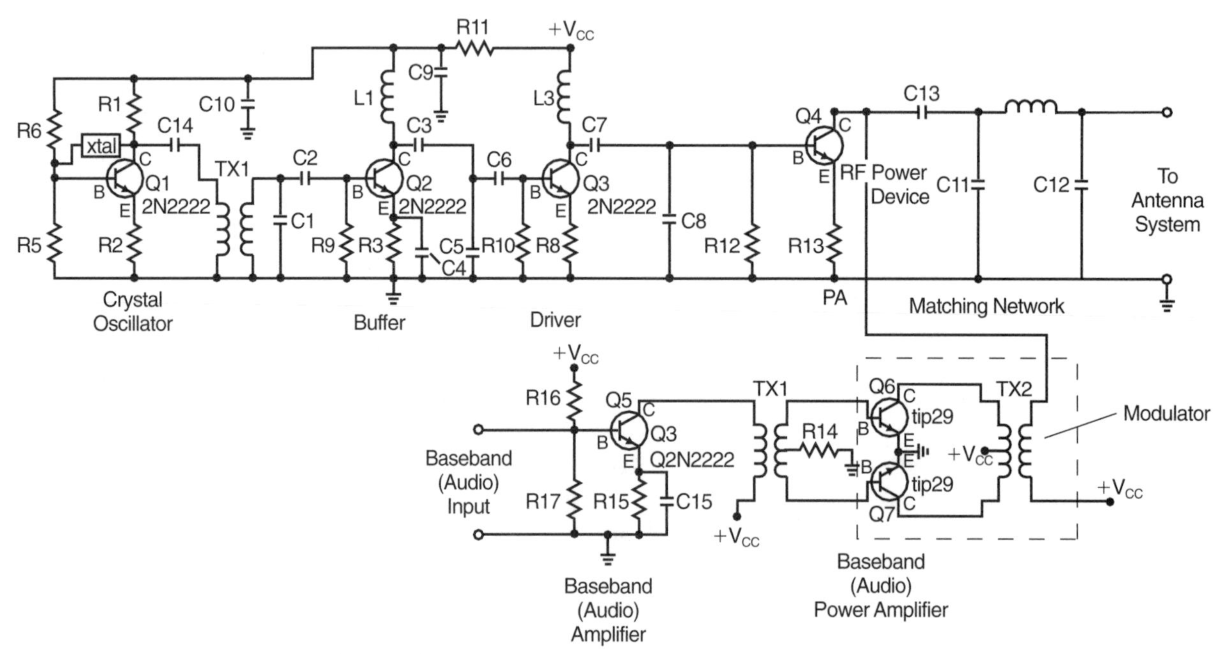


*Figure 7-3. AM transmitter schematic.*

The crystal oscillator is a Colpits oscillator built using a bipolar 2N2222 for Q1. The output of the oscillator is fed to the buffer stage through the transformer, TX1, between the two stages. The buffer and driver stages are constructed using a bipolar 2N2222 transistor for Q2 and Q3, and are class-C amplifiers. The resistors from the bases to ground provide a dc return path and the emitter resistors provide stabilizing negative feedback. The final power amplifier uses a bipolar transistor, Q4. This stage is coupled to the load or antenna through the LC network that provides the proper load to the transistor. Q5 is a baseband, linear audio pre-amplifier that raises the audio level sufficiently high to drive the power amplifier, which is constructed using two transistors, Q6 and Q7, interconnected as a linear push-pull audio power amplifier. The audio power amplifier is coupled to the final PA through the modulation transformer, TX2, which adds and subtracts voltage from the power supply to create the AM modulation and sidebands, similar to the PA previously described. Because the oscillator is a fixed-frequency, the transmitter is operating at one set frequency.

*Figure 7-4* shows a variation on the oscillator circuit that is used to produce a variable-frequency oscillator, which allows the transmitter to operate over a range of frequencies. Capacitor C4 is a mechanical capacitor made of metal plates that mesh with each other. By turning the capacitor shaft the frequency of the oscillator can be changed, hence changing the transmitter's carrier frequency. As more of the capacitor plates are merged, the capacitance increases. In practice, normal variable capacitors can change the frequency by about a 3 to 1 ratio — for example, from about 0.5Mhz to 1.5Mhz — while changing capacitance by a 10 to 1 ratio. This limitation is due to the construction of the capacitor. The oscillator's frequency will be the LC circuit resonant frequency given by:

$$f_r = \frac{1}{2\pi\sqrt{LC}}$$ where: $f_r$ = oscillator frequency in **Hz**
L = inductance in **henries**
C = capacitance in **farads**

### *Example 1: Calculating a Transmitter's Carrier Frequency*

Show that the carrier frequency of an oscillator changes by 3 to 1 when the capacitance used for the LC tuned circuit changes by 10 to 1. The inductance L = 1 mH when C changes from 25 pF to 250 pF.

***Solution:***

Since $f_r = \frac{1}{2\pi\sqrt{LC}}$ or $f_r^2 = \frac{1}{(2\pi)^2 LC}$

When L = 1 mH and C = 25 pF,

$LC = 1 \times 10^{-3} \times 25 \times 10^{-12} = 25 \times 10^{-15}$

and

$$f_r^2 = \frac{1}{6.28^2 \times 25 \times 10^{-15}} \cong \frac{1}{40 \times 25 \times 10^{-15}} = \frac{1}{1 \times 10^{-12}} = 1 \times 10^{12}$$

$$\therefore f_r = \sqrt{1 \times 10^{12}} = 1 \times 10^6 = 1 \text{ MHz}$$

When L = 1 MHz and C = 250 pF

$LC = 1 \times 10^{-3} \times 250 \times 10^{-12} = 250 \times 10^{-15} = 2.5 \times 10^{-13}$

and

$$f_r^2 = \frac{1}{6.28^2 \times 2.5 \times 10^{-13}} \cong \frac{1}{40 \times 2.5 \times 10^{-13}} = \frac{1}{100 \times 10^{-13}} = \frac{1}{10 \times 10^{-12}}$$

$$\therefore f_r = \frac{1}{\sqrt{10} \times 10^{-6}} = \frac{1}{3.18 \times 10^{-6}} = 0.318 \times 10^6 = 0.318 \text{ MHz}$$

The ratio of the two frequencies is 3:1

Check: Since $f_r = \frac{1}{2\pi\sqrt{LC}}$, if C increases by 10, LC increases by 10. The square root of 10 is 3.18; therefore, $f_r$ changes to about one-third of what it was.

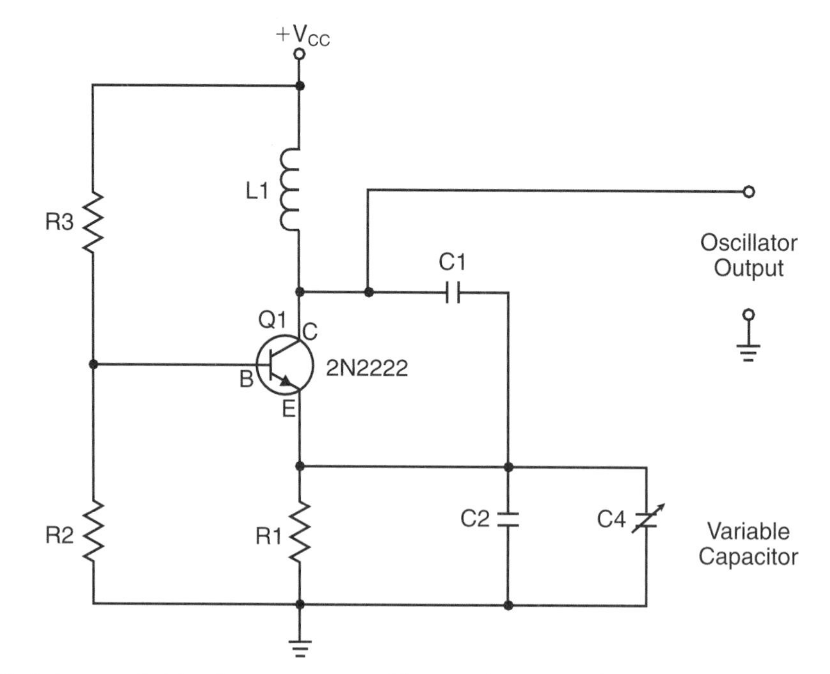


*Figure 7-4. Colpitts variable-frequency oscillator circuit built using a variable capacitor.*

A problem called "tracking error" occurs because the capacitor does not track in a linear fashion; that is, the oscillator's frequency does not change the same amount for each degree of rotation of the capacitor shaft. Manufacturers attempt to make the tuning capacitor's plates so they open and close in such a way as to provide linear tracking. Circuit designers have found that a better solution is to use specially-wound inductors that tune the oscillator by changing the inductance rather than the capacitance. The inductance is changed by inserting an iron core into and out of the coil form. *Figure 7-5* shows the variable capacitor and the variable inductor. Each tuning approach may have stability problems due to construction, mounting, and small physical changes in the components.

*Figure 7-5. Variable capacitor and variable inductor and their schematic symbols.*
*(Photos courtesy Gordon West Radio School.)*

## Frequency Synthesizers

As pointed out in Chapter 4, the problems of tracking and stability may be overcome by using a frequency synthesizer instead of an oscillator circuit. This allows the oscillator to have the stability of a crystal oscillator in a tunable oscillator. The block diagram of the synthesizer circuit used in Chapter 4 *(Figure 4-22)* is repeated here in *Figure 7-6.*

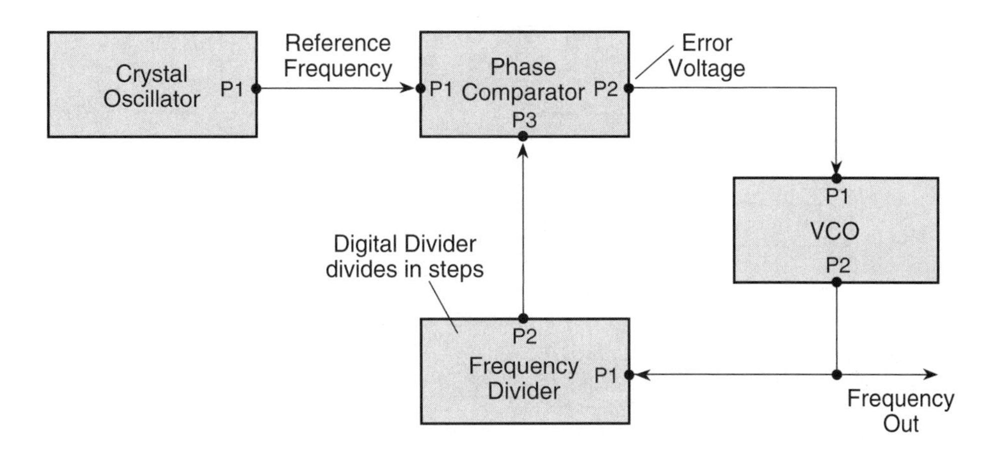


*Figure 7-6. Block diagram of synthesizer circuit.*

The circuit stays locked to a crystal oscillator by varying the output frequency of a VCO (voltage-controlled oscillator). The output frequency of the VCO is divided down digitally in steps and is one of the inputs to the phase comparator. The crystal oscillator produces a reference frequency that will be compared to the output frequency of the divider. If the two frequencies into the phase comparator are not equal, the output of the comparator is an error voltage. The circuit acts to reduce the error voltage to a net zero by adjusting the VCO frequency by a value that causes the output of the divider to equal the crystal reference frequency. The VCO will now be "locked" to the crystal oscillator. When the divider ratio is changed the VCO will change frequency. This allows the synthesizer, and thus the transmitter, to be tuned to a desired frequency. The frequency steps in the tuning depend on the digital divider and the oscillator reference frequency. If there are 128 tuning steps, the reference oscillator is 1/128th of the transmitter frequency. Each step will be equal to the reference oscillator frequency.

*Figure 7-7a* shows a synthesizer circuit using the MC145170 integrated circuit to implement the synthesizer. *Figure 7-7b* shows the circuit inside the IC package. Modern-day radio transmitters are tuned by using such synthesizers.

### *Example 2: Tuning of Transmitter Frequency*

What will the transmitter frequency be if the divider ratio is 32, 64, or 128, respectively, when the reference frequency is 400 kHz?

***Solution:***

Reference Frequency x Divider Ratio = Transmitter frequency

| Reference Frequency | Divider Ratio | Transmitter Frequency |
|---|---|---|
| 400 kHz | 32 | 12.8 MHz |
| 400 kHz | 64 | 25.6 MHz |
| 400 kHz | 128 | 51.2 MHz |

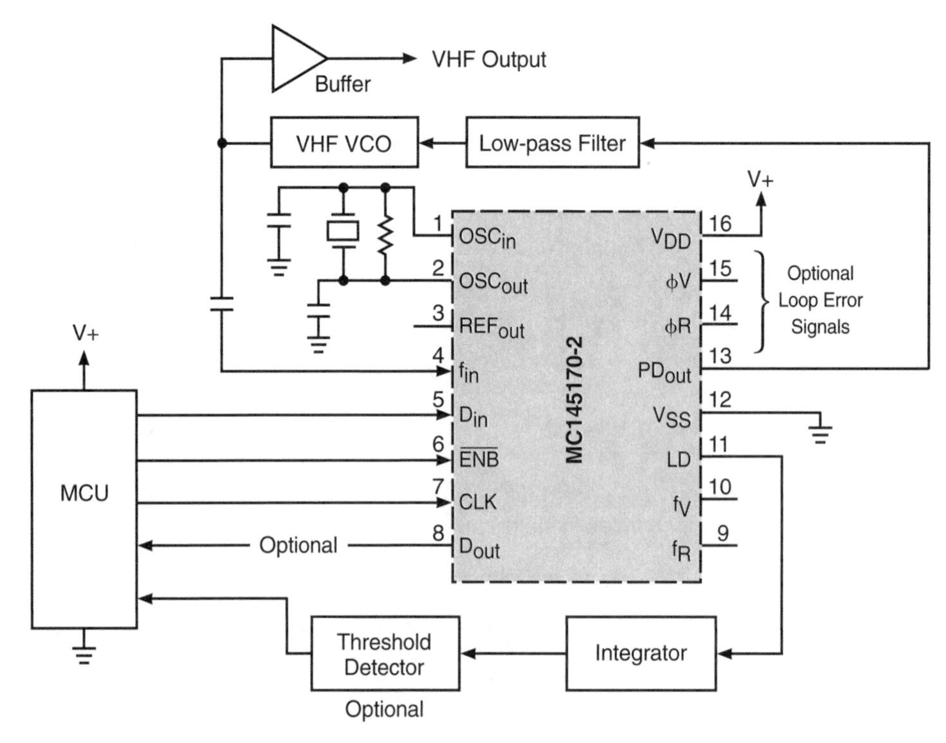

a. Synthesizer Circuit

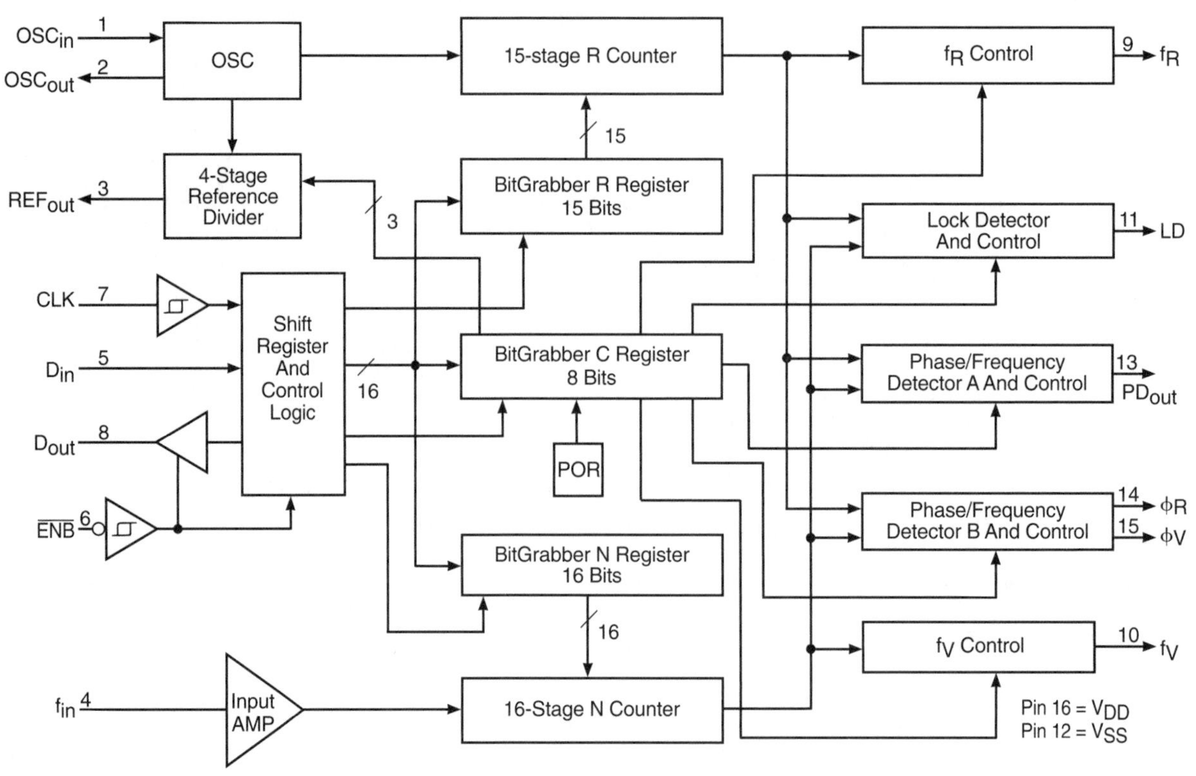

b. Circuit inside the MC145170-2

*Figure 7-7. Frequency synthesizer using the MC145170-2 IC.*
*(Courtesy of Motorola, Inc.)*

## Single-Sideband Transmitters

Single-sideband transmitters have many similarities to AM transmitters, except that the modulation is produced at low power levels. *Figure 7-8* shows the block diagram of an SSB transmitter. The oscillator stage is identical to the oscillator for the AM transmitter and may also be fixed or tunable. This oscillator determines the frequency at which the transmitter will transmit and, in modern-day transmitters, it is a frequency synthesizer.

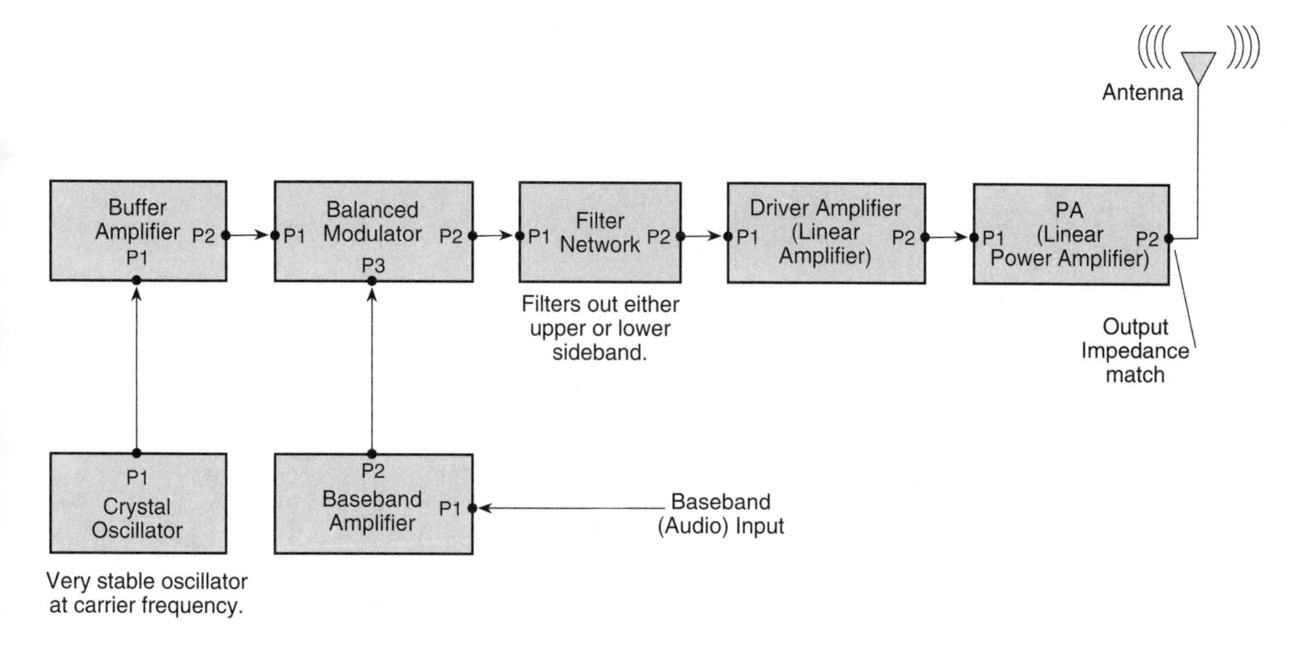


*Figure 7-8. Block diagram of an SSB transmitter.*

As in the AM transmitter, a buffer stage isolates the oscillator and the output signal from the buffer is fed to a balanced modulator. The baseband signal, again in this case an audio signal, also is fed to the balanced modulator and the oscillator (carrier) signal is modulated by the audio signal. The output from the balanced modulator will be the upper and lower sidebands with the carrier balanced out. When we discussed balanced modulators, the objective of the circuit is to suppress the carrier; therefore, the small amount of carrier signal still present depends on the effectiveness of the balancing. The output signal is a double sideband suppressed carrier (DSBSC) waveform.

The output signal from the balanced modulator is fed to a sharp filtering network that filters out either the lower or upper sideband. The remaining sideband is amplified by a driver stage that performs the same task as in the AM radio, except that it is a linear amplifier. More than one driver may be required, depending upon the final output power levels required. Linear amplifiers are required both for the driver and the final amplifier in order to preserve the modulated signal with no further distortion or unwanted modulation.

Once again, the final stage is the PA. In SSB, the modulation has already been accomplished at the balanced modulator; therefore, there is no requirement for modulation at the PA, as in the case of the AM transmitter. The active device used in the power amplifier circuit is chosen based on the transmitter power level required. And, as already indicated, the PA and the drivers are linear amplifiers. Both the driver and the power amplifier usually are tuned amplifiers operating in Class-AB mode.

*Figure 7-9* shows an entire schematic for an SSB transmitter. Comparing it to the AM transmitter in *Figure 7-3,* we see that the crystal oscillator supplying the very stable carrier frequency and the buffer amplifier are essentially the same. The transmitter is operating on one fixed frequency. The baseband amplifier, again in this case an audio amplifier, does not have to supply as much power for SSB as for the AM transmitter since the modulation is done at low level in the balanced modulator. The balanced modulator is a 1496 IC circuit similar to the mixer circuit of *Figure 6-10,* but balanced to suppress the SSB carrier and produce the double-sided output signal. A filter network within the balanced modulator block filters out the LSB or USB, depending upon the transmitter design. Its output is coupled to the driver stage. The driver stage raises the signal power to the level required to drive the PA, which then delivers the required power to the transmitting antenna. The PA circuits are tuned linear amplifiers operating in the Class-AB mode, and are tuned to the carrier frequency. The matching network performs the same function as the matching network in the circuit of *Figure 7-2.* Both amplifiers will have an efficiency of 50% to 60%. In most cases, the power output requirements of the driver and PA are less for SSB than for AM to achieve the same signal detection level, since only the single sideband is transmitted.

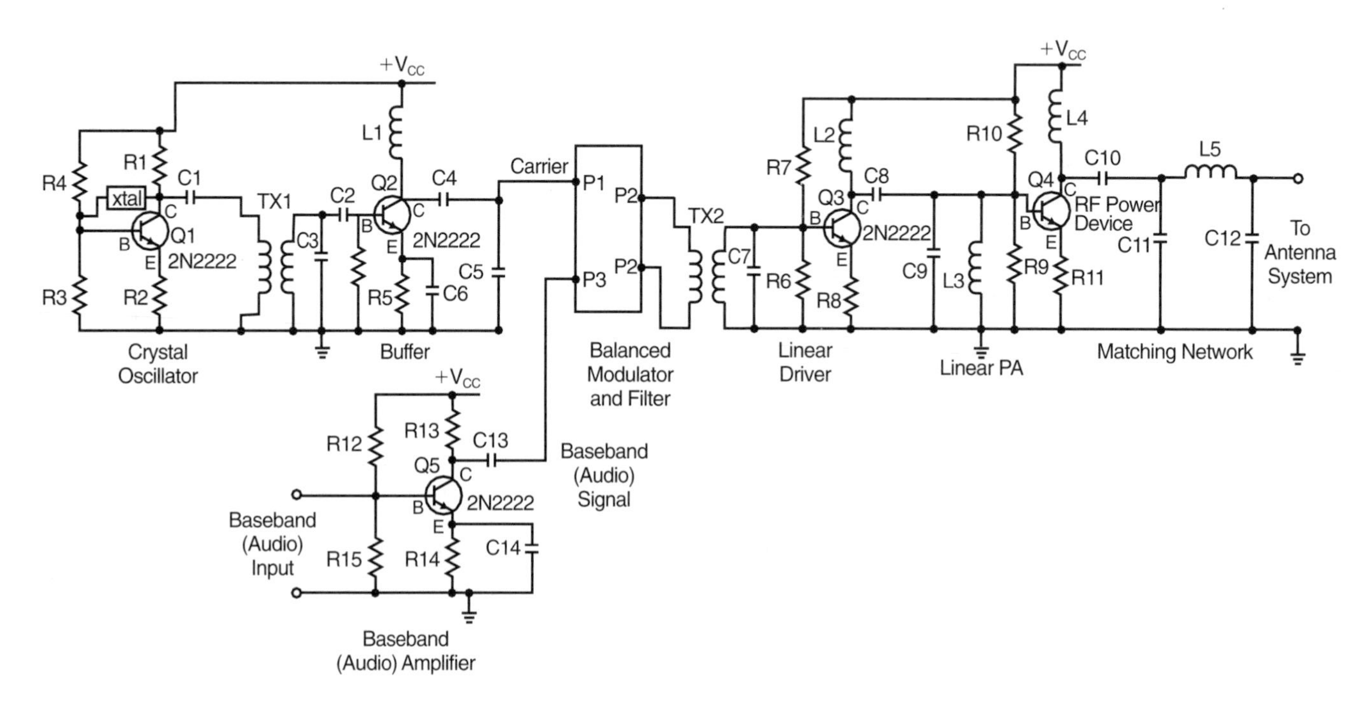


*Figure 7-9. Schematic of an SSB transmitter.*

## FM Transmitters

*Figure 7-10* is a block diagram of an FM transmitter. It is significantly different than an AM or SSB transmitter, but does have some stages that are very similar.

### Modulator

The modulation, of course, is frequency modulation that is produced at low power levels directly from the baseband signal input. FM modulation may be produced either by changing the frequency of an oscillator or by using a fixed oscillator and changing the phase angle. The frequency is changed directly by using a VCO (voltage-controlled oscillator) to generate the FM. The input baseband signal amplitude changes the frequency as the amplitude varies producing a frequency modulated output.

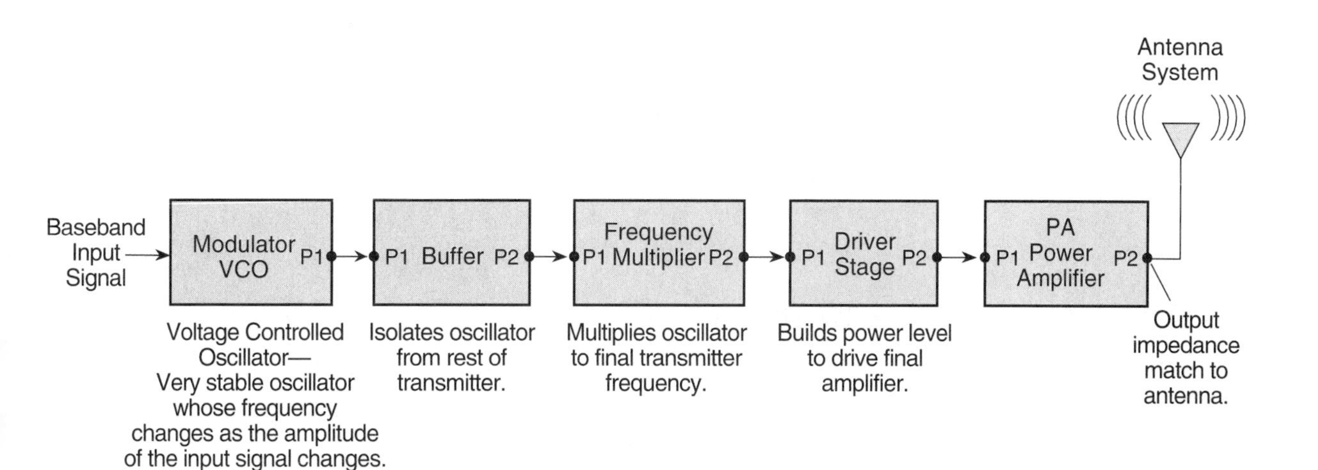


*Figure 7-10. Block diagram of an FM transmitter*

## Buffer Amplifier

The buffer amplifier performs the same task as in the AM and SSB transmitter. It isolates the oscillator (in this case, the modulator) from the rest of the transmitter circuits. It is a Class-C tuned amplifier. Class-C amplifiers can be used because all of the information is in the frequency of the FM signal, and any amplitude changes in the signal do not alter the information in the FM signal.

## Frequency Multiplier

This is a new stage different from any used in an AM or SSB transmitter. In this FM transmitter, the modulation is produced at a frequency lower than the actual transmitted frequency. The frequency multiplier is required to increase the signal frequency to the proper transmitting frequency. *Figure 7-11* is an example of a frequency multiplier circuit. A frequency multiplier is a tuned amplifier where the output is tuned to a harmonic frequency of the input. In *Figure 7-11,* the input signal has a frequency of $f_1$, but the output tuned circuit is tuned to $3f_1$. The frequency multiplier depends on the fact that signals contain harmonics, and most signals do because of distortion. As an example, in the circuit of *Figure 7-11,* if the input is tuned to 12 MHz and the output is tuned to 36 MHz, the circuit "selects" the 36-MHz signal as its output. This circuit is a frequency tripler. Frequency doublers and triplers are used in many applications.

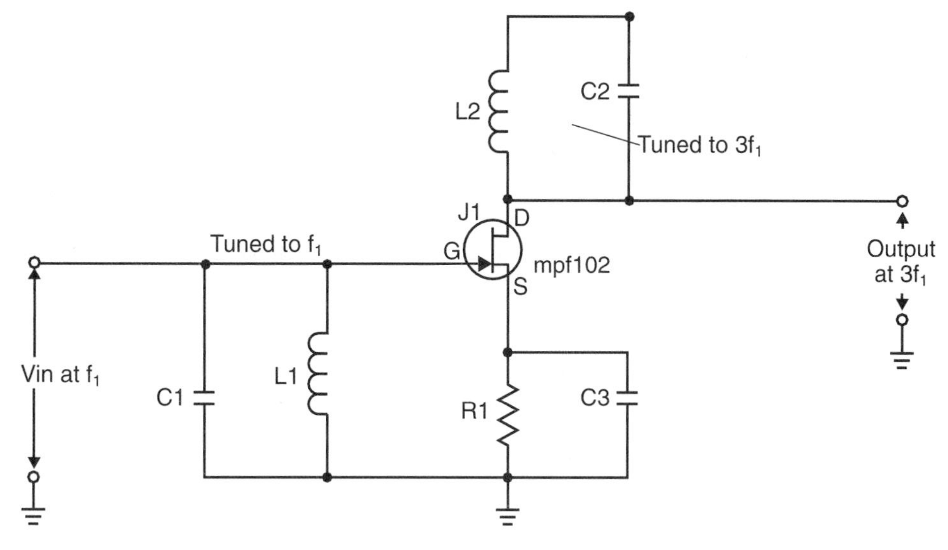


*Figure 7-11. Frequency multiplier circuit.*

## Driver and PA

The driver and PA serve the same purpose in an FM transmitter as in an AM or SSB transmitter. They build the signal power to the level required for the proper antenna transmission power. The PA and driver are again Class-C tuned amplifiers, and the PA must be able to deliver power into the transmission line and antenna efficiently.

## FM Transmitter — More Detail

*Figure 7-12* is a detailed schematic diagram of a simple FM transmitter. The modulator is a varactor VCO, like the one in *Figure 5-14.* It is called a crystal voltage-controlled oscillator. At first, this may seem like a contradiction since a crystal oscillator is stable and should not change frequency. However, when a capacitance is placed across a crystal, the frequency changes a small amount, but the basic stability of the oscillator remains. The varactor diode (voltage-variable capacitor) will change the oscillator frequency when the voltage across the diode is changed to produce FM modulation. The varactor diode is driven by the baseband modulating signal, which changes the capacitance of the diode in accordance with the modulating signal. The output of the oscillator is a frequency-modulated signal, but of low deviation because the crystal oscillator can only undergo small frequency changes.

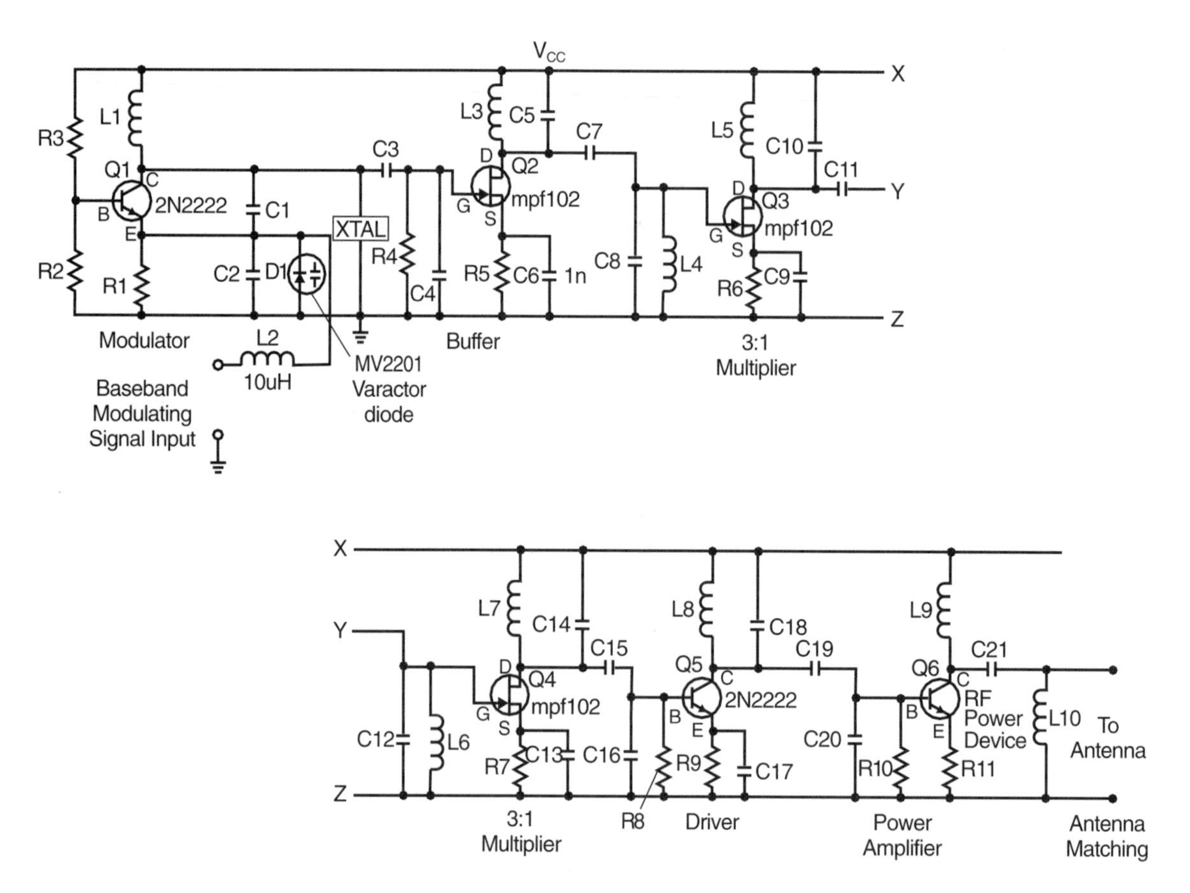

*Figure 7-12. Schematic of an FM transmitter.*

The buffer stage isolates the oscillator and also provides some amplification. Since the oscillator does not run at the required transmitting frequency, as previously described, a frequency multiplier stage is used to multiply the oscillator output to the transmitter frequency.

### *Example 3: Calculating Modulator Frequency & Deviation*

Calculate the modulator frequency and its deviation if the FM transmitter is to transmit at 140 MHz with a deviation of 5 kHz when the multiplier has a 9 times multiplication.

***Solution:***

$$\text{Modulator frequency} = \frac{\text{Transmitter frequency}}{9} = \frac{140\text{MHz}}{9} = 15.556\text{ MHz}$$

$$\text{Deviation of modulator} = \frac{\text{Deviation of FM Output}}{9} = \frac{5000\text{ Hz}}{9} = 555.56\text{ Hz}$$

In *Figure 7-12*, two multipliers are used, each with a multiplication of 3 for a total multiplication of 9. Q3 and Q4 are two class-C tripler tuned amplifiers. The driver amplifier and the power amplifiers, Q5 and Q6 respectively, are standard class-C amplifiers and simply raise the power level of the modulated signal. As mentioned previously, the power amplifier must couple the energy efficiently to the antenna system.

## Large Deviation FM

Other types of modulators are used in FM transmitters where large deviations are required. A reactance modulator, like that shown in *Figure 5-15,* is used frequently. *Figure 7-13* is a block diagram of a common arrangement. A baseband amplifier drives a reactance modulator, which changes the frequency of an LC oscillator. The output of the oscillator is coupled to a buffer and then a multiplier. Because the oscillator is LC controlled and may drift in frequency, a special automatic frequency control (AFC) is connected around the reactance modulator and the oscillator. The oscillator output frequency is compared to a crystal oscillator. The difference signal is used to adjust the reactance modulator to keep the modulator oscillator on frequency.

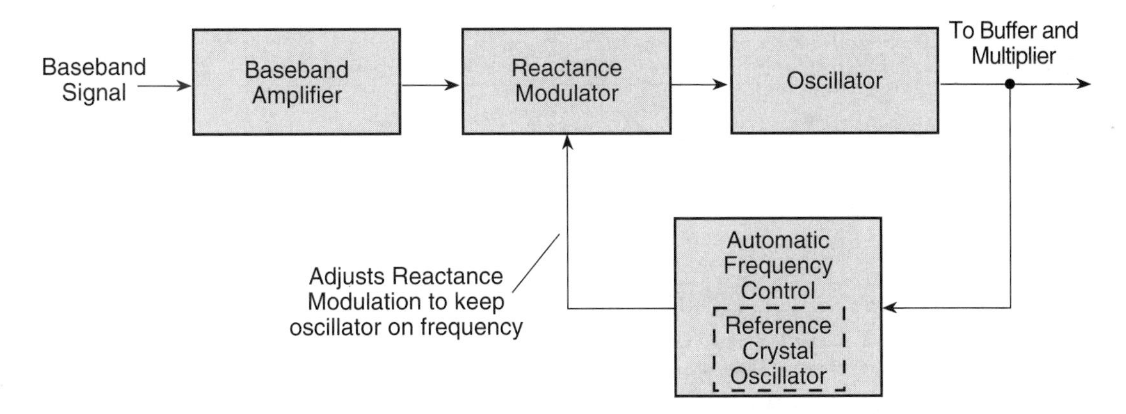


*Figure 7-13. Block diagram for large deviation FM modulation.*

## A Small System

*Figure 7-14* shows an FM transmitter made from a single integrated circuit (an MC2833) with added passive components. It produces a low-power FM signal that can be used in applications such as a cordless telephone. The basic chip functions as a VCO that is modulated by the audio signal. A transistor is used to provide a small amount of power gain and to drive a small antenna. It is a very-low-power transmitter useful over short distances for a variety of purposes — communications, of course, being the primary use. The various values of the external components are shown in *Figure 7-14* for specific output frequencies.

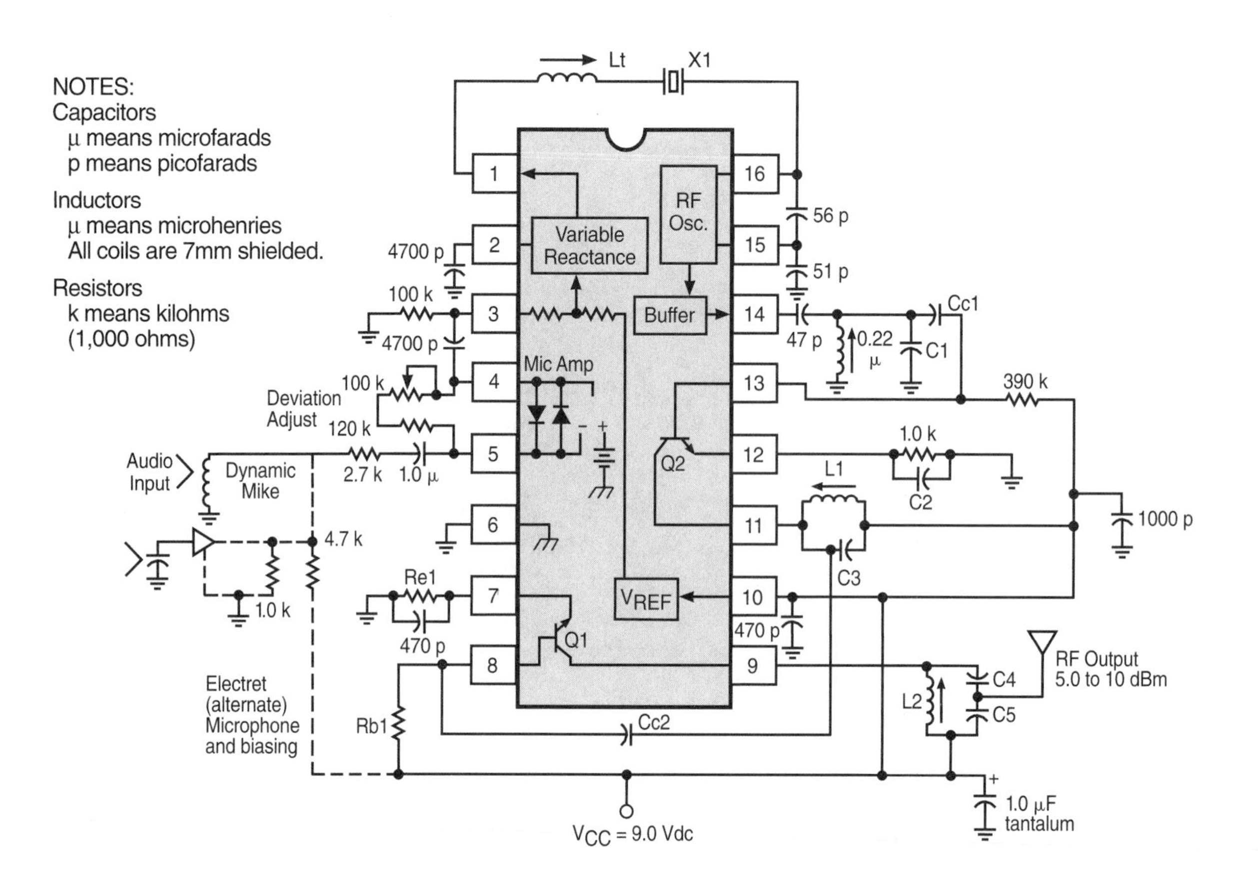


**Components vs. Output Frequency**

| Output RF | X1 (MHz) | Lt (µH) | L1 (µH) | L2 (µH) | Re1 | Rb1 | Cc1 | Cc2 | C1 | C2 | C3 | C4 | C5 |
|---|---|---|---|---|---|---|---|---|---|---|---|---|---|
| 49.7 MHz | 16.5667 | 3.3–4.7 | 0.22 | 0.22 | 330 | 390 k | 33 p | 33 p | 33 p | 470 p | 33 p | 47 p | 220 p |
| 76 MHz | 12.6000 | 5.1 | 0.22 | 0.22 | 150 | 300 k | 68 p | 10 p | 68 p | 470 p | 12 p | 20 p | 120 p |
| 144.6 MHz | 12.05 | 5.6 | 0.15 | 0.10 | 150 | 220 k | 47 p | 10 p | 68 p | 1000 p | 18 p | 12 p | 33p |

*Figure 7-14. Single-chip FM transmitter using an MC2833 IC.*
*(Courtesy of Motorola, Inc.)*

## Summary

In this Chapter, we have discussed transmitters — AM, SSB and FM. We examined block diagrams and detailed schematics to show how the transmitter system is made up of the basic functional blocks that were described in previous chapters. The transmitter is at the beginning of our communications system. In the next Chapter, we will focus on the end of the system as we discuss receivers.

# Quiz for Chapter 7

1. Why is frequency stability important in transmitters?
   a) Makes receivers smaller.
   b) Allow longer range transmission.
   c) Reduces or eliminates frequency drift.
   d) Increases power output.
2. What is spurious radiation?
   a) Harmonic frequencies that cause distortion in the transmitted signal.
   b) Radiation caused by sun spots that result in interference.
   c) Signals reflected back to the transmitter from the antenna.
   d) None of the above.
3. Why is a balanced modulator used to produce SSB?
   a) To get very high power.
   b) To eliminate the carrier.
   c) To eliminate one of the sidebands.
   d) To keep all of the sidebands.
4. Why are linear amplifiers required for power amplifiers in SSB?
   a) Class C amplifiers are too efficient.
   b) To make sure the SSB signal does not exceed the FCC's power limit.
   c) To preserve the shape of the SSB signal.
   d) None of the above.
5. Why are multiplier circuits used in FM transmitters?
   a) To increase the number of stages in a transmitter.
   b) To increase transmitter power out
   c) To make the receiver detector work better
   d) To allow the use of low-deviation modulators.
6. What is the purpose of a VFO in transmitters?
   a) To stay on the same frequency.
   b) To allow the transmitter to change frequencies.
   c) To transmit on more than one frequency at a time.
   d) Help eliminate frequency drift.
7. Why can both FM and AM transmitters use Class-C power amplifiers?
   a) AM and FM are the same types of modulation.
   b) FM and AM signals have no amplitude variations.
   c) FM has only frequency variation and AM is produced at an output class-C power amplifier.
   d) AM has only frequency variations and FM has no amplitude variations.
8. What is the purpose of buffer stages in transmitters?
   a) To eliminate the power amplifier.
   b) To decrease the drive to the PA.
   c) To increase the power in the sidebands.
   d) None of the above.
9. Why do modern transmitters often use frequency synthesizers?
   a) To allow easy frequency selection in the transmitter.
   b) To keep the transmitter signal stable in frequency.
   c) To make sure the transmitter can tune every possible frequency.
   d) All of the above.
10. What is the advantage of a crystal oscillator circuit used in transmitters?
    a) Very stable transmitter frequency.
    b) High transmitter output.
    c) Multiple frequency operation.
    d) High-frequency operation of transmitters.

**Answers:**
1 c, 2 a, 3 b, 4 c, 5 d,
6 b, 7 c, 8 d, 9 d, 10 a

## Questions & Problems for Chapter 7

1. The three most important characteristics of a radio transmitter are?
2. The driver stage in an AM transmitter usually is a ________________amplifier.
3. Why are frequency synthesizers used in transmitters instead of discrete component oscillators using variable capacitors or inductors?
4. What is a principle difference between AM and SSB transmitters?
5. In an FM transmitter, the amplitude of the baseband signal is used to change ___________ ___________ ___________ ___________ ___________ ___________ ___________ to produce the FM modulation.
6. What is the purpose of a frequency multiplier in an FM transmitter?
7. In what stage is the high-power level modulation produced for AM modulation?
8. If the crystal oscillator in a synthesizer is 320 kHz and the transmitter frequency is 81.9 MHz, what is the divider ratio?
9. What is the difference between the driver and the power amplifier of AM and SSB transmitters?
10. An FM transmitter with an output frequency of 108 MHz has three tripler stages as frequency multipliers. What will be the output frequency of the modulator VCO?

*(Answers on page 211.)*

# CHAPTER 8

# Receiving — Including Detection



Receivers — the destination end of our communications system — are the focus of this Chapter. The information transmitted by the transmitter that is sent over a transmission link ends up at the receiver. As with transmitters, the receivers we will be discussing are those used in a radio communications system in which the receiver is one of the key elements. Within the receiver is the capability of recovering the original information — the function we have defined as detection; therefore, detection is a major sub-topic of this Chapter. As with transmitters, we will find that a receiver is designed using the functions and circuits previously discussed that are put together to form a major system or sub-system.

Receivers, which are more complex than transmitters, require more design engineering to achieve optimum performance and to match them to their application. As an example, a high-frequency communications receiver has higher performance specifications than an AM commercial receiver.

## Receiver Principles

The basic purpose of a receiver is to receive an RF signal, amplify it, filter it to remove unwanted signals, and recover the desired baseband information. Since the RF signal usually comes from an antenna, its amplitude is very small, often on the order of a few microvolts. The receiver must be able to amplify the signal from that very small level up to usable levels of several volts or more. The antenna system at the receiver does not discriminate between the signal we want to receive and the many others that are present at the same time. The antenna presents all of the signals that excite it to the receiver. The receiver must select the required signal for amplification. Thus, two important specifications for a receiver are sensitivity and selectivity. Others are stability, dynamic range, and image rejection. Lack of performance in any of these areas will cause overall received performance to suffer.

### Sensitivity and Noise

Receiver sensitivity refers to the ability of a receiver to hear a "weak" signal with an acceptable signal-to-noise ratio. Sensitivity of a receiver usually is expressed in microvolts of signal at the input to the receiver that will produce a 10dB S/N ratio. The lower the number the more sensitive the receiver and the more gain the receiver must have to amplify a signal to the desired detection level.

Along with sensitivity we must include the specification termed *noise figure* (NF). Recall that noise is any unwanted signal added to the original information by the electronic circuits used, or by the transmission link environment. Environmental noise usually cannot be removed. The amplifiers in the receiver amplify the original signal as well as any noise added by the receiver circuits. The amount of noise added by the receiver is as important as the overall gain of the receiver, and may set the overall performance of the receiver with regard to weak signals. Generally speaking, the noise performance of the receiver becomes more critical the higher you go in frequency. For example, noise from the receiver is not a big problem in the AM broadcast band but becomes more important in the FM broadcast band, and is very important in the higher-frequency TV channels. Generally, background noise is very high up to frequencies of about 30 MHz, and may set system performance limits.

*Noise figure*, NF, is used to measure the noise performance of receivers. Noise figure is the ratio of the input S/N to the output S/N and is a measure of the noise introduced by the receiver. NF can be expressed as:

$$NF = 10 \log_{10} \times \frac{\text{input S/N}}{\text{output S/N}}$$

where NF = noise figure in **dB**
Input S/N = input signal-to-noise **ratio**
Output S/N = output signal-to-noise **ratio**
$\log_{10}$ = the logarithm to the **base 10**

As shown, NF, usually expressed in decibels, is 10 times the logarithm of the ratio of the *input-to-output signal-to-noise ratios*. A perfect receiver will have a NF of 0 dB, indicating that the receiver adds no noise to the received signal.

## Selectivity

Selectivity refers to the ability of the receiver to pick out or "select" the desired signal from all the unwanted signals present. It depends on the quality of frequency selection of the filter networks (sometimes crystal) and tuned LC circuits used in the receiver. Selectivity is important to reduce or eliminate interference from signals with frequencies close to the desired signal.

## Stability

Stability refers to the ability of the receiver to stay tuned to a particular incoming signal and not drift away from the desired signal. It is based on the ability of the electronic circuits — such as oscillators, filters, and tuned circuits — to remain stable at their specified frequency. A receiver appears to "drift" in frequency because its tuning oscillator changes frequency. If the change is large enough, the desired signal can drift out of the receiver's passband creating the possibility of no output signal from the IF, or that a new RF signal will drift in and produce a new IF signal. To obtain the original station signal, the receiver must be re-tuned, which may have to be done continuously if the drift is severe. As noted previously, phase-locked-loop synthesizers used as very stable oscillators that are linked to a crystal have contributed significantly to receiver stability.

Do not confuse the fading of incoming signals, especially those received from long distances, as receiver instability. In such a case, the receiver may have the ultimate stability, but the long-distance incoming signal fades due to loss of signal amplitude.

## Dynamic Range

Dynamic range is the ability of the receiver to handle both weak and strong signals without distortion. A receiver designed to handle weak signals may suffer from overload (heavy distortion) when it receives a strong signal that is close to the desired weak signal. Overloading a receiver will produce unwanted mixing that appears to be interference from signals at other frequencies. This is called intermodulation distortion and is most serious when you are trying to receive a weak signal in the presence of a strong interfering signal.

## Heterodyning — A Great Design Aid

Early receivers, called tuned radio frequency (TRF) receivers, were designed with very high-gain over a broad range, but were very unstable. Today, modern receiver designs using the superheterodyne principal and IF amplifiers that are designed for one frequency — with steep-sided, narrow-frequency tuned circuit filters — can have high gain, good stability, and excellent sensitivity and selectivity.

## Image Rejection

Recall the discussion in Chapter 6 on image frequencies, which occur at a station frequency plus twice the IF frequency. Image rejection is a measure of a receiver's capability to reject image frequencies. It is enhanced first by careful selection of the IF frequency, and second by the design of the LC tuned circuits and filters that select the frequencies in the receiver. Receiver front end selectivity is a must for good image rejection.

# AM Receivers

## Generic

*Figure 8-1a* shows the block diagram of a "generic" AM receiver. The first three blocks are the RF amplifier, mixer, and oscillator. They form what is known as the "front end" of the receiver.

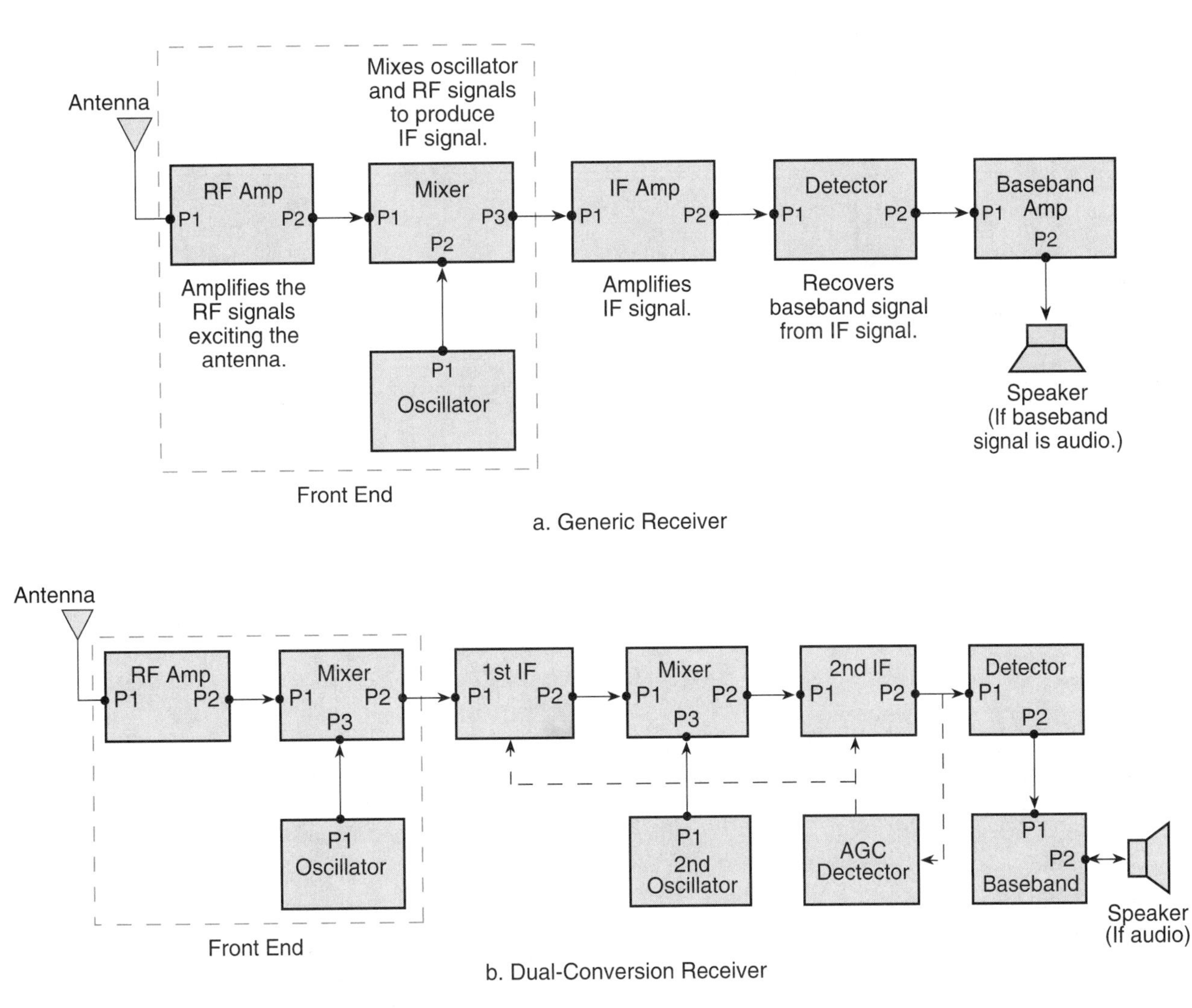


*Figure 8-1. Block diagrams of generic and dual-conversion AM receivers.*

## Front End

The front end is very important in setting the noise figure of the receiver. In particular, the RF amplifier should be a low-noise design because it amplifies all the low-level signals from the antenna — including noise — with a significant amount of gain. As a result, the RF amplifier has a great effect on the noise performance of the receiver. A low-noise mixer also influences the noise performance of the receiver. The better the low-noise design of the RF amplifier the better the mixer design must be in terms of noise. If the RF amplifier and mixer both have good low-noise characteristics, then the receiver can be designed with enough amplification to be very sensitive.

## IF Amp

The next block is titled "IF Amp." It may be a single stage or a number of stages, usually made up of almost identical circuits. The series of IF amplifiers (called an "IF strip") are used to amplify the desired signal and to provide the bulk of the receiver's selectivity. Often, the IF strip will have crystal or ceramic filters to provide even shaper selectivity characteristics. If the front end of the receiver is very low noise, then the IF amplifiers may determine the receiver's noise figure.

*Figure 8-2* is a schematic of two IF amplifiers using mpf102 JFETs as RF amplifiers (recall that these JFETs operate in the depletion mode). Each stage is an LC-tuned amplifier like the one discussed in *Figure 4-9*. The two stages are coupled together using a crystal filter that aids greatly in providing selectivity because of the steep sides of the filter passband. The filter passband of the amplifiers is shown in *Figure 8-2b*. Note the effect of the crystal filter in making the sides of the passband much sharper. The output of the second IF amplifier is coupled to the next block — the detector.

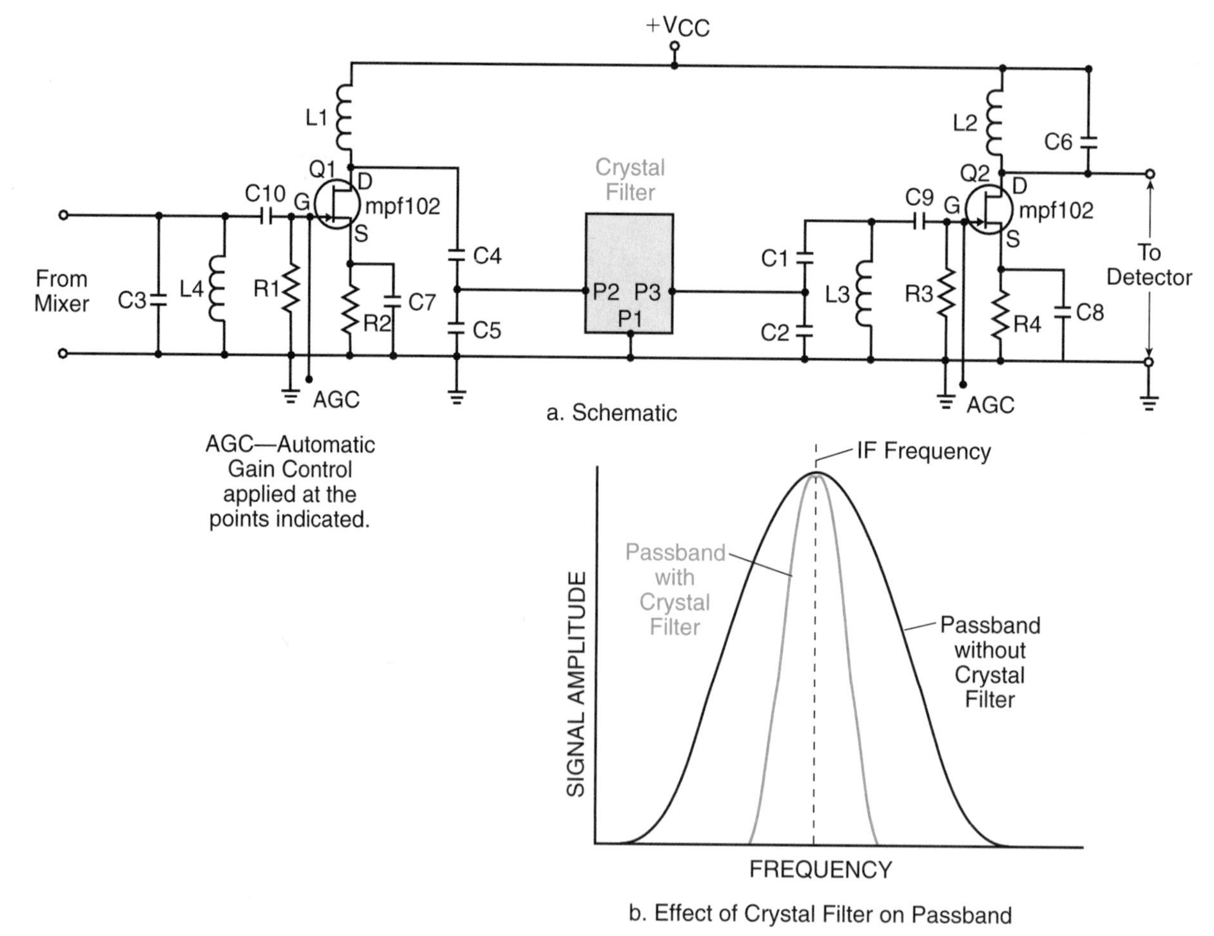


*Figure 8-2. Cascaded IF amplifiers. The active devices are JFETs. The value of most components depends upon the frequency of operation.*

## Detector

More will be said about the detector function and the detector circuit a bit later, but for now let's repeat what was said previously — detection is a function that provides a receiver with the capability of recovering the original information transmitted by the transmitter and transferred over the transmission link.

## Baseband Amplifier

The output of the detector is coupled to the baseband amplifier. The baseband amplifier raises the detected signal to the required power levels. If the baseband signal is audio, this circuit is usually a push-pull amplifier, like *Figure 4-16b*.

## Dual-Conversion Receiver

*Figure 8-1b* is a dual-conversion receiver block diagram. This receiver design, which has two mixers and two IF strips, improves the image rejection and achieves excellent selectivity. It has the first IF at a frequency higher than the highest receiving frequency. This improves the image rejection, but the receiver's selectivity suffers. Thus, a second IF at a lower frequency is used to improve selectivity. In some cases, high-performance receivers are designed with three IFs and are known as triple-conversion receivers.

### *Automatic Gain Control*

In the dual-conversion receiver of *Figure 8-1b,* there is a block shown as an AGC detector. This AGC detector is used to develop a dc voltage that is proportional to the signal strength. The dc voltage is then fed back to the IF amplifier stages as shown in *Figure 8-2* to control the overall receiver gain. This is called automatic gain control (AGC) because it keeps the output of the receiver at a nearly-constant level even though the input RF signal levels vary widely. If the input signal increases, the gain is decreased; if the input signal decreases, the gain is increased.

# Detection

All receivers contain some form of detection circuitry as part of their design. The detector circuit is used to "unmodulate" the signal and produce the baseband signal from the modulated signal. Detection translates the signal in the frequency spectrum from the carrier frequency back down to the baseband frequency. *Figure 8-3* illustrates the reverse frequency shift caused by detection.

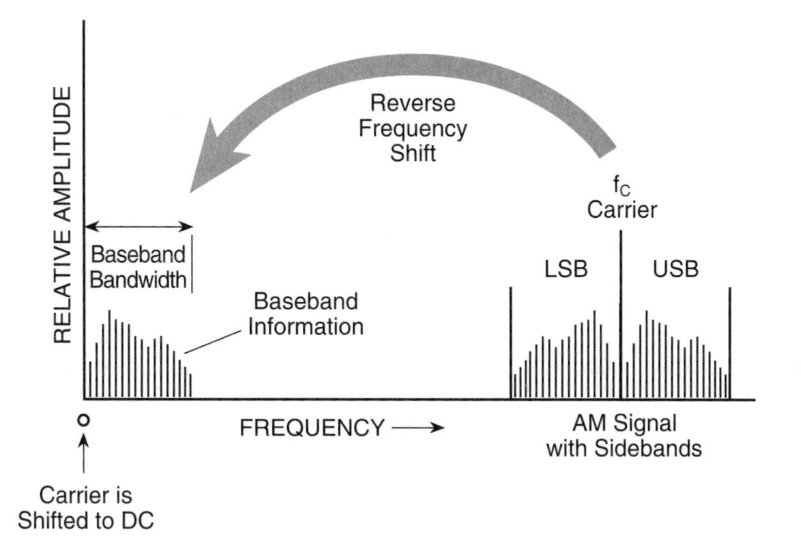


*Figure 8-3. Frequency translation during detection of the baseband signal.*

## AM Detection

*Figure 8-4* is a typical detector circuit used in AM radios. It is a very simple circuit and is one of the reasons that AM is a popular form of modulation. Recall from Chapter 3 that the diode will allow current only in one direction, in this case, toward the resistor and capacitor. When the anode is 0.7 volts more positive than the cathode, the diode will conduct current and charge the capacitor, C1, to the positive input voltage. As the input voltage increases, the charge (and voltage) on the capacitor increases. If the input voltage is such that the anode is negative with respect to the cathode, the diode fails to conduct current and the charging stops. The diode acts as a half-wave rectifier.

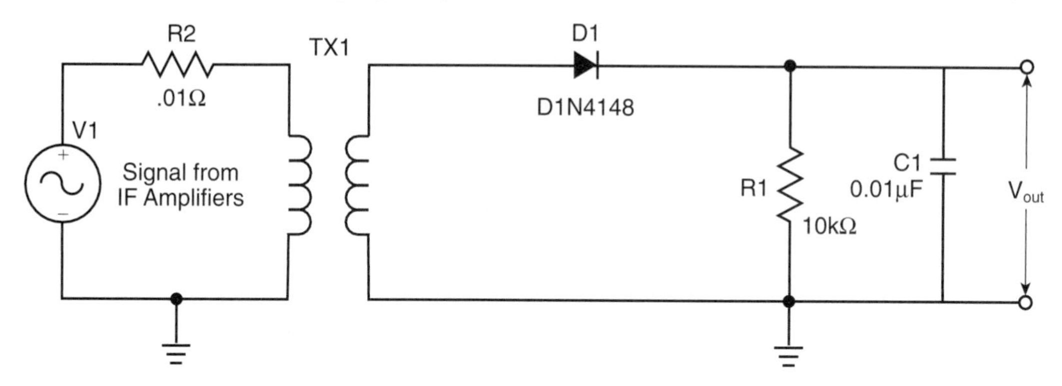


*Figure 8-4. Typical AM detector circuit.*

## *No Modulation*

In *Figure 8-5a,* a carrier (no modulation, thus no sidebands) is applied to the detector circuit. *Figure 8-5b* shows the output. Note that on the first positive half cycle of the

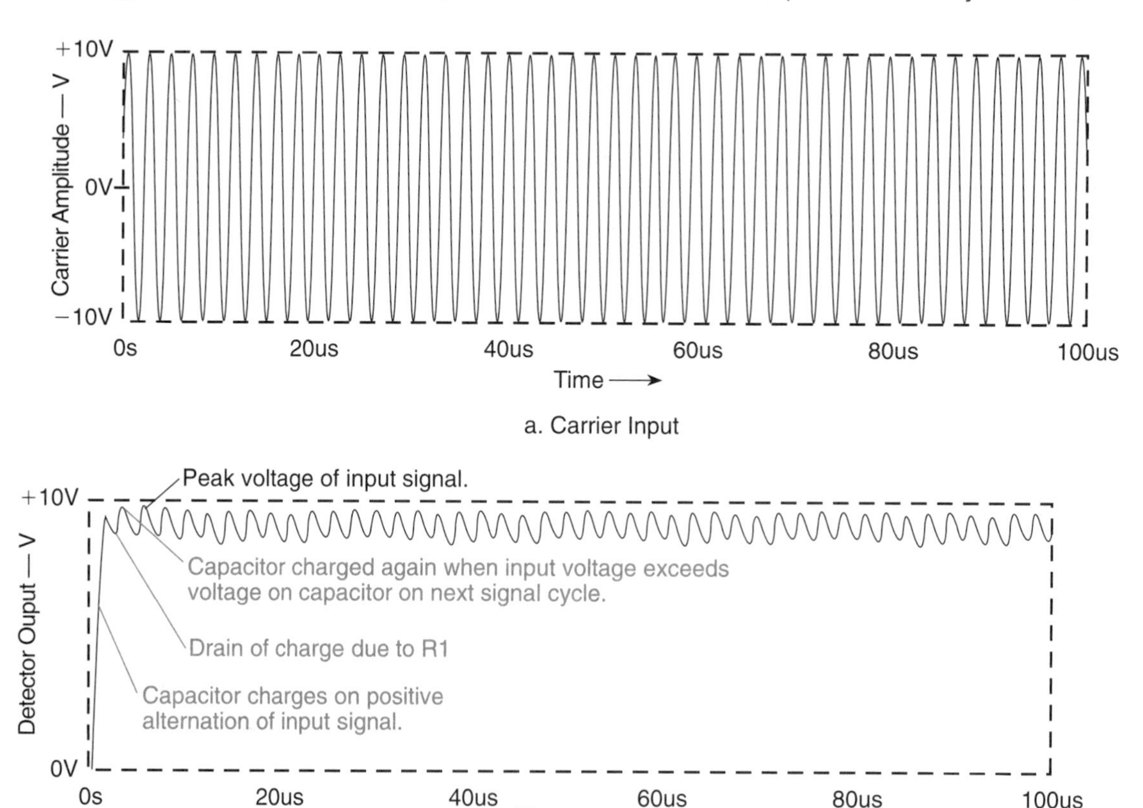

*Figure 8-5. Computer simulation of detector waveforms with no modulation.*

carrier, the detector output across the capacitor is charged by the diode conducting as the carrier signal makes the anode more positive than the cathode. The charge on the capacitor follows the positive carrier voltage. As soon as the carrier voltage starts in a negative direction, the anode of the diode becomes negative with respect to the cathode because the capacitor holds the positive voltage to which it was charged. The diode no longer conducts. As a result, the diode conducts on the positive peaks of the carrier until the capacitor is fully charged. The resistor across the capacitor drains charge from the capacitor and reduces its voltage. The resistor value is set so it will not drain much charge between carrier peaks; therefore, the capacitor will be charged by the carrier peaks, discharged by the resistor until the next positive peak, then recharged on the next carrier peak. With just the carrier applied, the voltage across the capacitor will be a DC voltage that follows the carrier peak amplitude, but due to the resistor drain, has a very small amount of "ripple" on it. With this type of detector operation, the circuit is usually referred to as an envelope detector.

## Modulation Added

In *Figure 8-6,* the AM modulation is a square wave. *Figure 8-6a* is the modulated carrier, and *Figure 8-6b* is the detector output. The modulation in this case is a square-wave signal that increases the amplitude in fast-rising steps. It is not the typical modulation, but is used in this example to demonstrate the action of the detector circuit. It is a slowly-varying ac square wave that is much lower in frequency than the carrier. The

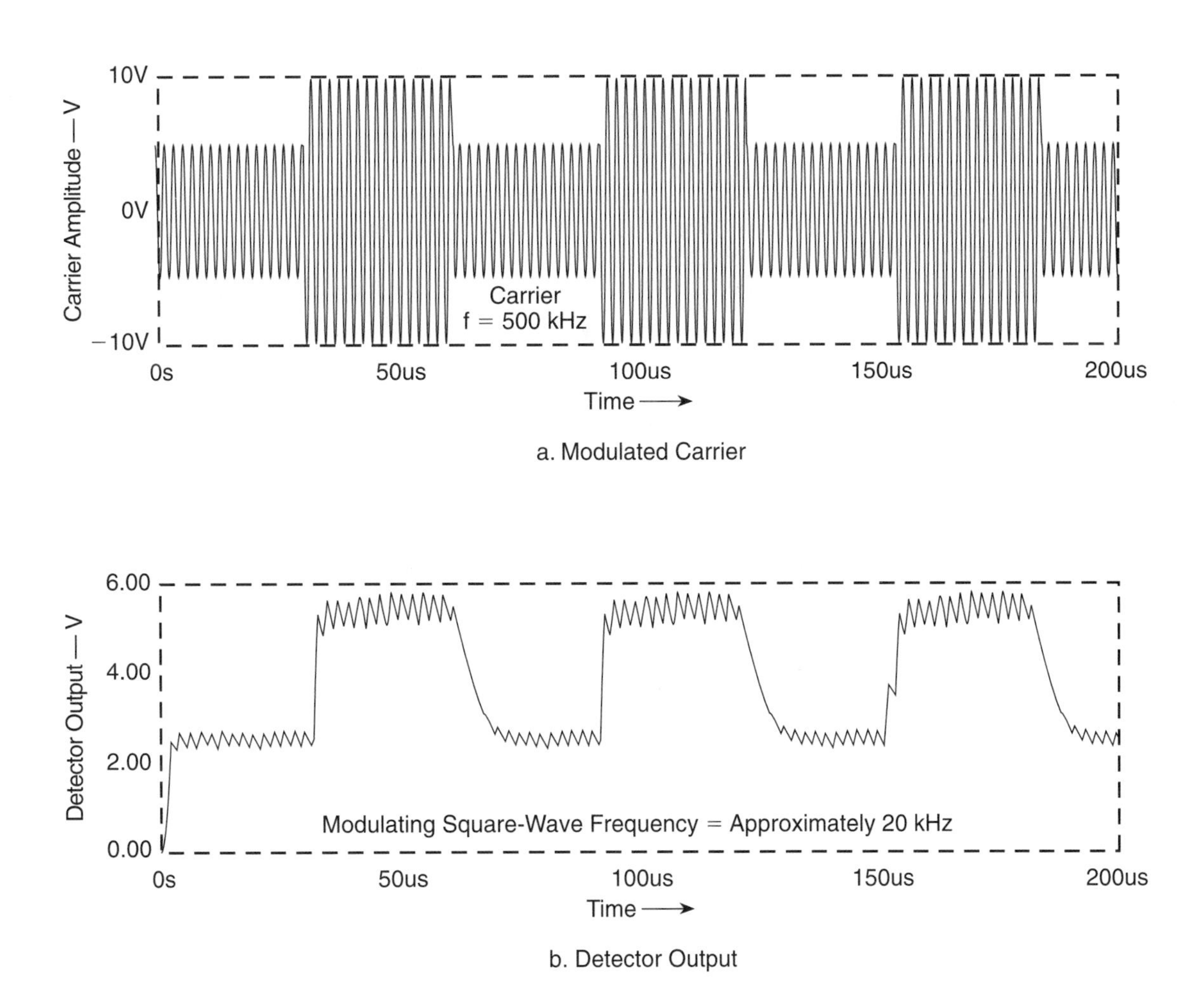


*Figure 8-6. AM detector action when modulation is a square wave.*

carrier frequency is 500 kHz and the modulating square-wave is approximately 20 kHz. The amplitude of the voltage across the capacitor will change to keep up with the peak value of the composite signal (carrier plus sidebands). Note that the voltage across the capacitor adjusts, within a few carrier cycles, to the new peak value of the modulated signal. The values of the resistor and capacitor in the detector circuit are chosen to be a low-pass filter. They allow the lower-frequency modulation square-wave voltage to charge the capacitor and change as the peak value of the modulated carrier changes but filter out the high-frequency carrier down to a small ripple. Note that only a positive signal voltage is output. The negative alternations have been eliminated by the detector circuit's operation.

### *Example 1: Calculating Resistor Value*

When a resistor is in parallel with a capacitor, the discharging curve from an initial voltage across the capacitor is as shown. The value of a resistor in ohms times a capacitor in farads is called an RC time constant (in seconds). In one RC time constant, the capacitor will discharge 63% of its initial value, leaving a 37% charge. In 5RC, the capacitor is considered totally discharged.

What value of resistor should be used in the detector circuit of *Figure 8-4* to prevent the capacitor from discharging too much voltage between peaks of the incoming IF signal of 500 kHz? The capacitor has a value of 0.01 microfarads.

***Solution:***

The time between peaks (the period) of a 500 kHz input signal is

$$\frac{1}{500 \times 10^3} = \frac{1}{5 \times 10^5} = 2 \times 10^{-6}, \text{ or 2 microseconds.}$$

To make the ripple between peaks (the discharge of the capacitor) small, the RC time constant should be 10 times the period of the input waveform. Therefore:

$$RC = 20 \times 10^{-6}, \text{ and since } C = 0.01 \times 10^{-6}$$

$$R \times 0.01 \times 10^{-6} = 20 \times 10^{-6}$$

$$R = \frac{20 \times 10^{-6}}{0.01 \times 10^{-6}} = \frac{20}{1 \times 10^{-2}} = 20 \times 10^2 = 2 \text{ kilohms}$$

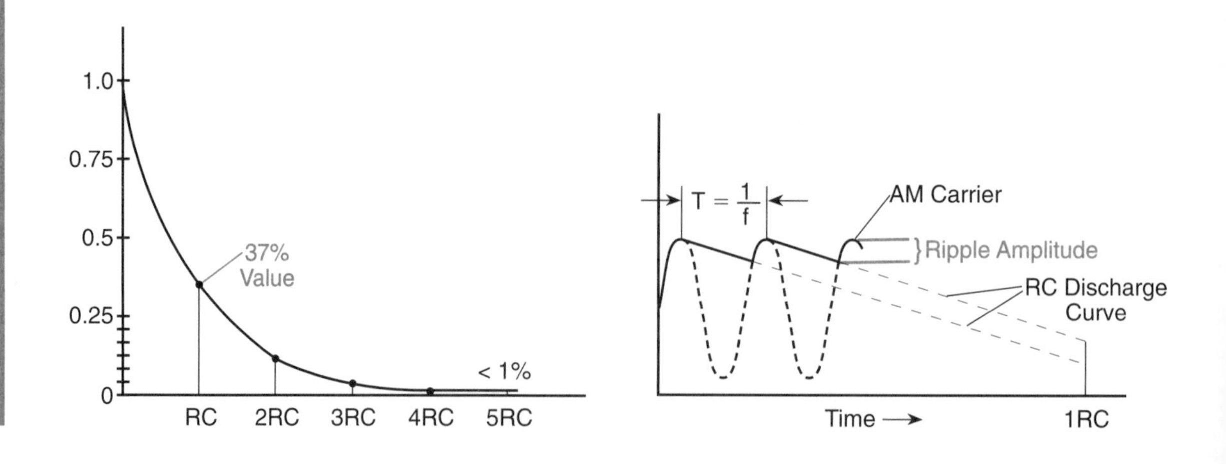

## *A Computer Simulation*

*Figure 8-7* shows a computer simulation of the detection of an AM signal. Recall that in AM modulation the peak amplitude of the composite signal is changed in accordance with the amplitude of the baseband signal. Since the AM detector circuit will "follow" the peak value of the composite signal, the voltage across the capacitor will be the baseband signal. *Figure 8-7a* shows an RF carrier modulated with a single sine-wave baseband signal. *Figure 8-7b* is the detector output that is coupled to the baseband amplifier for further amplification. Note the envelope detection and that no carrier is left in the signal — it has been filtered out by the RC network.

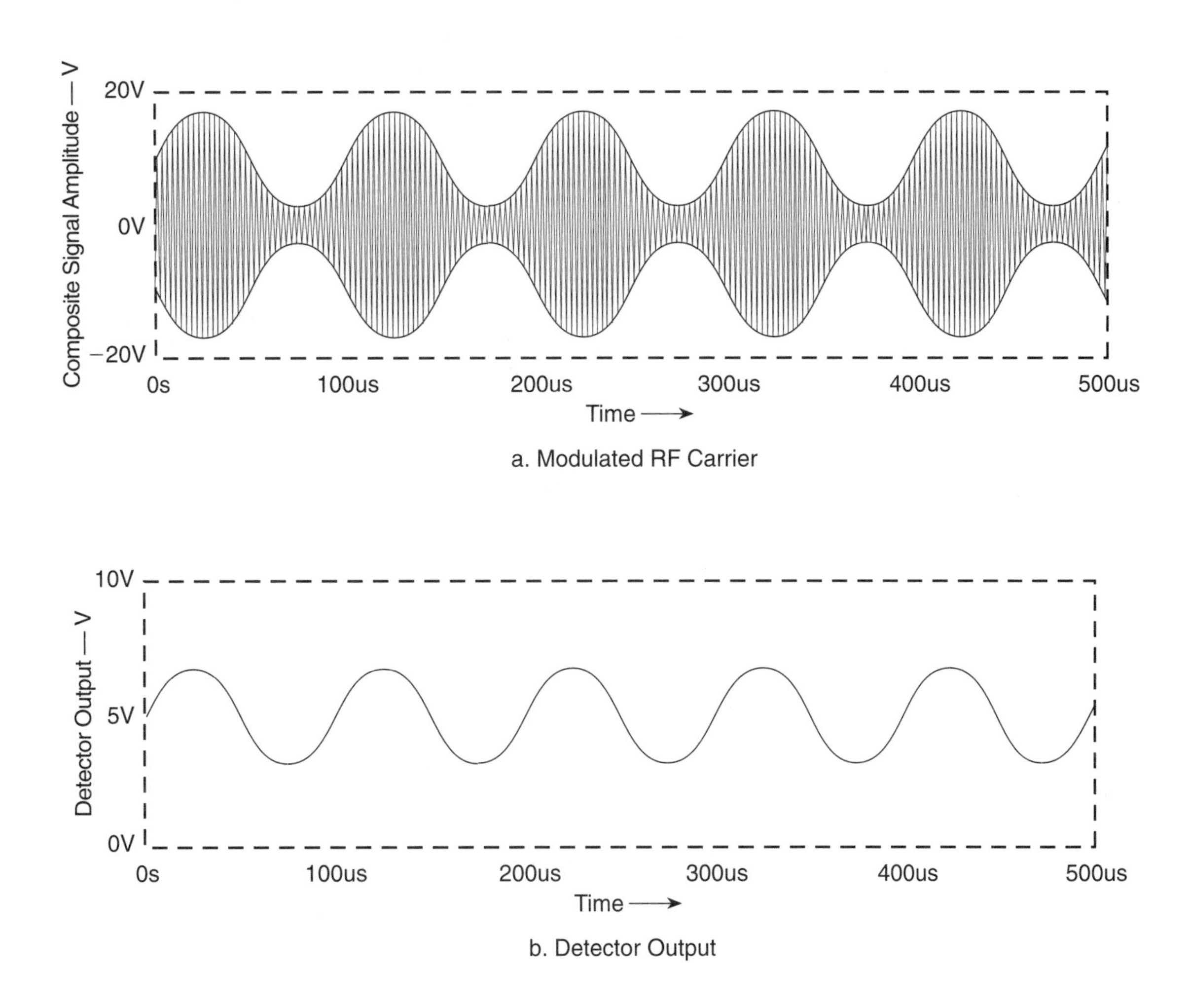


*Figure 8-7. Detection of an AM RF signal.*

One disadvantage of this type of detector is the amount of voltage required to drive the diode circuit. Since the diode requires 0.7 volts to start working, the voltage applied to the detector must have a fairly large value. Usually 10 to 15 volts peak is sufficient to give adequate performance. Note that composite signal amplitudes below 0.7 volts peak will not be detected.

## SSB Detection

Receiving a single-sideband transmission requires a different kind of detection scheme than AM. Since the carrier has been removed from the SSB signal, there is no carrier to reference for detection. To provide proper detection, the carrier must be inserted at the receiver and mixed with the SSB signal. The effect of the mixing is to provide a frequency translation down to the original baseband signal. This type of detection is called product detection, and a circuit that performs the function is shown in *Figure 8-8.* The circuit uses a mpf102 JFET, which operates in the depletion mode, that has the IF signal applied to the gate through an LC-tuned circuit formed by L1 and C3, C4. The carrier signal is generated by a BFO (beat-frequency oscillator) and inserted on the source through the LC tuned circuit of L2 and C1, C2. The circuit operates similar to the mixer of *Figure 6-6,* and is biased to provide the peak value composite signal required to detect and recover the baseband signal.

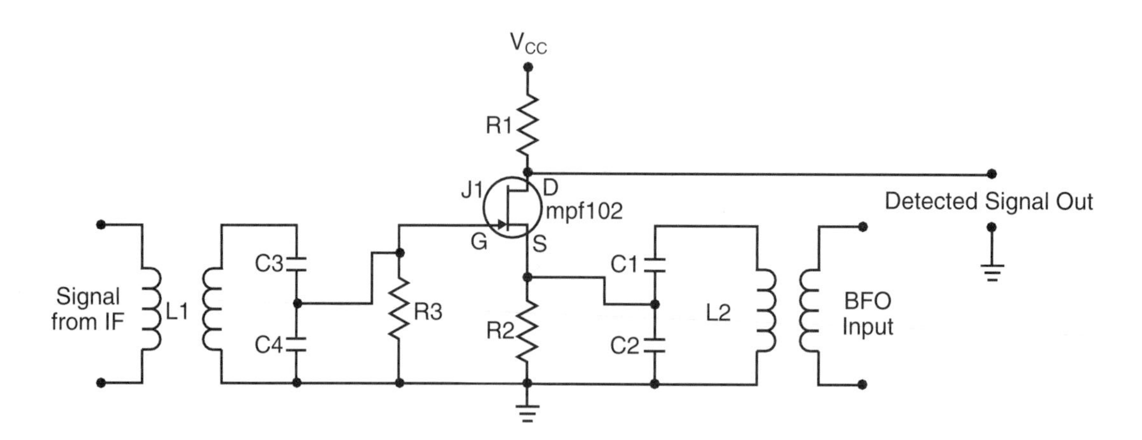


*Figure 8-8. Product detector circuit used for SSB.*

The advantage of product detection is that it can take place at lower signal levels than envelope detection. The disadvantage is that if the inserted carrier from the BFO is not at the same frequency as the original carrier, errors occur.

Product detection also is used for the detection of Morse code (CW) modulation. In CW detection, the inserted carrier is used to produce a difference signal in the audio range. By adjusting the BFO-inserted frequency to an audible beat frequency, the tone used to identify the dits and dahs of the Morse code can be selected to be most pleasing to the code operator. An 800-Hz tone is one that is commonly used.

## FM Detection

### Discriminator

Detecting an FM signal is more complex than detecting an AM signal. Since the information is contained in the carrier's frequency changes (not the amplitude), it is the frequency deviation of the carrier that must be detected to recover the baseband signal. *Figure 8-9* illustrates the technique.

To understand the technique, remember that the modulating signal at the transmitter changed the frequency in proportion to the amplitude of the signal. This is reviewed in *Figure 8-9.* The dotted line sine wave represents the modulating signal at the transmitter. When the modulating signal is at zero amplitude, as at **A**, there will be no frequency deviation. On the voltage vs. frequency plot of *Figure 8-9,* the point **A** is at the zero point on the frequency change axis (no modulation) and at the zero volts point of the voltage axis. When the modulating signal is at its most positive amplitude (point

**B**), the frequency will have changed to point **B** on the positive frequency axis. Point **C** on the modulating signal produces a point **C** frequency change on the frequency axis. Point **D** frequency change on the negative frequency change axis results due to the most negative change of the modulating signal. A corresponding negative frequency change occurs at point **E** for the negative amplitude of point **E**.

At the FM detector in the receiver, the detector output is to be a reproduction of the modulating signal. We can see from the detector transfer characteristic of *Figure 8-9,* that a frequency deviation to point **B** on the positive frequency change axis will produce a detector positive output voltage of $V_1$. A frequency deviation to point **C** will produce a positive voltage of $V_2$. A frequency deviation to point **E** will produce a negative voltage of $V_3$, and to point **D** will produce a negative voltage of $V_4$. If the frequency changes are plotted against time as they were made at the transmitter by the modulating signal, the output of the detector will be a reproduction of the original information signal that modulated the transmitter signal.

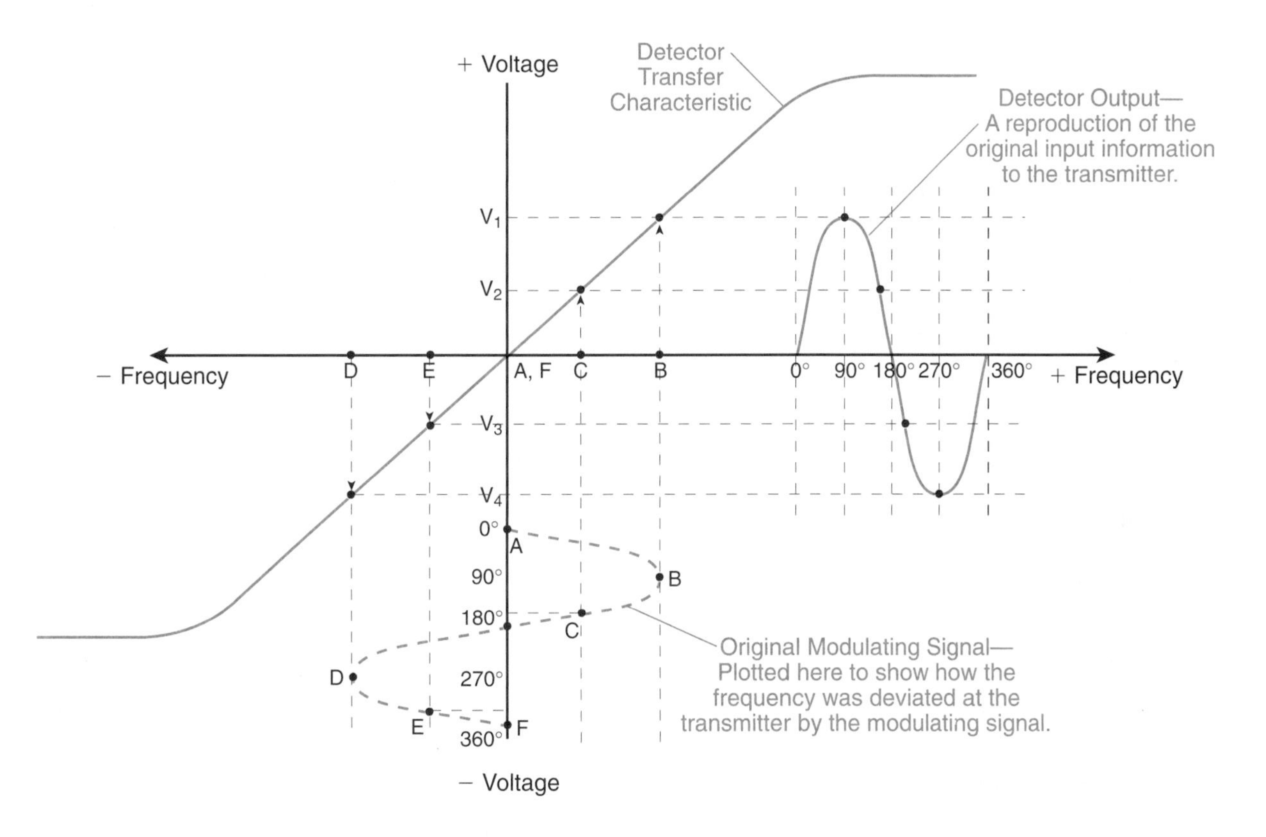


*Figure 8-9. FM detection technique.*

In actual circuits, if the input frequency is at the carrier frequency (no modulation), the detector output is not zero but some offset dc voltage representing the carrier frequency. When the carrier is deviated (frequency changed) by a modulating signal amplitude level, the detector corresponds with an equivalent voltage output either above or below the offset voltage. This type of detector is called a *discriminator*. A discriminator may be viewed as a frequency-to-voltage converter, since the instantaneous output voltage of the detector is determined by the frequency of the carrier.

## Slope Detector

One of the simplest discriminator circuits is a slope detector. Slope detection is implemented as shown in *Figure 8-10.* An IF amplifier is designed so that the FM carrier is on the side (slope) of the IF's tuned circuit response. As the FM deviates the carrier

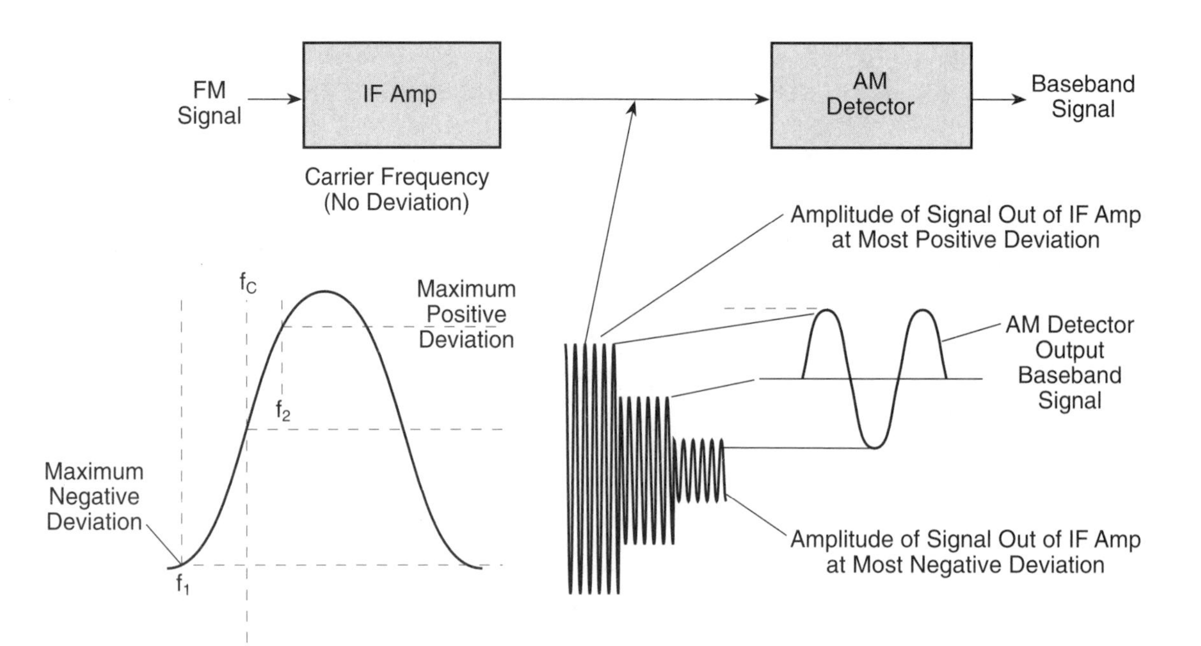


*Figure 8-10. Block diagram of FM slope detector.*

frequency, the amplitude of the deviated signal changes. The amplitude-varying frequency signal is coupled to an AM detector, which reproduces the original baseband signal. To prevent any amplitude variations in the FM signal from causing distortion, limiters are placed in front of FM detectors. The limiter function will be described when an FM receiver is discussed a little later in this Chapter.

## PLL Discriminator

Today, one of the most commonly-used FM discriminators uses a PLL (phase-lock-loop) circuit, shown in block diagram form in *Figure 8-11.* The modulated carrier is compared in frequency against a VCO (voltage-controlled oscillator). The VCO is pulled to the instantaneous IF amplifier signal frequency by the filtered output of the phase comparator. As the signal frequency changes due to the FM, the error voltage of the phase comparator and filter will change, changing the VCO to keep the VCO at the same instantaneous frequency as the IF amplifier signal. The error voltage, which is used to recover the baseband signal, is filtered and fed to baseband amplifiers to raise the power level of the recovered signal. The PLL "tracks" the deviated carrier signal, and as long as the loop stays locked, the baseband information is recovered.

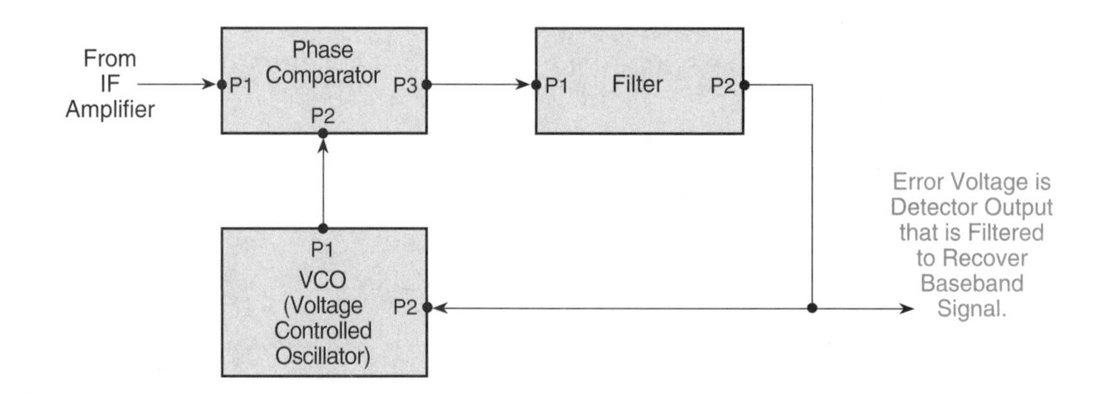


*Figure 8-11. Block diagram of a PLL FM discriminator.*

Early FM receivers used discriminators such as the Foster-Seely discriminator, or a ratio detector. Their circuits are fairly complex and used tuned circuits and diodes to recover the baseband signals. The circuits needed matched components and careful alignment to operate properly. Modern receivers use integrated circuit PLL discriminators because the IC provides closely-matched components manufactured at the same time within a single IC package, thus providing excellent circuit performance, reduced component cost, and lower assembly cost.

## AM Receiver Details

As mentioned previously, a receiver is a major sub-system that will have many of the circuits that we discussed in this and previous chapters combined together to implement the system. *Figure 8-12* is a schematic diagram of a basic AM receiver. The front end of the receiver is made up of Q1, the RF amplifier stage, Q2, the mixer stage, and Q3, the oscillator stage. Q1 is a JFET tuned amplifier, like *Figure 4-9;* Q2 is a JFET mixer, like *Figure 6-6;* and Q3 is a variable-frequency oscillator, like *Figure 7-4.* By changing the variable capacitor, C13, in the oscillator, the receiver is tuned to the desired signal. In many receivers, this oscillator is replaced with a PLL synthesizer circuit like *Figure 7-6.* The tuned circuit at the input to Q1 couples the RF energy from the antenna to the RF amplifier, and also supplies some selectivity. The output of the mixer Q2 is tuned to the IF frequency and the signal is fed to the IF amplifiers Q4 and Q5, whose circuits are the same as *Figure 8-2.*

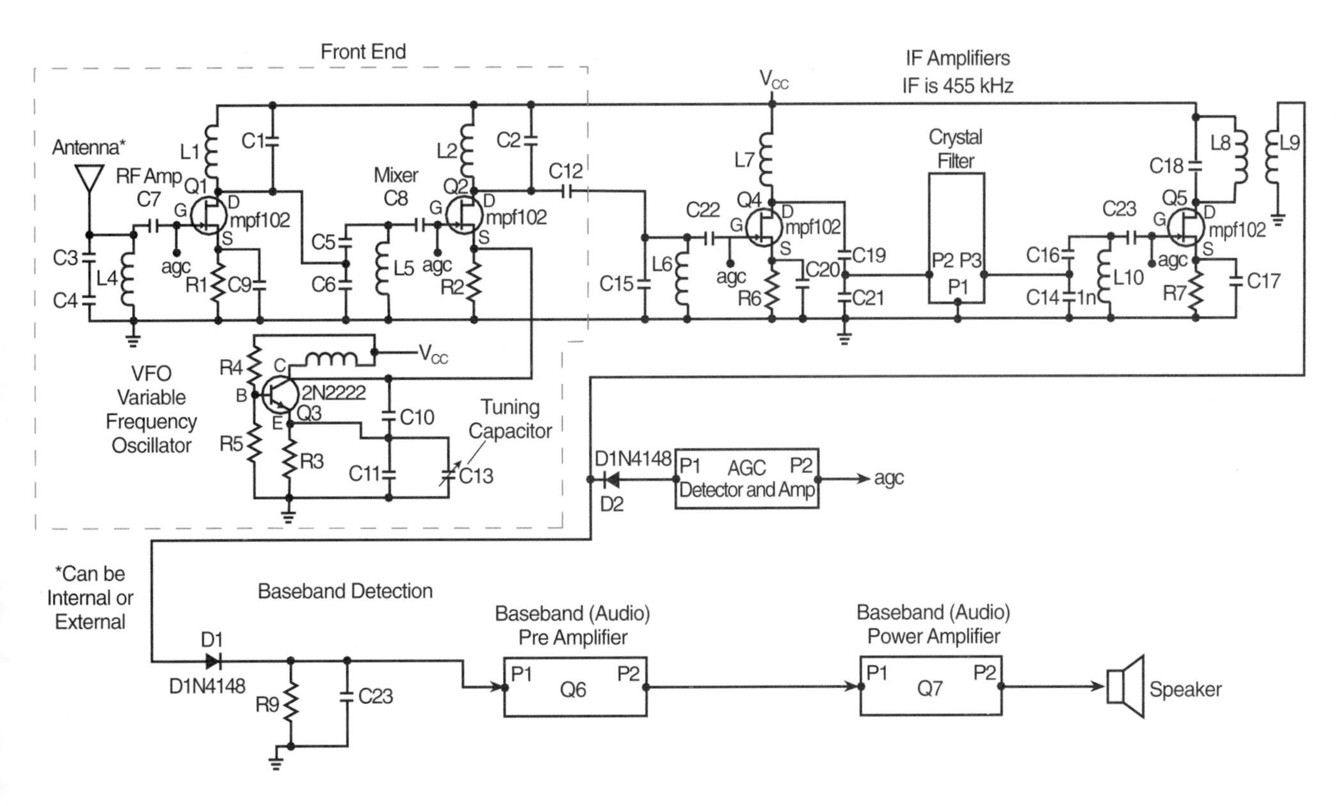


*Figure 8-12. Detailed schematic diagram of an AM receiver*

The IF amplifiers, Q4 and Q5, share a crystal filter circuit between them to provide a sharp selectivity curve. All of the tuned circuits associated with Q4 and Q5 are tuned to the IF center frequency. The diode circuit, D1, is the detector for the AM receiver and its output will drive the baseband amplifier, in this case an audio amplifier. The detected signal is fed to Q6, which is a low-level audio amplifier (preamplifier) like *Figure 4-5,* that is coupled to the audio power amplifier Q7. Q7 is a transformer-coupled amplifier, like *Figure 4-16b.* The diode circuit, D2, is the AGC detector, which develops a voltage proportional to the strength of the signal from the station receiver. Its output changes the bias on the amplifier stages which changes their gains, thus regulating the signal level at the detector. With strong signals, the gain is reduced so the receiver amplifier stages do not overload and cause distortion.

## SSB Receiver Details

*Figure 8-13* is a schematic diagram of an SSB receiver. It is very similar to the AM receiver of *Figure 8-12*. The main difference is the detector, which is not a simple diode circuit, but a product detector with a BFO like *Figure 8-8*. The front end is essentially the same as the front end of the AM receiver, as are the IF amplifiers, except the component values are adjusted for the frequency range desired. The crystal band-pass filter used for the SSB receiver will be narrower than the AM filter (3 kHz rather than 6 to 12 kHz). If the SSB receiver is used for CW reception, the filter passband can be as narrow as 500 Hz. The baseband preamplifier and power amplifier are the same as for the AM receiver, so they are shown only as blocks, as is the AGC detector.

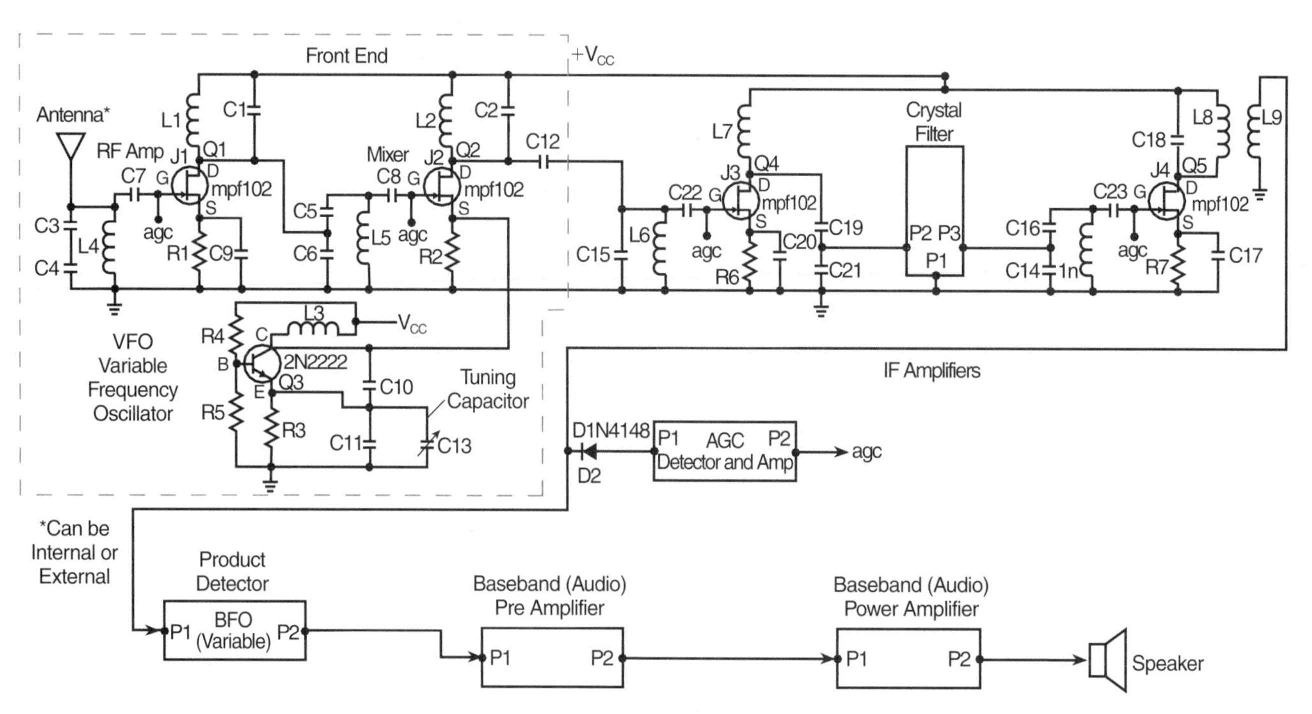


*Figure 8-13. Detailed schematic of an SSB receiver.*

## FM Receiver

*Figure 8-14* is a block diagram of an FM receiver. The basic diagram is similar to the AM receiver of *Figure 8-12*, but FM receivers have a few different functions. First, the detector is a discriminator. Second, before the detector there is a function called a limiter. The limiter will provide immunity from a noise signal added to the FM signal. This is especially true of environmental noise. And third, after the discriminator there is an added function called the squelch circuit. This prevents the listener from hearing a constant noise output when no signals are present. Both the limiter and the squelch functions improve the performance of the FM receiver for the listener.

### Limiter

The limiter does just what its name implies — it limits the amplitude of its input signal to an output with a constant minimum value. This eliminates most amplitude variations that may be caused by noise. The limiting function is shown in the block diagram of

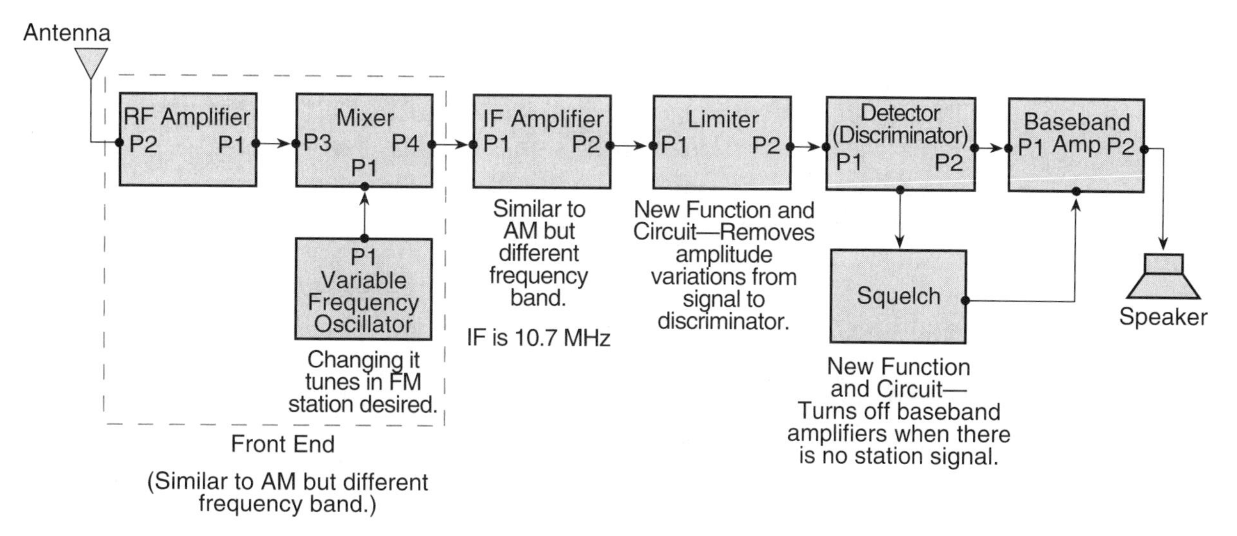


*Figure 8-14. FM receiver block diagram.*

*Figure 8-15a.* It is placed in the FM receiver just before the FM discriminator. An input signal to the limiter contains the desired frequency changes due to FM modulation, but it also contains amplitude variations that can cause noise in the receiver output. The limiter eliminates the amplitude variations and outputs a signal to the discriminator that still contains all the FM information, void of any amplitude variations. *Figure 8-15b* shows the block diagram of two IC op-amps, operated as inverting, overdriven amplifiers, that provide the limiting function.

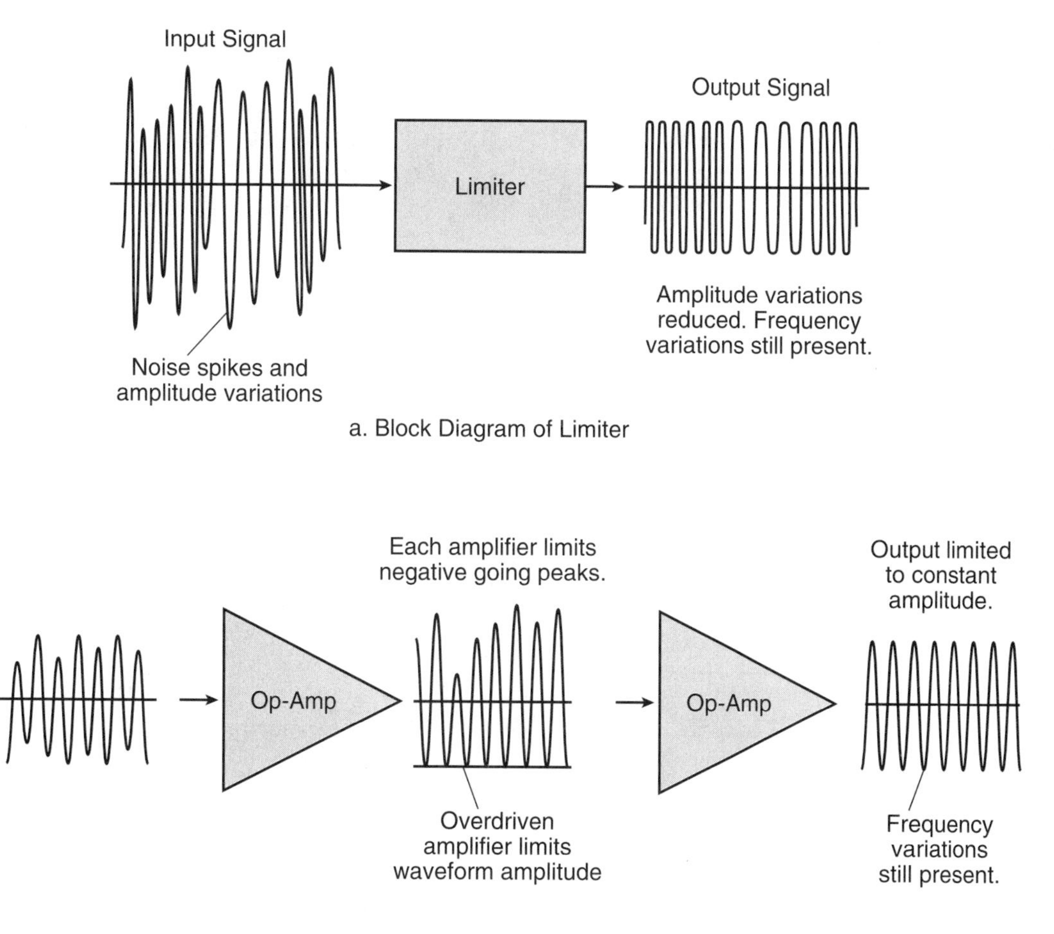


*Figure 8-15. FM limiter function.*

Due to the way the FM modulation and its detection in the FM receiver occur, the receiver will normally have a great deal of noise at the output when there is no signal present. The squelch circuit detects the lack of signal and turns off the audio baseband amplifier stages so no noise is heard. With an RF signal present (carrier detected), the audio stages are turned back on so the detected signal can be heard.

The receiver front end and IF amplifiers are similar to AM, but are designed to operate in the FM band (55 to 108 MHz), rather than in the AM band (550 to 1700 kHz). The IF frequency for the FM commercial entertainment receiver is 10.7 MHz , while it is 455 kHz for the AM receiver.

## FM Receiver Detail

*Figure 8-16* is the schematic diagram for an FM receiver. It is a somewhat unusual schematic because most of its circuitry is contained in integrated circuits. The front end circuits are still discrete component circuitry because of all of the carefully-designed tuned circuits for tuning and filtering, but the rest of the circuitry is ICs. The front end uses similar components to the circuits for the AM and SSB receivers, except for the frequency band. The receiver is tuned either with a PLL synthesizer or with a similar VFO (variable frequency oscillator), which will most likely be tuned mechanically. *Figure* 8-16 includes a simple VFO circuit with an NPN bipolar 2N2222 (Q3) as the oscillator transistor. The Q1 circuit is an RF amplifier and the Q2 circuit is the mixer. The IF strip and detector is implemented with an LM1965 FM IF-system integrated circuit. This IC includes the IF amplifiers, limiter, discriminator, squelch circuit, AGC circuit, and audio pre-amplifier. As stated previously, the limiter strips off all amplitude information that may be present before the detector. There also is an added auxiliary function — the IC includes circuitry to drive a signal-strength meter. Thus, the receiver gives its operator an indication of the relative strength of the station signals received.

The LM383 IC that provides audio amplifiers to increase the audio signal to the required power level is shown in *Figure 8-17.* The amplifiers will have a bandwidth tailored to the baseband signal. Complementary bipolar circuitry is used and a complementary-symmetry power amplifier is used, similar to the one discussed in Chapter 4.

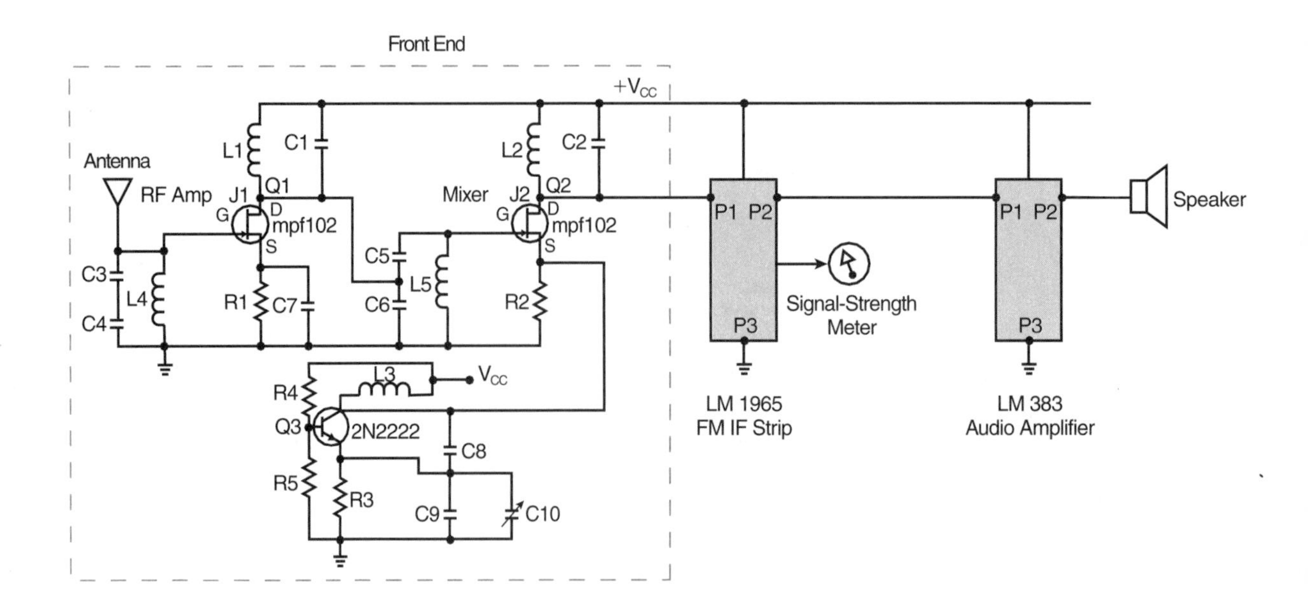

*Figure 8-16. FM receivers with discrete component front end and ICs for IF strip and audio power amplifier.*

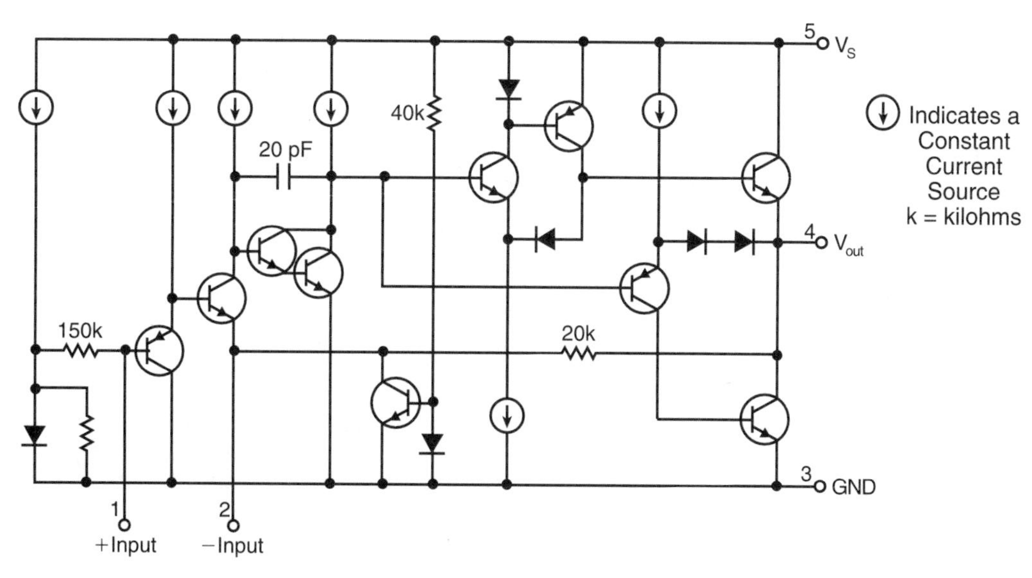


*Figure 8-17. Detailed schematic of the LM383 audio power amplifier.*
*(Courtesy National Semiconductor)*

## Radio on a Chip

The LM1868 is an integrated circuit that provides much of the circuitry for an AM/FM radio receiver. A block diagram of the circuit functions contained in the IC and the schematic of the external components that are interconnected to make up a complete AM/FM radio are shown in *Figure 8-18.* The IC provides most of the active devices required, while all of the tuned circuits and bypassing capacitors are external. A separate, discrete-component FM front end is used. The Q1 circuit again is an RF amplifier and the Q2 circuit is a mixer. The AM and FM front ends, IF section, and detection are each

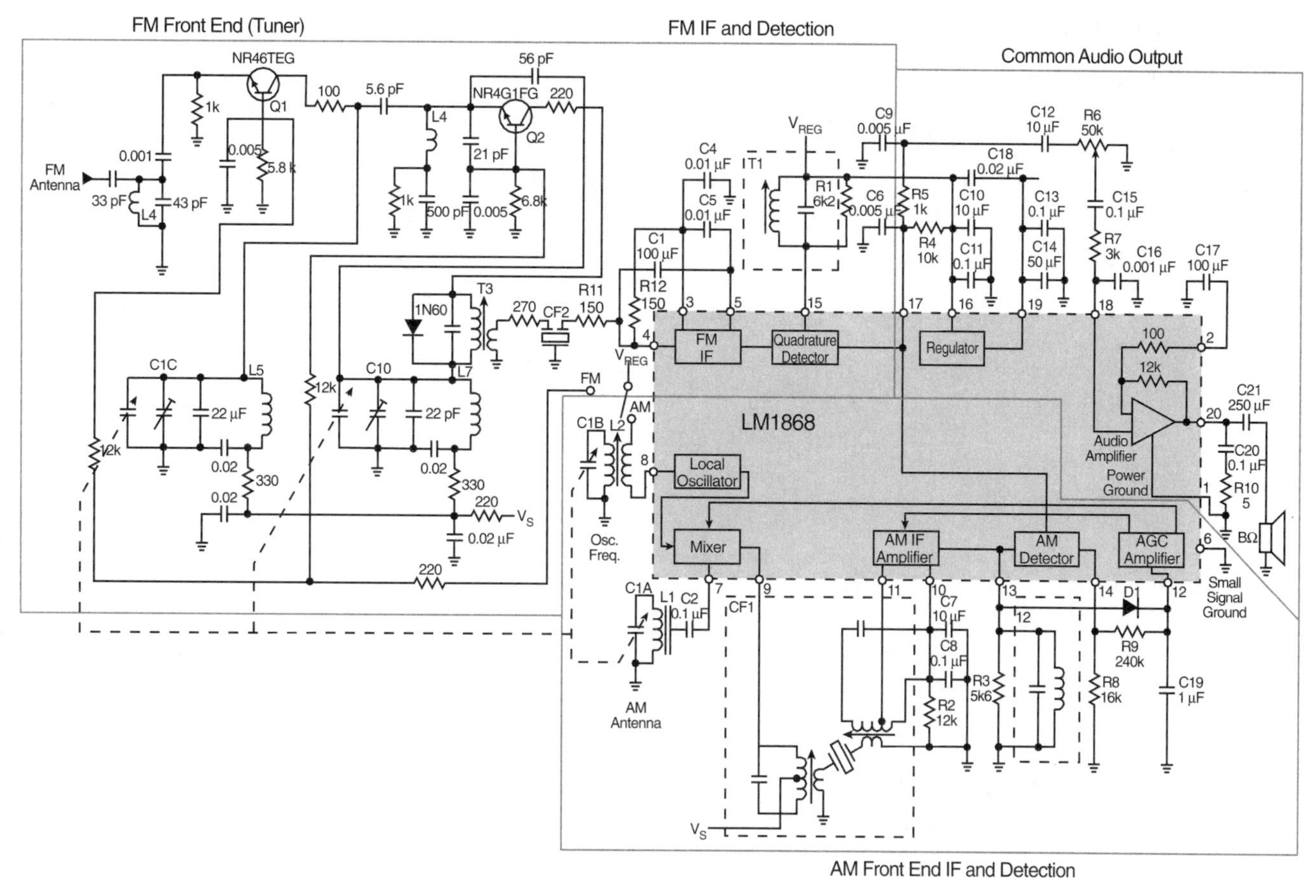


*Figure 8-18. LM1868 IC used as AM/FM radio.*
*(Courtesy National Semiconductor)*

identified in *Figure 8-18.* There is a common audio output section used for both AM and FM. The selection of AM or FM operation is through the application of a DC voltage to the input circuitry of the respective front end circuits. The 20-pin package is shown in *Figure 8-19,* and the function of each pin on the package is identified.The radio is tuned by a multi-ganged capacitor for both AM and FM. This application demonstrates the size reduction and assembly cost savings produced by integrated circuits, which provide uniform matched characteristics of active devices and integrated components.

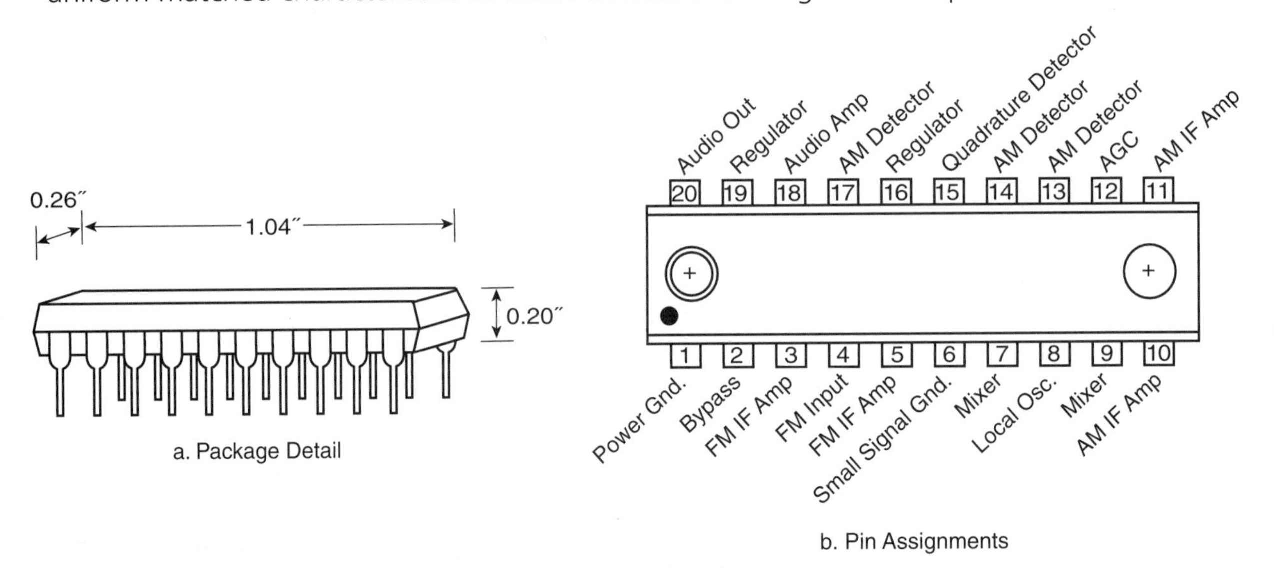


*Figure 8-19. LM1868 20-pin package and terminal functions.*
*(Courtesy National Semiconductor.)*

## Summary

In this Chapter, we have discussed receivers — AM, SSB and FM. As with transmitters, we examined block diagrams and detailed schematics to show how receiver systems are made up of the basic functions that were described in previous chapters. Our transmission link for both of those chapters has been wireless radio signals with the transmitter at one end and the receiver at the other. In the next Chapter, we will examine the transmission link in more detail.

# Quiz for Chapter 8

1. What is the "front end" of a receiver?
   a) The circuits closest to the dial.
   b) The RF amplifier and mixer stages.
   c) The RF amplifier, mixer, and first local oscillator.
   d) The antenna matching circuits.
2. What is quieting with regard to FM receivers?
   a) The lack of noise when no signal is being received.
   b) All of the noise caused by the carrier.
   c) The lack of noise when a carrier is present.
   d) Having a good volume control.
3. Explain the process of detection for AM signals?
   a) Finding out what the carrier frequency is.
   b) Recovering the carrier.
   c) Recovering the baseband information from the composite signal.
   d) All of the above.
4. What is the IF strip in a receiver?
   a) Amplifiers that operate at the carrier frequency.
   b) Amplifiers used to increase the voltage of the local oscillator.
   c) The intermediate frequency amplifiers which have most of the receiver's gain.
   d) Part of the detector circuit.
5. Explain sensitivity with regard to receivers?
   a) The ability to "hear" a weak signal.
   b) How stable the receiver is in frequency.
   c) How well the receiver operates in extreme environmental conditions.
   d) The ability to reject unwanted signals.
6. What are image frequencies?
   a) An unwanted signal close to the desired signal.
   b) Another signal that enhances the wanted signal.
   c) An interfering signal with a frequency that is twice the IF frequency above the wanted signal frequency.
   d) Interference due to receiver overload.
7. What is the purpose of the mixer circuit in receivers?
   a) Pick up the weakest possible signal.
   b) Translate the wanted signal to the IF frequency
   c) Translate the wanted signal to a randomly-selected frequency.
   d) Remove noise from a wanted signal.
8. What is meant by selectivity in receivers?
   a) The smallest detectable signal.
   b) Having a bandwidth just wide enough for the wanted signal.
   c) Being able to tune the receiver to receive only the signal you want.
   d) None of the above.
9. Why do FM receivers limit the input signal before detection?
   a) To make sure all of the amplitude information gets through.
   b) To eliminate amplitude variations.
   c) To eliminate frequency variations.
   d) To protect the detector from current overload.
10. An FM detector will...
    a) Give a voltage output that is proportional to the frequency of the input signal.
    b) Reject all sideband information.
    c) Reject all carrier information.
    d) Have no output of any kind when only the carrier is present.

**Answers:**
1 c, 2 c, 3 c, 4 c, 5 a, 6 c,
7 b, 8 c, 9 b, 10 a

## Questions & Problems for Chapter 8

1. _______________ is used to measure noise performance in receivers.
2. What is the purpose of an automatic gain control (AGC) circuit?
3. What device is used to insert the carrier frequency in an SSB receiver?
4. A dual-conversion receiver uses two _______________ and two _______________ to improve image rejection and achieve excellent selectivity.
5. What is the purpose of a limiter in an FM receiver?
6. What are the possible oscillator frequencies for a receiver with an IF of 2.5MHz receiving a 6.25 MHz signal?
7. What is the image frequency of a receiver with a 455-kHz IF listening to a 12-MHz signal?
8. An AM signal modulated with 1kHz and 3kHz is received by a receiver. What are the audio output signals?
9. If a dual-conversion receiver is receiving a 10-MHz signal and the IF frequencies are 4.7 MHz and 1.7 MHz, what do the oscillator frequencies have to be to tune-in the signal?
10. When a PLL is used for an FM discriminator, the _______________ voltage represents the output.

*(Answers on page 212.)*

# CHAPTER 9
# Transmission Links



Recall that a communication system is made up of a source (transmitter) and a destination (receiver), with a transmission link between the two. We have concentrated on wireless transmission in describing a radio transmitter and receiver. In this chapter, we will add more detail to the wireless transmission link, as well as discuss the other means of transferring energy over transmission lines used as transmission links.

## Completing the Wireless Transmission Link

Each wireless transmitter and receiver requires an antenna. At the transmitter, energy is transferred from the final power amplifier to the antenna, which then radiates energy in the form of electromagnetic waves into the air to produce the wireless transmission. At the receiver, as the radiated electromagnetic energy passes through the antenna, a voltage is induced into the antenna that produces signal current in the input amplifiers of the receiver. The magnitude of the induced voltage depends on the intensity of the electromagnetic energy at the receiving antenna. The antenna at each end is a vital part of the wireless transmission link.

## Antennas

The simplest antenna is a single piece of wire. When there is current in a wire due to an applied voltage, a magnetic field is produced around the wire. Thus, there are two fields produced — one by the voltage, the second by the current. If the voltage is ac, it alternates in amplitude and polarity at the frequency of the ac voltage. As the voltage varies, the current varies in amplitude and direction. The two fields developed around the wire are shown in *Figure 9-1* — the electric field in *Figure 9-1a* and the magnetic field in *Figure 9-1b.* Note that the two fields are at right angles to each other, as shown in *Figure 9-1c* and *Figure 9-1d,* and change direction as the voltage changes polarity and the current changes direction.

### A Transmitting Antenna

Let's look again at *Figure 9-1,* especially *Figure 9-1b.* Notice that there are two field parts shown — a near field and a far field. There is a critical distance from the wire that separates the two. For the near-field part, if the current stops in the wire, the electromagnetic field collapses and returns to the wire; however, in the far-field part, the electromagnetic field doesn't return. It radiates into space. The far-field radiated by the transmitting antenna is the part that is used to transfer the information in a wireless transmission link.

The electromagnetic field radiates through space in a direction that is at right angles to both the contained electric and magnetic fields, which we have said previously are at

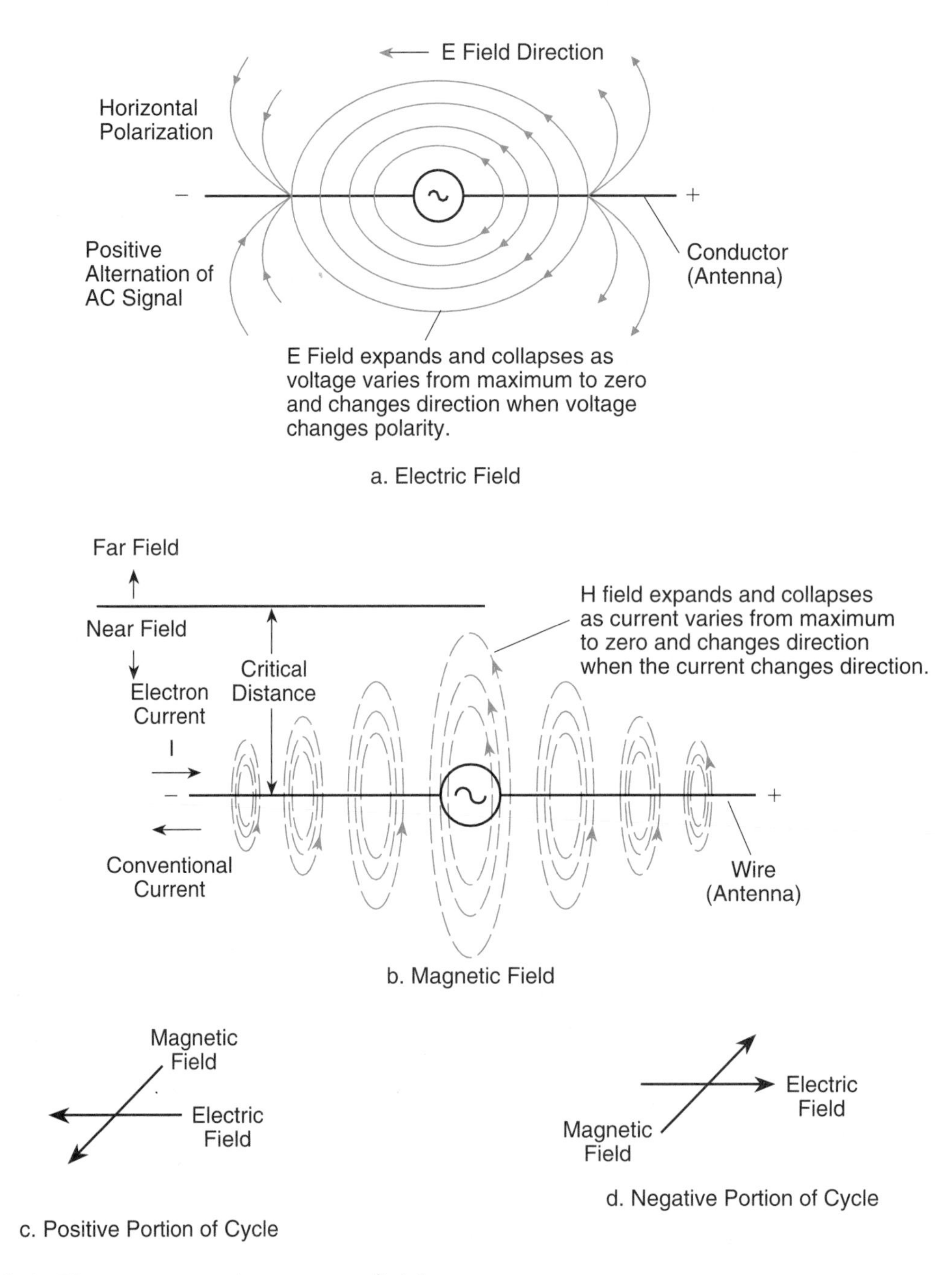


*Figure 9-1. Electromagnetic antenna fields.*
*(Source: Antennas, A.J. Evans, K.E. Britain, ©1998, Master Publishing, Inc., Lincolnwood, IL)*

right angles to each other. A visual representation of the wave travel is shown in *Figure 9-2.* As the electric (E) field varies in amplitude and polarity, the current varies and thus the magnetic (H) field varies. When there is zero current, the magnetic field will be zero. In *Figure 9-2,* point **X** is a position in space, and if one were at that position and the electromagnetic wave passed by, you would see the variations in electric and magnetic fields as shown.

## A Receiving Antenna

To help understand the way the antenna operates at the receiver, look at *Figure 9-3.* The magnetic field variations of *Figure 9-1b* are shown again in *Figure 9-3.* If another wire is placed close to the wire with the changing magnetic field around it such that the changing magnetic field cuts through the wire, a voltage will be induced in the second wire. This is the same principle as that used in transformers. If the second wire is in the far field and is considered to be a receiving antenna as at position **X** of *Figure 9-2,* and

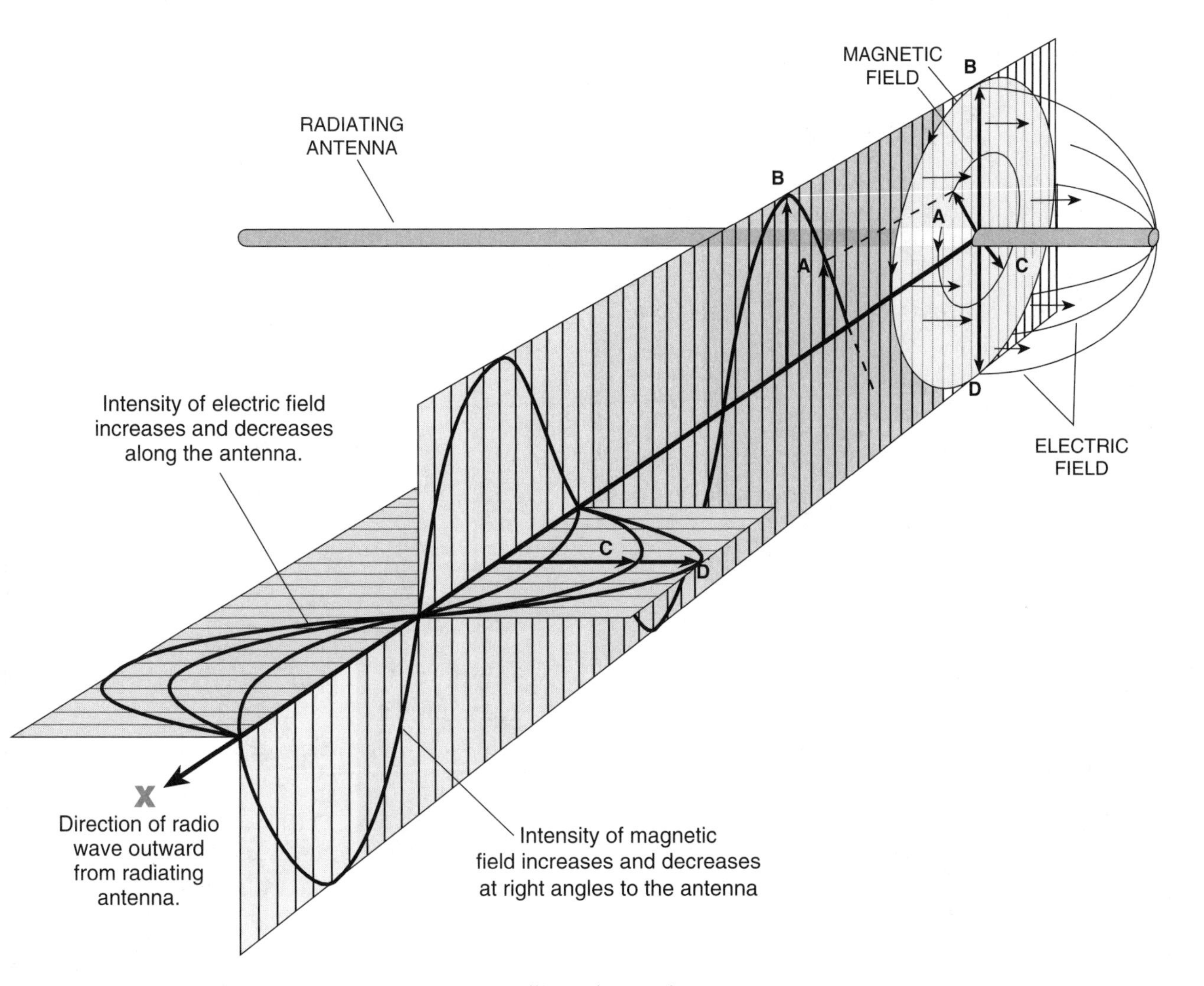


*Figure 9-2. Electomagnetic wave travelling through space.*
*(Source: Basic Electronics, G. McWhorter, A.J. Evans, ©1994, Master Publishing, Inc., Lincolnwood, IL)*

the radiated electromagnetic wave passes by, the variations (changes) in magnetic field will induce a voltage in the receiving antenna. The induced voltages in the receiving antenna cause current variations in the input amplifier circuits of the receiver. As we discussed in Chapter 8, the receiver then amplifies and detects the signals received by the antenna from the wireless transmission link.

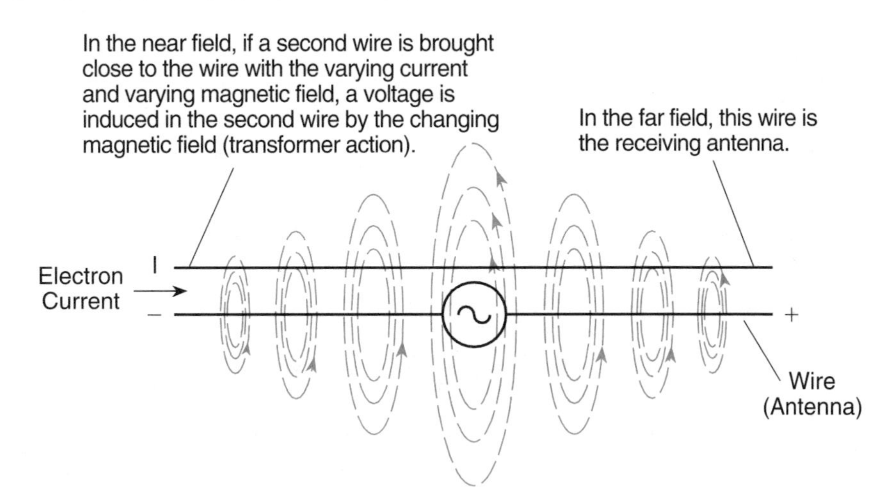


*Figure 9-3. Magnetic field variations induce voltages in wires in the near field or in the far field. When the wire is in the far field it is a receiving antenna.*
*(Source: Basic Electronics, G. McWhorter, A.J. Evans, ©1994, Master Publishing, Inc., Lincolnwood, IL)*

# Antenna Design

## Antenna Characteristics

Antennas are specialized devices that are matched to their application. It is common practice to discuss antennas as if they were transmitting antennas, but they work equally well as transmitting or receiving antennas, except that the transmitting antenna must be designed to handle the required transmitter power output. In addition, the antenna impedance is "matched" to the transmission line feeding the antenna, which is coupled to the transmitter to provide efficient transfer of transmitter power output to radiated energy. The impedance is set at an appropriate level for the transmitter, transmission line, and antenna. Here are some characteristics of antennas that are important:

1. Polarization
2. Element Size
3. Radiation Pattern
4. Gain and Directivity
5. Impedance

All of the above will contribute to the overall strength of the signal radiated in the far field. Let's discuss each of these characteristics separately.

### *Polarization*

Antennas are commonly classified as having vertical or horizontal polarization. This refers to the way the electromagnetic waves are oriented as they are radiated. The *direction of the electric field* determines the polarization. A vertical antenna will most likely have a vertical electric field, and thus it has vertical polarization. Horizontally-polarized antennas have a horizontal electric field. The polarization of transmitting and receiving antennas should be the same for efficient transfer of energy. *Figure 9-4* summarizes the concept.

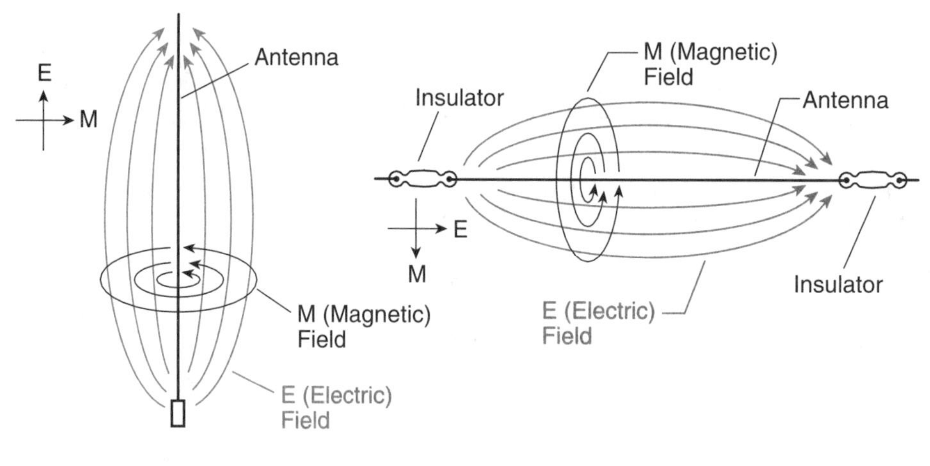


*Figure 9-4. Polarization of antennas. An antenna's polarization is determined by the direction of the E (Electric) field.*
*(Source: GROL Plus, F. Maia, G. West, ©1996, Master Publishing, Inc., Lincolnwood, IL)*

There are circularly-polarized antennas where the E and H fields are not only at right angles to each other but their orientation rotates so that the signal can be received equally well by either vertically- or horizontally-polarized antennas. Antennas normally have a driven element to which power is delivered by the feed line. The orientation of the electric field in the driven element will determine the antenna's polarization.

### *Example 1: Determining the Polarization of Antennas*

Four antennas are shown. Determine their polarization.

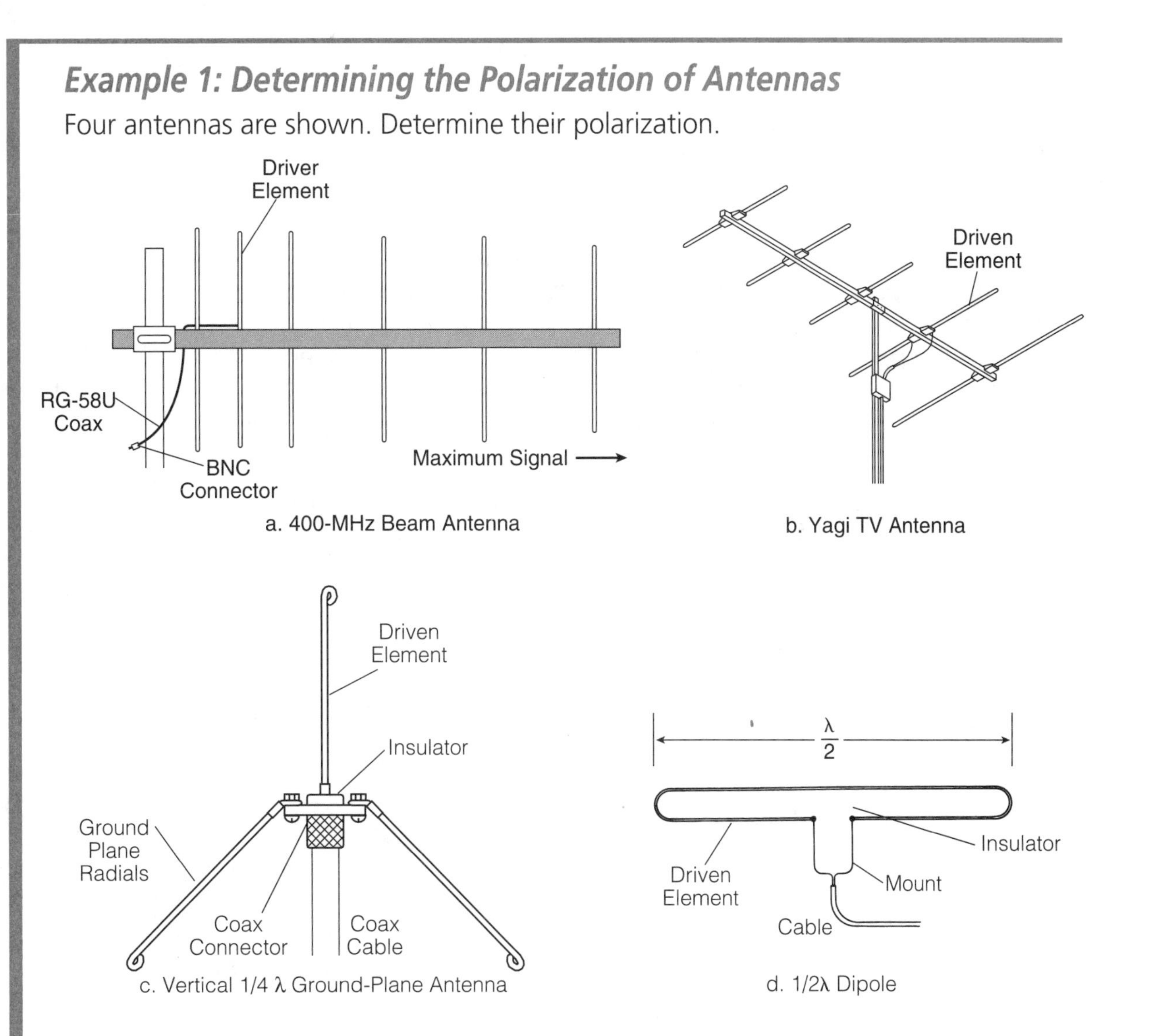


a. 400-MHz Beam Antenna

b. Yagi TV Antenna

c. Vertical 1/4 λ Ground-Plane Antenna

d. 1/2λ Dipole

#### *Solution*

Find the driven element of the antenna and the direction of the electric field to determine the antenna's polarization.

A. The driven element is vertical; therefore, electric field is vertical for vertical polarization.

B. The driven element is horizontal; therefore, electric field is horizontal for horizontal polarization.

C. Vertical electric field, vertical polarization.

D. Horizontal electric field, horizontal polarization.

## Element Size

Recall that wavelength is defined as the distance the electromagnetic wave travels in space in the time required to complete one cycle of the transmitted signal. In equation form it is (remember, the Greek letter λ represents wavelength):

$$\text{Wavelength } (\lambda) \text{ in meters} = \frac{\text{Velocity (v) in meters per sec}}{\text{frequency (f) in cycles per sec}}$$

$$\text{Distance} = \text{velocity} \times \text{time} = \text{velocity} \times \frac{1}{f} = \frac{v}{f}$$

In order for antennas to be efficient radiators of electromagnetic waves, their size (or length) must be some submultiple or multiple of the wavelength of the signal being radiated. Usually, a quarter of a wavelength is the minimum and one-half of a wavelength is a very common antenna length.

## *Radiation Patterns*

For a particular antenna design, the energy will be radiated in a particular pattern. The radiation pattern is a plot of the field strength of the radiated energy at various directions and/or elevation angles from the antenna. For example, *Figure 9-5* shows the radiation pattern of a half-wavelength horizontal dipole antenna (called a Hertz antenna). The maximum field strength is at right angles to the antenna, with little or no energy radiated from the ends. In a dipole, the transmitter power is fed in the middle of the element.

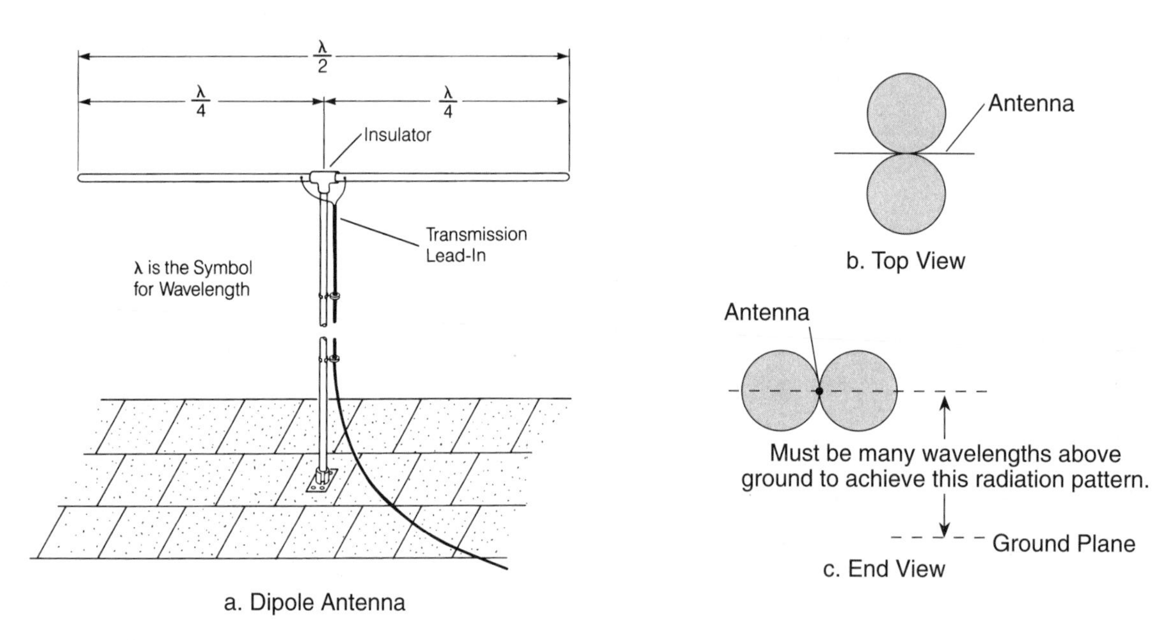

*Figure 9-5. Radiation pattern of a half-wavelength dipole antenna.*
*(Source: Antennas, A.J. Evans, K.E. Britain, ©1998, Master Publishing, Inc., Lincolnwood, IL)*

Broadcast station antennas are designed specifically to have particular radiation patterns over specific areas surrounding the station. This helps define the station's service area.

Some antenna designs have a pattern with a preferred direction of radiation. There are two patterns of interest used to evaluate the direction of radiation — the horizontal and the vertical. The horizontal pattern is viewed from above, as though you were looking down on the antenna, as shown in *Figure 9-5b.* The vertical pattern is viewed from the end, as shown in *Figure 9-5c,* or sometimes from the side as though you were beside the antenna. One should also keep in mind that the horizontal and vertical patterns depend upon the angle of the radiation from a ground plane. For horizontal antennas, this angle depends on the height of the antenna above ground. *Figure 9-6* shows how the vertical radiation of the dipole antenna of *Figure 9-5* varies as the height above ground is varied.

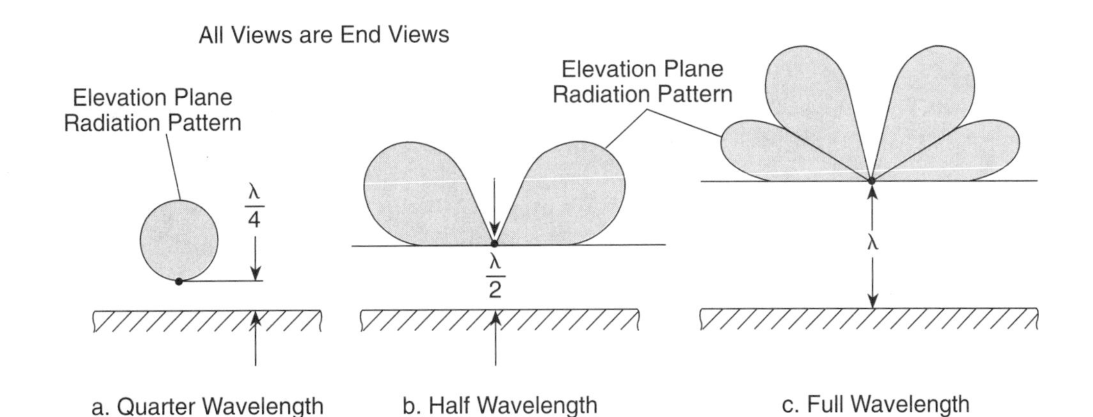


*Figure 9-6. Variation of radiation patterns as height above ground is varied.*
*(Source: Antennas, A.J. Evans, K.E. Britain, ©1998, Master Publishing, Inc., Lincolnwood, IL)*

## Gain and Directivity

### *Isotropic Radiator*

A theoretical antenna that works equally well in all directions is called an isotropic radiator. Its pattern is spherical around the antenna, as shown in *Figure 9-7,* and no matter where you are the field strength from the antenna will be the same. In practice, it is impossible to build an isotropic radiator, but it is used as a theoretical standard of comparison for antenna systems.

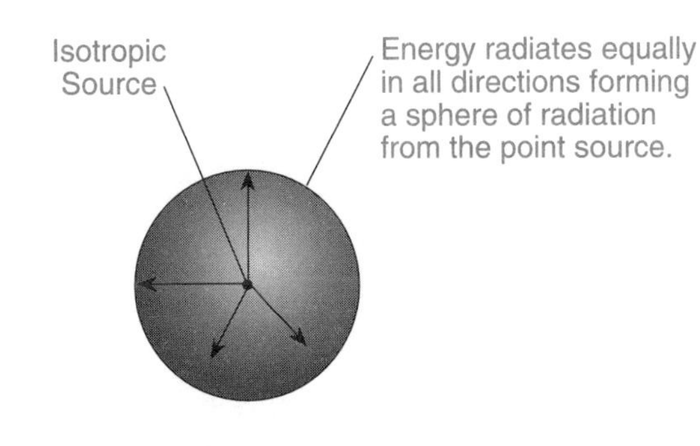

*Figure 9-7. Isotropic radiator pattern.*

### *Antenna Gain*

Antennas are said to exhibit power gain. Actually, the antenna simply radiates energy in a preferred direction and the "gain" is the fact that more signal is radiated in the preferred direction than that radiated by an isotropic radiator. This means that the antenna will work better in some directions than it will in others. However, this does not mean that the antenna will radiate more power than is coupled to it from the transmitter. For example, if 100 watts is coupled to an antenna then slightly less than 100 watts will be radiated, because all antennas have some power loss due to heating.

Let's look at an example of antenna "gain." *Figure 9-8* shows an antenna that radiates four times the signal strength in the most favored direction as that expected from the same power into an isotropic radiator. Antenna gain, like amplifier power gain, is measured in decibels. The antenna is specified to have 6 dB gain (3 dB for each doubling of power) over the isotropic radiator. This usually is stated as 6 dBi gain because it is referenced to the isotropic. In the least-favored direction, the antenna does not work as well as the isotropic radiator. Radiating energy greater than the isotropic radiator in the preferred direction gives rise to the term "effective radiated power" (ERP). ERP is the amount of power you would need to couple into an isotropic radiator to have the same signal strength in the preferred direction as the antenna in question.

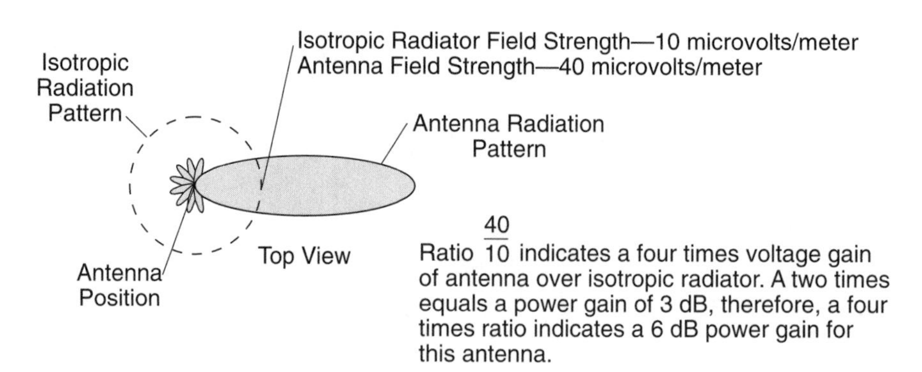


*Figure 9-8. "Gain" of an antenna.*

The isotropic radiator is characterized in "free space" rather than over a real ground. As shown in *Figure 9-6,* when antennas are mounted over real ground the position has a strong influence on the radiation pattern. For horizontally-polarized antennas over ground, there is a theoretical 6-dB gain over an isotropic radiator due to ground effects. This directional pattern effect is illustrated in *Figure 9-8.* The dipole, due to its directional characteristics, has about 2.4-dB gain over isotropic. Consequently, by mounting it at least a half-wavelength above ground, the basic dipole antenna can have a gain of about 8 to 8.4 dB over an isotropic radiator.

## Directivity

Directivity is a term that means that an antenna has been designed to concentrate its radiation pattern in a preferred direction. *Figure 9-9* shows a common TV antenna designed for this purpose. *Figure 9-9a* is the directional radiation pattern directed at the TV station when mounting the antenna, and *Figure 9-9b* shows the antenna construction.

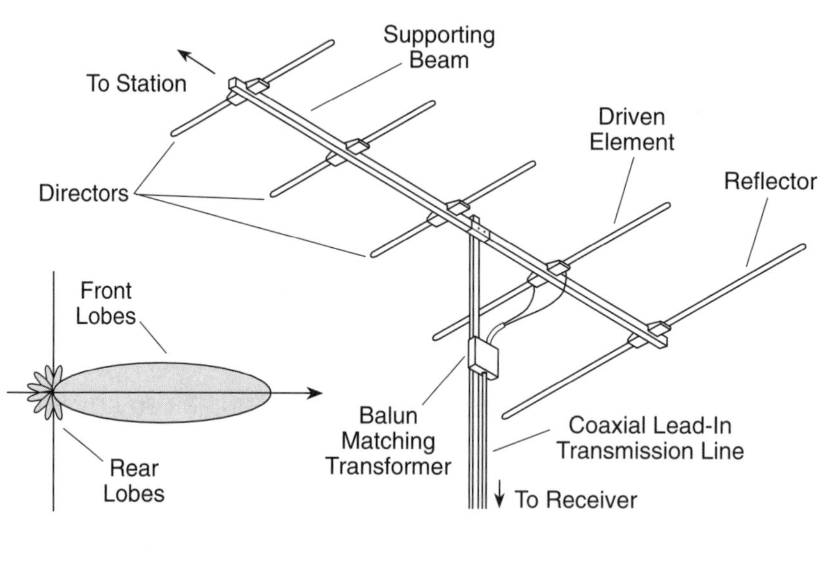


*Figure 9-9. A Yagi-Uda directional beam antenna.*
*(Source: Antennas, A.J. Evans, K.E. Britain, ©1998, Master Publishing, Inc., Lincolnwood, IL)*

Recall that an antenna has the same characteristics when used as a transmitting or receiving antenna. We will describe the operation of the antenna of *Figure 9-9b* as a transmitting antenna. It is called a multi-element antenna because it uses extra elements in addition to the driven element, which is the element actually coupled to and driven by the transmitter. By adding the extra elements, it is possible to increase the amount of signal radiated in a specific direction. This antenna is called a Yagi-Uda antenna after its inventors. The extra elements are called parasitic elements because they are not connected to the feed line from the transmitter but are close enough to the driven element to have voltages induced in them by the radiation from the driven element. The currents resulting from the induced voltages will produce radiation from the parasitic elements and, if the antenna is arranged correctly mechanically, the parasitic radiation will be in phase with the driven element radiation and the total radiation will be focused in a preferred direction. This beam forming will cause the antenna to have a gain using one parasitic element of 4 dB, and a second parasitic element will add another 1.5 to 2 dB of gain. It is important to note that power is not increased, just the amount of power being radiated in one specific direction.

There is another way to achieve gain with added elements that are not parasitic. If the added elements are fed at different phase angles by using different lengths of feed line, the resultant antenna — called a phased array — will have a directional pattern. Phased arrays are often used at medium and high frequencies. The most common example is one used for an AM commercial broadcast station to set the station's radiation pattern. *Figure 9-10* is an example of a four-element vertical phased array. The physical construction is shown in *Figure 9-10a* and *Figure 9-10b* shows its horizontal radiation pattern.

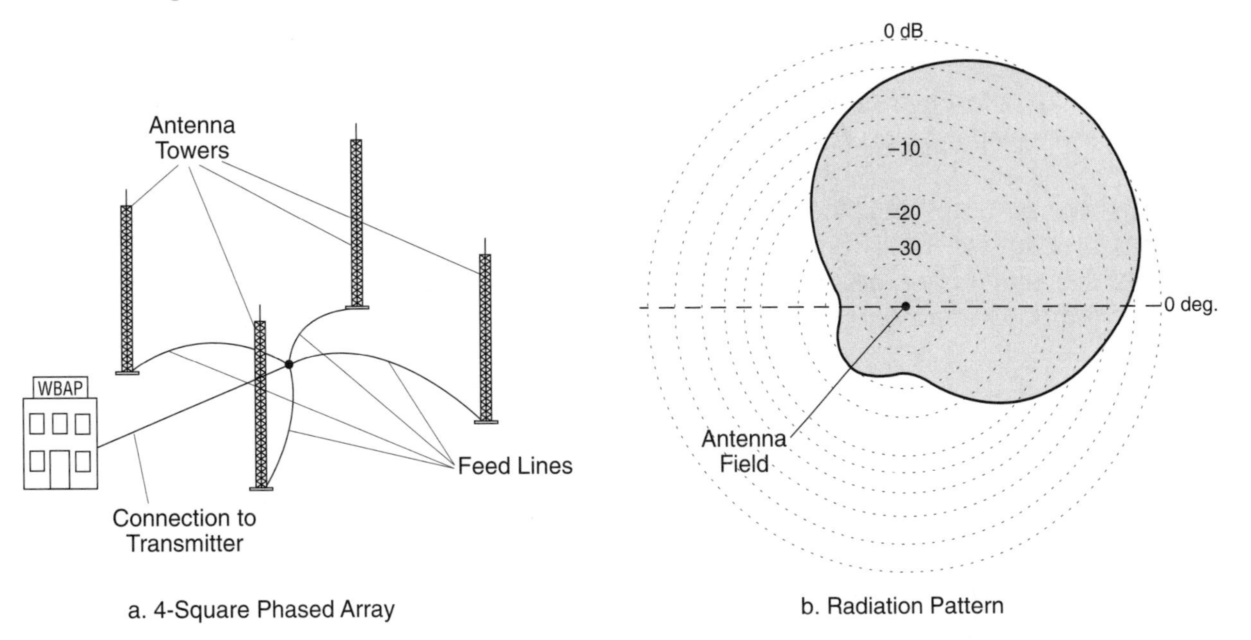


*Figure 9-10. Four-element phased-array like those used for commercial AM broadcasting.*

## A Common Vertical Antenna

Since we have discussed radiation patterns and their effect, let's cover a special type of vertical antenna. It is used commonly as an automobile radio antenna, and *Figure 9-11* shows how the antenna is mounted. The antenna shown is one-quarter wavelength long. The unusual thing about this antenna is that the body of the automobile serves as the ground plane and an image one-quarter wavelength antenna is formed in the ground plane. This type antenna is called a "ground plane" or Marconi antenna after its inventor. The ground acts as the other half of the antenna to make it a one-half wavelength antenna. The better the ground the better the quarter-wavelength vertical will perform. The radiation pattern is shown in *Figure 9-11b*. It indicates that the antenna radiates equally well in all directions, but tends to have a fairly low-angle vertical lobe. Vertical antennas find application throughout the radio spectrum. The automobile radio antenna usually is made as long as practical, and can approach a quarter-wavelength for the commercial FM bands. The input impedance of the quarter-wavelength antenna is about 37 ohms at resonance and generally it is coupled to the transmitter with coaxial cable.

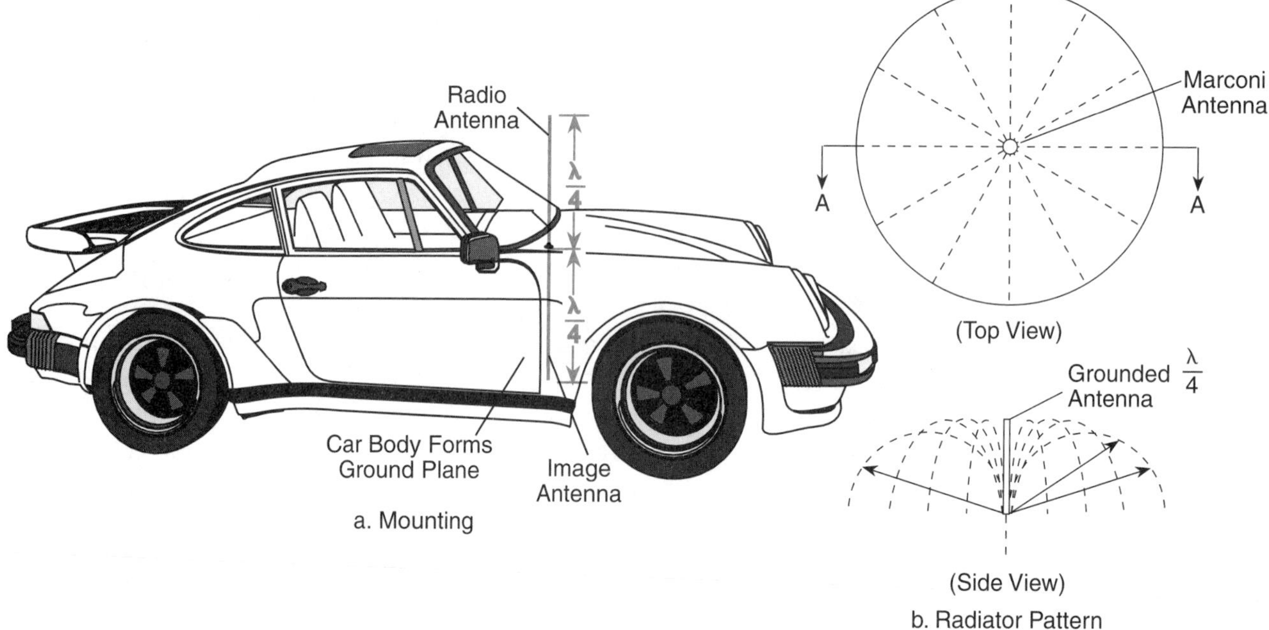


*Figure 9-11. A 1/4 wavelength UHF/VHF vertical mounted to an automobile. The car body serves as the ground plane.*

There are many variations of the 1/4-wavelength vertical. The most common is a 5/8-wavelength vertical. This antenna has a lower angle of radiation and does not depend as much upon ground as the 1/4-wavelength antenna. The 5/8-wavelength antenna finds wide use in mobile applications at VHF and UHF frequencies.

## Field Strength

Field strength is a measure of the intensity of the signal at a particular location and is expressed in terms of the strength of the electric field in volts-per-meter; that is, how many volts will the radiating field induce in an antenna that is one meter long. Since receiving antennas are great distances from the transmitter, the field strength is expressed in millivolts or microvolts per meter. The field strength at the receiving antenna is determined by the amount of power transmitted, the transmitting antenna gain, the distance from transmitter to receiver, and signal losses.

## Impedance

Just like other components in electronic circuits, antennas have impedance. The impedance is determined by the physical design of the antenna and is characterized as minimum at the resonant frequency for which the antenna is designed. TV antennas often have an impedance of 300 ohms because of the construction of the driven element. The characteristic impedance of the twin-lead transmission line that is commonly used to couple the TV to the antenna is matched to this. Many coax cables used for transmission lines have characteristic impedances of 50 ohms and 75 ohms, and since they are used regularly to couple receiving antennas to receiving systems, antennas are designed to have impedances that match these line impedances.

# Propagation

## The Paths

Discussion of the wireless transmission link is not complete without talking about the path that the radio signal takes from transmitter to receiver — the propagation of the signal. *Figure 9-12* shows the four paths — a direct line-of-sight path, a ground-wave path, a ground-reflected-wave path, and a sky-wave path.

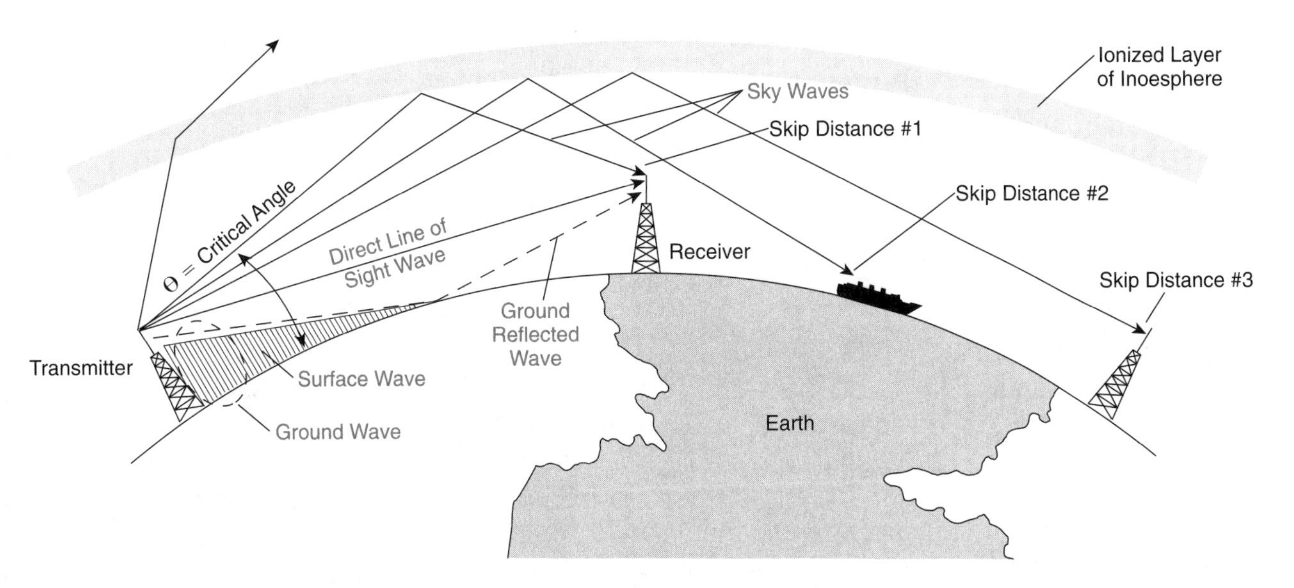


*Figure 9-12. Propagation of radio waves.*

A line-of-sight or direct-wave propagation, which is the first of the three ground-wave propagations, is just as the name implies — it is the reception of a signal at the receiver in a propagation path that is direct and where the receiver antenna "can be seen" from the transmitter antenna. In line-of-sight propagation, the transmitter antenna usually is located at the highest point in the radiating area in order to achieve the greatest propagation distance.

The second ground-wave propagation is called the surface wave because it "hugs" the surface of the Earth. This type of signal propagation is used by navigation systems (Loran) and by many military service communications. There is a third ground-wave propagation called a reflected-ground-wave that can be a "skip" type of propagation that depends on ground surface conditions and on frequency. Ground waves travel best over surfaces that conduct electricity, such as seawater, swamps, bogs, and the like.

The fourth propagation is called sky-wave propagation. It occurs when radio waves from the transmitter travel upward into the sky and are reflected back to Earth by the ionized layers that form in the ionosphere. As shown in *Figure 9-12,* the skip distance can be great, and depends on the angle of the radiated wave and the frequency. The angle must be less than or equal to the critical angle theta (θ), and the signals usually are between the frequencies of 500 kHz and 30 MHz, although some sky waves have been known to occur up to 100 MHz.

## Propagation Characteristics

### *Direct Waves*

From *Figure 9-12,* it is easy to see why an antenna well below the horizon does not receive direct waves. Line-of-sight is the path for communications of all signal frequencies above about 30 MHz and, for the most part — especially for commercial services —the range of reliable operation is determined by the height of the antenna above ground.

Occasionally other forms of propagation occur, as when a second TV channel seems to appear for a while, interferes with the one you are viewing, and then disappears. This is due to "atmospheric ducting" caused by an inversion layer in the upper atmosphere that acts as a duct and keeps the signal energy channeled or "ducting" near the Earth's surface. This can result in long-distance communications, but is not a reliable means of communication and in many cases is a nuisance.

### *Ground Surface Waves*

At low-frequencies below 500 kHz, a radio signal will appear to "bend" over the Earth's surface to offer a reliable means of communications. However, the amount of bandwidth is limited and large antennas and high transmitted power usually is required for this means to be successful.

### *Sky Waves*

At frequencies from around 500 kHz to about 30 MHz, long-range skip propagation can take place. As stated previously, the radiated energy is "reflected" (actually refracted) from ionized layers in the ionosphere. In addition, it can be reflected again from the Earth's surface so that the signal will "bounce" in several hops and provide communications on a world-wide basis. This is not the most reliable form of communications since the ionosphere is in constant change and varies with the time of day, day of year, and over the duration of the 11-year "solar flux" cycle. The stronger the solar flux the stronger the ionization and often the better the communications.

As pointed out previously, the sky-wave path is dependent on the critical angle of the radiated energy and its frequency. The success of a sky wave radiated at less than the critical angle is given by a measure called the Maximum Usable Frequency (MUF). MUF is used to characterize propagation conditions at any given time of day between any two given points on the surface of the Earth. If the radiated signal is at frequencies below the MUF, there is a reasonable chance for communications to take place. Signals with frequencies above the MUF pass through the ionosphere and are lost. At frequencies well below the MUF, (called the Lowest Usable Frequency, or LUF) the opposite effect takes place — the signals will be absorbed by the ionosphere and not reflected. Thus, at any given time, there will be a band of frequencies between the MUF and LUF over which sky-wave propagation will occur between any two widely-separated points on the Earth's surface. High sunspot numbers cause the MUF to go up and generally make it easier to communicate over long distances.

The MUF goes up in daylight and goes down in darkness, thus the required signal frequency is higher in the daytime and can be lower at night. These variations in the MUF are quite apparent at night for AM broadcast stations. Stations from as far away as several thousand miles can sometimes be heard due to the sky-wave propagation of their signals, which are reflected by the ionosphere during the nighttime hours. This is why some AM broadcast stations are not allowed on the air at night. Such stations broadcast on the same frequency as other stations that are allowed to broadcast 24 hours-a-day, which are called "clear-channel" stations.

Occasionally, there is some "skip" propagation in the VHF and UHF regions, but these paths are not a reliable means of communications.

## Satellite Transmission Links

Because of the variations in sky-wave communications due to atmospheric conditions, much of the long-haul communications has been taken over by a much more reliable wireless transmission link — one using satellites orbiting above the Earth. *Figure 9-13* illustrates how geo-stationary satellites are used by cable networks to transmit TV signals to large areas of the United States. The TV signals are transmitted on an "up-link" to the satellite, which is positioned in space so that it stays in the same parked location above the Earth's surface. The satellite then rebroadcasts the TV signal on a "down-link" frequency to the cable network's distribution substations. Down-link frequencies range from 3.7 to 4.2 GHz on the C-band, and from 10.7 to 12.2 GHz on the Ku-band. The local cable company then distributes the signals it receives over a coax-cable transmission line network to all of its home, school and business customers. A similar system is used in the newer, small-dish, direct-broadcast, DBS satellite TV systems, except the reception is at each location where there is a small-dish antenna.



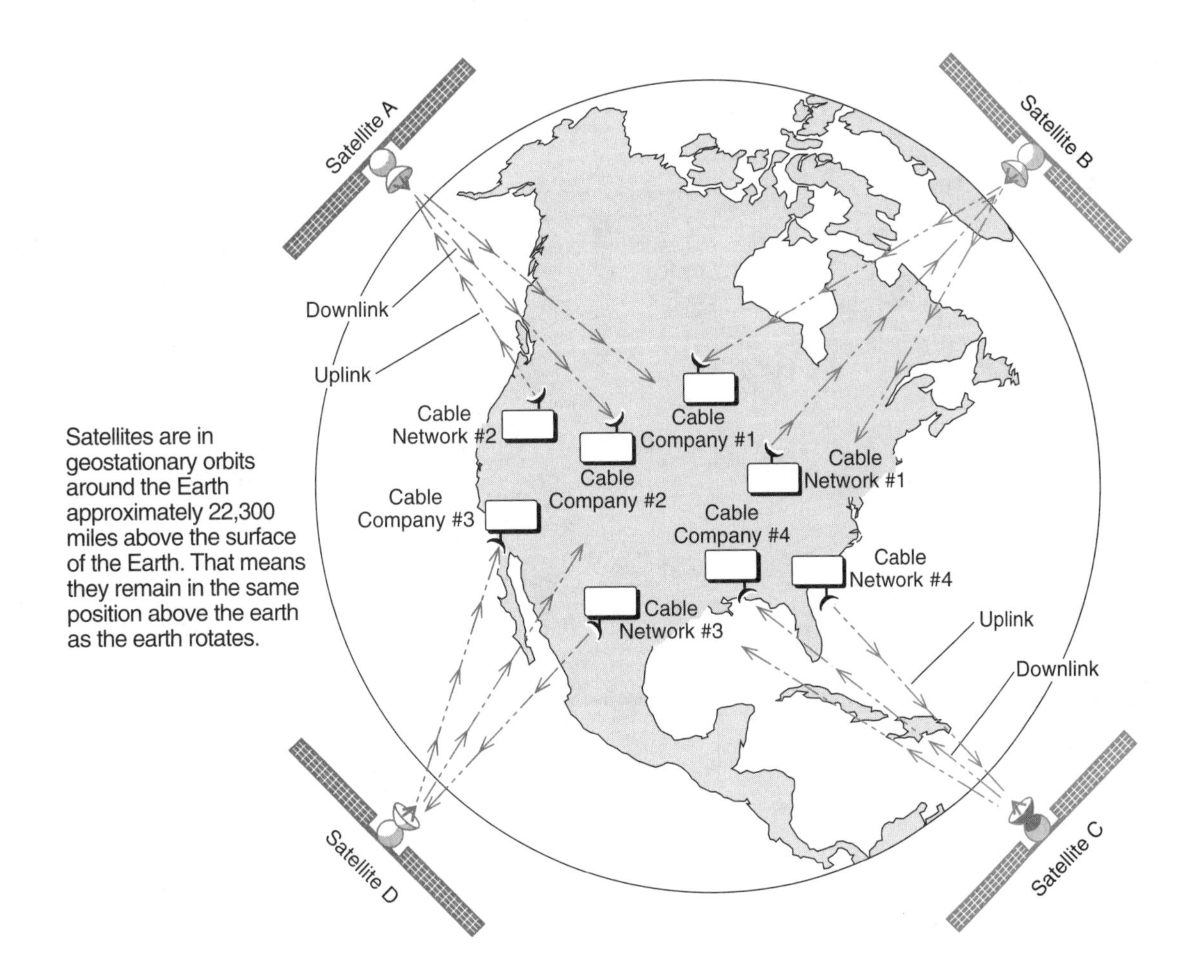


*Figure 9-13. Example of satellite transmission links used by cable TV networks.*
*(Source: Installing TV Video Systems, L.W. Pena, B.A. Pena, ©1996, Master Publishing, Inc., Lincolnwood, IL)*

## Transmission Lines

The wireless transmission links that we have discussed normally are used to transfer energy over long distances with very low efficiency. In our communication systems, there often is a need to transfer energy with high efficiency and/or at high power levels. We stated in Chapter 1 that this is done with transmission lines. Transmission lines may be individual or multiple wires, or specially-constructed cables. Or they may be special structures, such as waveguides for microwave signals, or fiber optic cables for light-wave transmissions.

In communication systems, there are eight principal types of transmission lines or special structures, as shown in *Figure 9-14:*

1. Single wire
2. Twisted pair
3. Two-wire shielded pair
4. TV twin-lead line
5. Open-wire ladder line
6. Coaxial cable
7. Waveguides
8. Fiber-optic cable

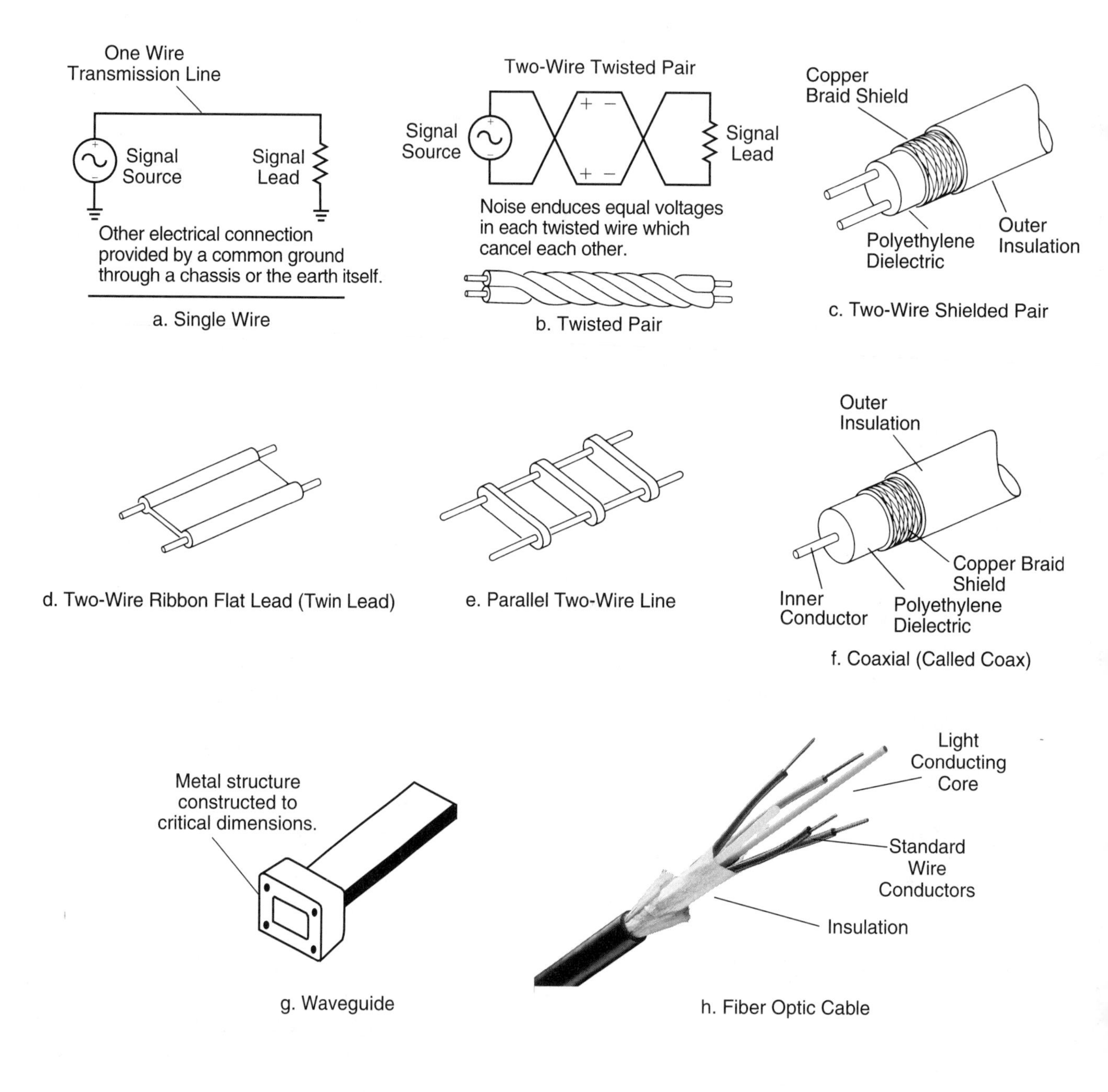


*Figure 9-14. Types of transmission lines. (Photo of fiber optic cable courtesy of Siecor.)*

## Single Wire Lines

Single wire transmission lines, as shown in *Figure 9-14a,* are really a two-wire system, but half of the connection is provided by a common ground line, a metal chassis containing a circuit, the frame of a piece of equipment, or the good Earth itself. Initial telephone systems used the Earth as a common connection for the return signal path. Single wire connections for many electronic and electrical circuits are not used as transmission line connections *per se*. They are used at such low frequencies, with such short lengths, and with such a variety of impedances, that they are treated as lumped circuit components that contain some resistance, inductance and/or capacitance, and that provide circuit paths for voltages and currents.

## Twisted Pair

As circuits were required to operate at higher frequencies and in noisier environments, twisted pair wiring (*Figure 9-14b*) was introduced. Here, the wire providing the return signal path was included and both wires twisted so that any noise that induces stray voltages would do so equally in both twisted wires and cancel out the noise.

## Two-Wire Shielded Cable

As circuits became more sensitive and outside noise became more difficult to handle, the two wires were shielded (*Figure 9-14c*) into a two-wire shielded cable. The shield is connected to ground and conducts any stray noise voltages to ground.

## TV Twin-lead Line

When TV became a popular household appliance in the 1950s, there was a need for a true transmission line with relatively low signal loss at a reasonable price. Thus, twin-lead TV transmission line (*Figure 9-14d*) was invented. It contains two parallel conductors separated and held in position with a molded plastic covering. It has a characteristic impedance of 300 ohms.

## Open-Wire Lines

As broadcast stations demanded more transmitter power output and lower losses when coupling the transmitter to their elaborate antenna arrays, parallel open-wire transmission line, sometimes called "ladder line" (*Figure 9-14e*), was used to provide the power-handling capability with low loss. Two parallel conductors of the appropriate size are spaced for the desired characteristic impedance by insulators forming the transmission line. Many lines have characteristic impedances of 300 to 600 ohms. These lines are subject to interaction with nearby objects and are dangerous to touch when transmission occurs, so they must be installed carefully in isolated locations.

## Coax Cable

Coaxial cable was invented to provide a transmission line that had relatively low loss per foot, excellent shielding from outside signals, durability for hard wear, and pliability for all varieties of applications at a wide range of frequencies. As shown in *Figure 9-14f*, it has a center conductor surrounded by a mesh copper or aluminum shielding, or both. The center conductor is separated from the mesh shielding by a low-loss insulator. The shielding has an outer jacket of heavy plastic to protect the cable from exterior damage.

Coaxial cables have higher losses than open wire lines but are not influenced by nearby objects and are easier to install. Commonly used RG-type coaxial cables include RG-58 and RG-8, and RG-59 and RG-6. The RG-58 and RG-8 lines have 50-ohm characteristic impedances, while RG-59 and RG-6 have a 75-ohm impedance. There are many types of coaxial cable, and *Figure 9-15* shows typical specifications for various types. Each differs in size and in their line losses. The losses increase with frequency. Normally, the larger the lines physically the lower the losses. The losses are given in dB per 100 feet and usually are specified as a function of frequency. At VHF and UHF frequencies, special coax transmission lines, called hard lines, are used. These coaxial cables have a large inner conductor and a hard outer conductor. Such lines are often used in commercial operations with large antenna towers where long coaxial runs are required. Hard lines have very-low loss per 100 feet compared to the RG type and can operate up to frequencies of 1500 MHz.

**Coax Cable Type, Size and Loss per 100 Feet**

| Coax Type | Size | Loss at HF 100 MHz | Loss at UHF 400 MHz |
|---|---|---|---|
| RG-6 | Large | 2.3 dB | 4.7 dB |
| RG-59 | Medium | 2.9 dB | 5.9 dB |
| RG-58U | Small | 4.3 dB | 9.4 dB |
| RG-8X | Medium | 3.7 dB | 8.0 dB |
| RG-8U | Large | 1.9 dB | 4.1 dB |
| RG-213 | Large | 1.9 dB | 4.5 dB |
| Hardline | Large, Rigid | 0.5 dB | 1.5 dB |

*Figure 9-15. Coax specifications.*

## Waveguides

The need for higher-frequency operation, driven by radar applications in particular, created the need for waveguide transmission lines (*Figure 9-14g*). Coax cables were too lossy, and signals at GHz frequencies behave differently. As frequency increases, RF energy no longer is conducted throughout a conductor — it confines itself to the skin surface of the conductor. Thus, a waveguide is a hollow metal pipe that directs microwave electromagnetic signals within its interior (only skin deep), and forms a tuned cavity for the energy as it propagates through this special type of transmission link. A waveguide has very-low loss and is very efficient in its transmission link duties. Its physical size is very critical to the successful operation of a system at a chosen frequency.

## Fiber Optics

Wire transmission lines are quickly being replaced by fiber optic cables. Fiber optic cables have a great advantage over copper cable because of their ability to handle larger bandwidth (much larger than coaxial cable), their freedom from interference, and their very compact size. Fiber optic cables required careful handling but provide very reliable wide-band communications.

A fiber-optic cable, as shown in *Figure 9-14h,* has a core of glass fibers — it is a light "pipe" down which light is transmitted. Most of the light is confined to the inside of the cable with a minimum of leakage. At the transmitter end, a light "modulator" is used to couple the information as light into the fiber. Some transmitters use special light emitting diodes (LEDs) for short distances but, for longer distances, transmitters use special solid-state lasers. At the receiving end, various light-sensing devices — for example, a phototransistor — are used to detect the transmitted signal.

Fiber optic cables are used extensively in the telephone industry because one cable has the capability to transfer thousands more channels when compared to older, copper transmission lines. Fiber optic cables are designed into systems where wide bandwidth and extremely high-frequency operation are keys to the system's success. And, in the future, the use of glass fiber will allow the delivery of wide bandwidth signals over cable networks.

## Transmission Line Basics

We have found in our discussions that transmission lines that transfer signal energy must do it with efficiency — meaning low losses — and must handle a variety of signal power levels. At the same time, wide ranges of frequency operation are required. Losses are a function of frequency and the higher the frequency the higher the losses.

To understand how to use transmission lines properly to accomplish these objectives, let's take a closer look at some transmission line basics. Transmission lines are not like a circuit with discrete component resistors, capacitors and inductors. Transmission lines have what are called "distributed parameters," which are stated as so much "resistance per foot," or so much "inductance per foot," or so much "capacitance per foot." We have spoken about the characteristic impedance of a transmission line. Each transmission line is designed to have a particular characteristic impedance, $Z_O$. Its value is based on the physical size of the conductors used, the separation distance of the conductors, and the dielectric of the insulation separating the conductors. It is very important to understand that the characteristic impedance is the same everywhere along the length of the transmission line.

For radio frequency transmission lines,

$$Z_0 = \sqrt{L/C}$$

Where $Z_0$ = characteristic impedance in **ohms**
$L$ = inductance per foot in **henries**
$C$ = capacitance per foot in **farads**

For two-wire lines,

$$L = 0.281 \log_{10} \frac{b}{a} \text{ microhenries per foot}$$

$$C = \frac{3.677}{\log^{10} \frac{b}{a}} \text{ micromicrofarads per foot}$$

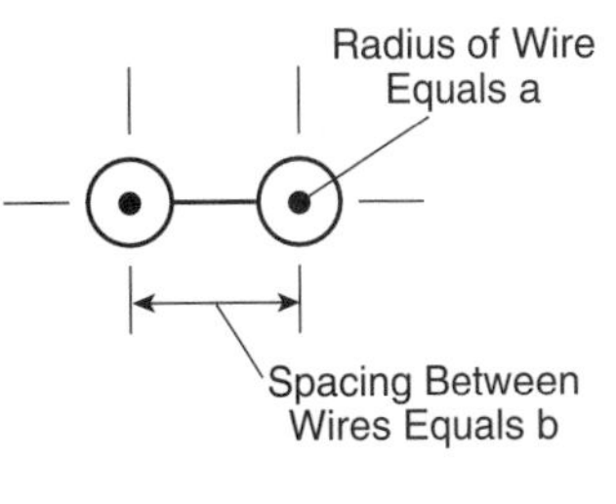


For coaxial lines,

$$L = 0.140 \log_{10} \frac{b}{a} \text{ microhenries per foot}$$

$$C = \frac{7.354K}{\log_{10} \frac{b}{a}} \text{ micromicrofarads per foot}$$

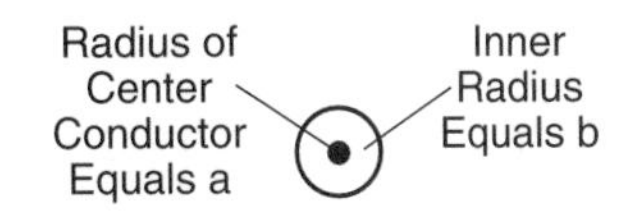


In the above L and C equations, for the two-wire lines, b is the spacing between the wire centers and a is the radius of the wire; and for the coax line, b is the inner radius of the outer conductor and a is the radius of the center conductor. K is the dielectric constant of the insulator separating the conductors. K = 1 for air. a and b are in centimeters.

As a result, for the two-wire lines,

$$Z_0 = 276 \log_{10} \frac{b}{a} \text{ ohms}$$

and for coaxial lines,

$$Z_0 = \frac{138}{\sqrt{K}} \log_{10} \frac{b}{a} \text{ ohms}$$

The characteristic impedance appears as a resistance for the lossless line.

## Why is Characteristic Impedance So Important?

Let's look at *Figure 9-16* to help us understand why the characteristic impedance, $Z_0$, of a transmission line is so important for the efficient transfer of energy. *Figure 9-16a* shows a transmitter coupled to an antenna with a transmission line that has a characteristic impedance of $Z_0$. The transmitter's output impedance is $Z_T$ and the antenna input impedance is $Z_A$, the load impedance that the antenna presents to the transmission line.

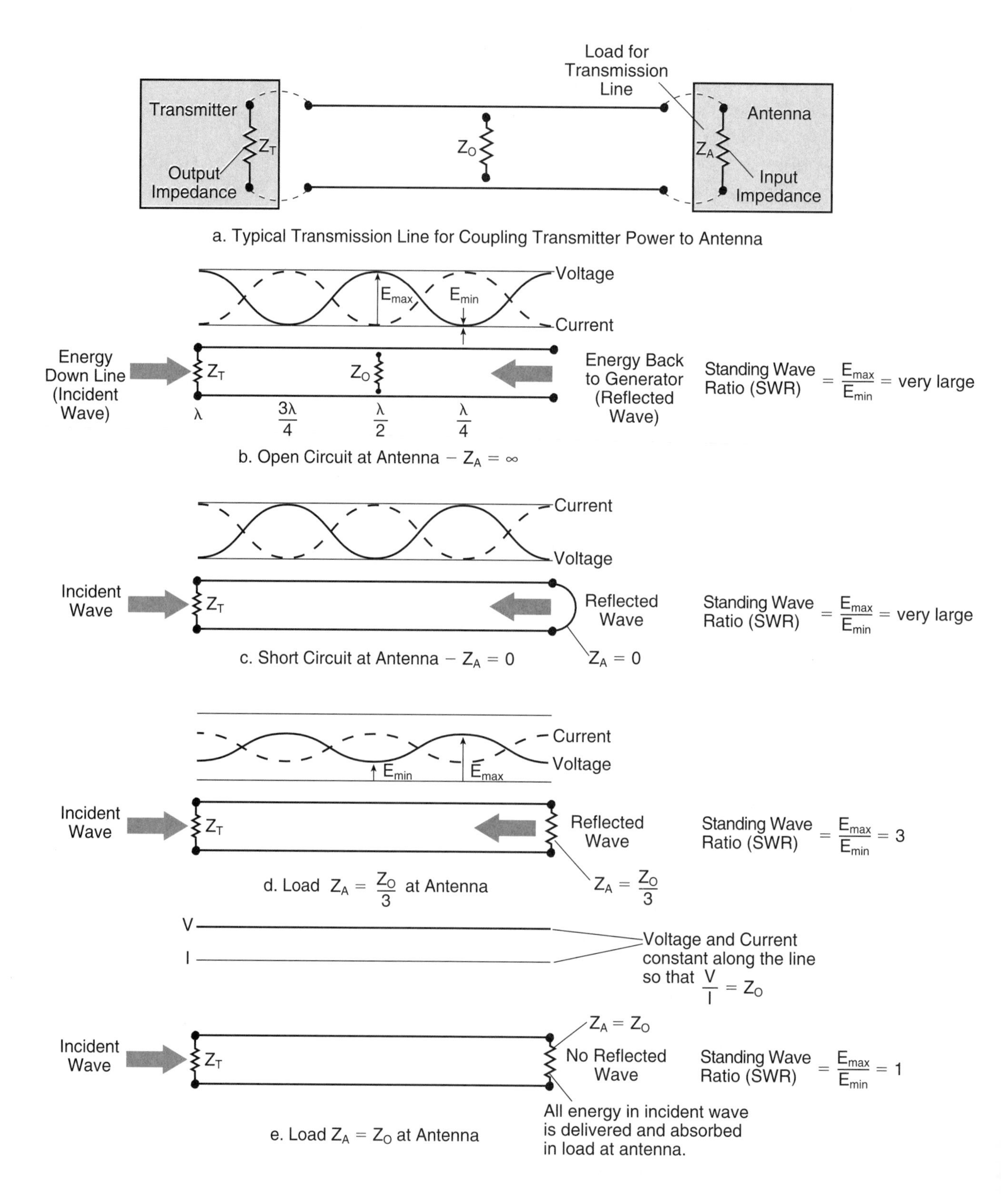


*Figure 9-16. Coupling transmitter and antenna with transmission line.*

Let's assume that $Z_A$ is infinity — in other words, the load on the transmission line at the antenna is an open circuit. *Figure 9-16b* is a diagram of the interconnection and shows the voltages and currents that exist on the transmission line. Since $Z_A$ is an open circuit, the current at the end of the line is zero and the voltage is maximum. Since the wave of energy sent from the transmitter has no load to absorb the energy, it is reflected back to the transmitter. The voltage and current waveforms shown are the combined resultant of the wave from the transmitter (called the incident wave) and the reflected wave from the load end. In this case, the total wave is reflected back and no signal is transmitted.

## Standing-Wave Ratio

The ratio of the maximum amplitude, $E_{max}$, of the resultant voltage waveform on the line to the minimum amplitude, $E_{min}$, is called the *Standing-Wave Ratio*, SWR, or more correctly the *Voltage Standing-Wave Ratio*, VSWR. In the case of an open load, since there is no energy absorbed at the load and all energy delivered is absorbed as losses in the line or absorbed by the transmitter's output resistance, the VSWR is very high (in the order of 20 or greater). VSWR is a measure of how well energy is being transferred over a transmission line to the load. VSWRs below 1.5 indicate quite efficient transfer of energy (VSWR = 1 is perfect), and VSWRs above 2 begin to indicate poor matching.

Now let's look at *Figure 9-16c.* Here the $Z_A$ is a short circuit. That means the voltage is zero at the end of the line and the current is maximum. Again, no energy is absorbed by the load. It is all reflected back down the line. The voltage and current waveforms vary in minimums and maximums every quarter wavelength, as in the open-circuit case, but are 180° out of phase from the open-circuit case. The standing-wave ratio again is very large and no energy is transferred.

*Example 2* illustrates how to calculate the VSWR on a transmission line if the load impedance (all resistive) and the transmission line $Z_0$ are known. $\rho$ (the Greek letter rho) which is the reflection coefficient (the ratio of reflected to incident voltage), can be calculated from $Z_l$ and $Z_0$. From $\rho$ the VSWR may then be calculated. If the $Z_l$ is not resistive then the magnitude of $\rho$ must be used to calculate the VSWR.

## A Matched Case — $Z_A = Z_0$

*Figure 9-16e* is the case where all the energy in the incident wave is absorbed by the load. There is no reflected wave because $Z_A = Z_0$. The antenna load is matched to the characteristic impedance of the transmission line and maximum transfer of power from the transmitter to the antenna occurs. The standing-wave ratio is 1 because the voltage and current in the line are constant and are such that they equal $Z_0$ everywhere along the line. Thus, we see how important it is to match the load impedance to the transmission line $Z_0$. With $Z_A = Z_0$, there will be some small losses in the transmission line, but if there is a mismatch, energy will be reflected from the load back to the transmitter and large losses on the line will be dissipated as heat.

## Some Other Observations

Note that on a transmission line, the characteristics of voltage and current repeat themselves every half-wavelength. Every quarter-wavelength, the voltage and current change phase. If the voltage is large and the current is small, one-quarter wavelength away they are reversed. Every half-wavelength, the impedance is repeated, and the load appears to be the same. Such properties of transmission lines can be used to reflect particular impedances and match load impedances to source impedances, or to use transmission lines for tuning. As an example, a quarter-wave stub of a waveguide will reflect an infinite impedance (open circuit) to the source when it has a short circuit as a load or, if it has an open circuit as a load, it will reflect a short circuit at the source. In addition, inductive and capacitive loads can be reflected on transmission lines if the load impedance has reactive components.

### *Example 2: Calculating Reflection Factor and VSWR*

Reflection factor $\rho = \dfrac{\dfrac{Z_L}{Z_0} - 1}{\dfrac{Z_L}{Z_0} + 1}$ where: $Z_L$ = load impedance in **ohms**
where: $Z_0$ = characteristic impedance in **ohms**

and $\rho = \dfrac{\text{sigma} - 1}{\text{sigma} + 1}$ and $\text{sigma} = \dfrac{E_{max}}{E_{min}}$ = VSWR therefore, $\rho = \dfrac{\text{VSWR} - 1}{\text{VSWR} + 1}$

Calculate VSWR when $Z_A = \dfrac{Z_0}{3}$ as in *Figure 9-16d.* $Z_A$ is the load impedance of the antenna.

***Solution:***

Since $Z_A = Z_L = \dfrac{Z_0}{3}$,

then $\rho = \dfrac{\dfrac{1}{3} - 1}{\dfrac{1}{3} + 1} = \dfrac{\dfrac{2}{3}}{\dfrac{4}{3}} = \dfrac{2}{4} = 0.5$

with $\rho = 0.5$

$0.5 = \dfrac{\text{VSWR} - 1}{\text{VSWR} + 1}$

0.5 VSWR + 0.5 = VSWR - 1

Transposing,

1 + 0.5 = VSWR - 0.5 VSWR

1.5 = 0.5 VSWR

3 = VSWR

$\therefore$ VSWR = 3

## Impedance Matching

A word about coupling the transmitter and the antenna to the transmission line for efficient power transfer. Two examples are shown in *Figure 9-17.* The first (*Figure 9-17a*) is the use of an antenna tuner to match the impedance of the transmitting antenna to the transmission line feeding it. The tuner is placed right at the antenna for best results. Another impedance match may be necessary at the transmitter to match the transmitter to the transmission line and adjust the system so that there is maximum available power transfer from the transmitter to the antenna. Difficulties arise when the impedances are not truly resistive and high SWRs will result with reactive (inductance or capacitance) loads. An SWR bridge is shown in *Figure 9-17a*. It is used to measure SWR so the impedances can be adjusted.

The second (*Figure 9-17b*) technique is to use transformers to match impedance levels. A transformer is used at each end of a TV coax 50-ohm transmission line to match the 300-ohm antenna to the 50-ohm transmission line and the 50-ohm transmission line to the 300-ohm input to the TV. The same principle of maximum available power transfer is used, but the emphasis here is not power transfer, but maximum signal strength because of the low-level signals.

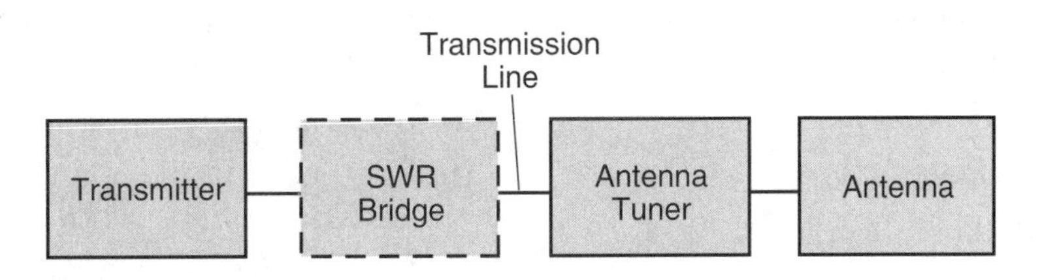


a. Antenna Tuner Matches Antenna to Transmitter Transmission Line.

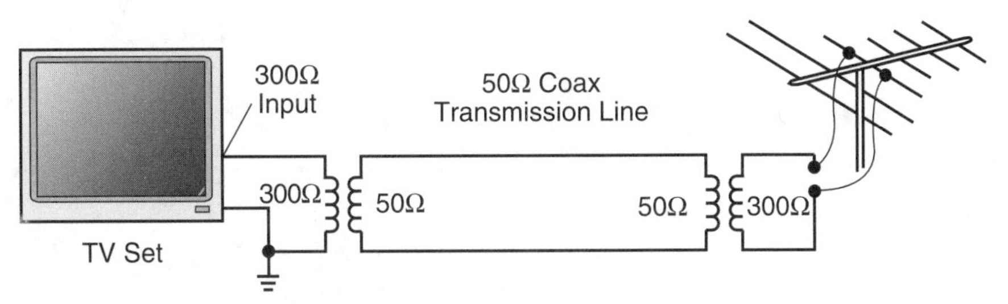


b. Transformers Used to Match Impedances of TV Antenna Input

*Figure 9-17. Impedance matching.*

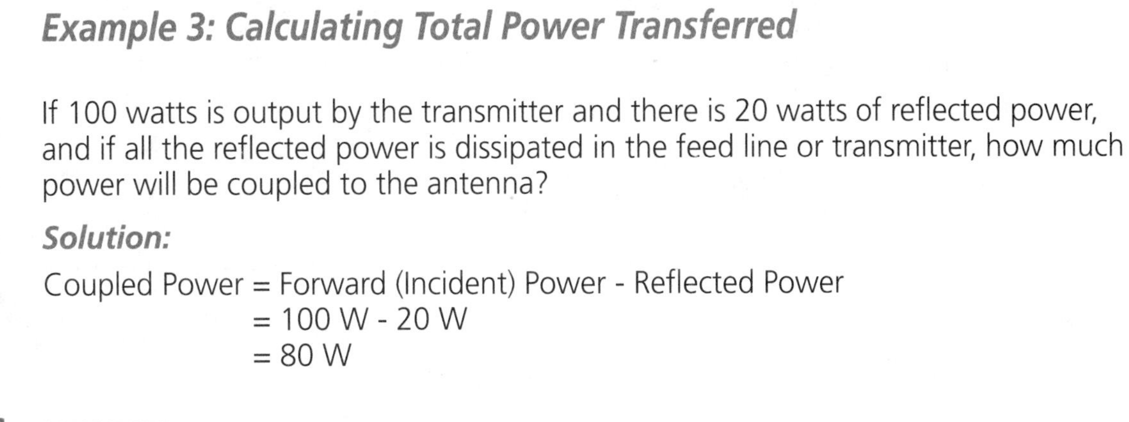

### *Example 3: Calculating Total Power Transferred*

If 100 watts is output by the transmitter and there is 20 watts of reflected power, and if all the reflected power is dissipated in the feed line or transmitter, how much power will be coupled to the antenna?

***Solution:***

Coupled Power = Forward (Incident) Power - Reflected Power
= 100 W - 20 W
= 80 W

## Summary

In this Chapter, we have discussed antennas and transmission lines to complete the transmission link. We discussed the characteristics of antennas and their operation and their importance to the radio communications systems. We discussed the types of transmission lines and some transmission line basics to show that, for maximum power transfer, the transmitter's output impedance and the load must be matched to the transmission line characteristic impedance.

In the next Chapter, we will discuss how the communication functions are implemented in IC form.

# Quiz for Chapter 9

1. What two fields are radiated from a transmitting antenna?
   a) Positive and negative.
   b) Horizontal and vertical.
   c) Magnetic and Electric.
   d) Circular and linear.
2. What is the polarization of a vertical antenna?
   a) Vertical.
   b) Horizontal.
   c) Circular.
   d) Negative.
3. What is an isotropic radiator?
   a) A heating device to keep ice off of antennas.
   b) A theoretical antenna that works equally well in all directions.
   c) An insulated antenna for SSB.
   d) None of the above.
4. What is meant by antenna gain?
   a) The ability to increase the amplitude of the transmitted signal.
   b) The ability to increase the voltage of a received signal.
   c) The ability to send or receive a signal in a preferred direction.
   d) The ability to switch between horizontal and vertical polarization.
5. What are the propagation paths for transmitting antennas?
   a) Ground wave and line-of-sight.
   b) Sky wave.
   c) Reflected ground wave.
   d) All of the above.
6. What is the purpose of an antenna tuner?
   a) To match the impedance of the antenna to the transmission line that feeds it.
   b) To pre-tune signals before they are sent to the receiver.
   c) To match the frequency of the transmitter to the frequency of the antenna.
   d) None of the above.
7. Why do some antennas, such as a Yagi, have gain?
   a) They have power amplifiers.
   b) They use parasitic elements to increase signal strength.
   c) They are mounted high off the ground to avoid interference.
   d) The signals are fed with an impedance matching device.
8. What factors influence the radiation pattern of an antenna?
   a) The coupling of the transmission line to the antenna.
   b) Reflection of the signal off of the ionosphere.
   c) Antenna length and height above the ground plane.
   d) None of the above.
9. Why should transmission lines be kept as short as possible?
   a) To maintain signal quality and strength.
   b) To keep costs of construction low.
   c) To prevent signal losses.
   d) a and c.
10. What is the advantage of fiber optic cables?
    a) They handle a large bandwidth of signals.
    b) Their freedom from interference.
    c) Their compact size.
    d) All of the above.

**Answers:**
1 c, 2 a, 3 b, 4 c, 5 d, 6 a, 7 b, 8 c, 9 d, 10 d

## Questions & Problems for Chapter 9



1. How is electromagnetic energy sent from a transmitting antenna transferred to a receiving antenna?
2. The direction of the _________ field determines the polarization of an antenna.
3. What is another name for a vertical antenna, and a dipole antenna.
4. The length of an antenna usually is a submultiple of the wavelength to be sent or received. What are two common antenna wavelengths?
5. What causes sky-wave propagation to vary from year to year?
6. What is MUF?
7. If a satellite is in a geostationary orbit, how far above the Earth's surface is it's orbit?
8. Why is the characteristic impedance of a transmission line so important?
9. What does a VSWR of 1.5 or less indicate?
10. If the antenna impedance is 0.5 $Z_0$, what is the VSWR? Check your answer by using $\text{sigma} = \frac{1+\rho}{1-\rho}$.
11. If the incident power from the transmitter is 1000 watts and the coupled power to the antenna is 850 watts, what is the reflected power?
12. From the table of Coax Specifications in *Figure 9-15,* what is the difference in losses of 1000 ft. of RG-8X and RG-8U coaxial cable at 400 MHz?

*(Answers on page 212.)*

# CHAPTER 10
# Analog Integrated Circuits

In this chapter, we will detail what the term "integrated circuit" means, explain how ICs are made, and discuss the factors that make analog integrated circuits different from digital integrated circuits.

In earlier chapters, the following circuits were identified as integrated circuits (ICs):

- Op-Amp: a universal amplifier circuit used for lower-frequency applications whose gain can be controlled by the ratio of an external feedback resistor and an input resistor.
- MC1496: a balanced modulator or mixer.
- CA3080: a modulator made from an amplifier whose gain can be controlled by a control voltage.
- PLL: a phase-locked loop can be used as a frequency synthesizer and in frequency selection applications.
- IF, Detector and Audio: the ICs necessary to provide IF amplifiers, the detector, and the audio amplifiers for a radio receiver.
- In Chapter 11, we will discuss these IC's: A/D and D/A — ICs that convert analog signals to digital (A/D) and digital signals to analog (D/A).

Other very-common analog ICs are:

- Voltage regulators.
- Comparators.
- Timers.
- All processing circuits for TV.
- Line drivers and receivers.
- Input buffers.
- Current sources.
- Specialized circuits for telecommunications.

## What is an IC?

An integrated circuit is a miniature electronic circuit produced on and within a single piece of semiconductor material, usually silicon. All of the circuit components — the active and passive devices — are manufactured at the same time both on and within a small chip of silicon material. Since all of the components — diodes, transistors, resistors, capacitors — are within a solid piece of silicon, they are called *monolithic ICs*. Early in their development era, ICs were called "solid-state" circuits.

## About ICs

Integrated circuits have become the major building blocks of electronic systems because they provide the following benefits when compared to circuits built using discrete components:

1. They are a low-cost means of building complete circuits in a single manufacturing process, since all of the devices are formed in steps at the same time on or within a single piece of semiconductor material, almost always silicon. However, for very-high-frequency applications, gallium arsenide ICs are used.
2. They have uniform circuit parameters across a wafer and, in many cases, important matched-device characteristics.
3. They provide small active devices for low power and high-frequency operation.

Integrated circuits evolved from printed circuit and transistor technology. The same manufacturing facilities used to fabricate transistors are used to produce integrated circuits. ICs are increasingly more innovative, more powerful, smaller in size, and less expensive to make. Both bipolar and MOS transistors are at the center of integrated circuit designs, and there is an emphasis on packing more and more components into smaller-sized chips. For example, when ICs were first developed in the 1950s, a chip might hold 10 transistors. Today, an ultra-large IC (which might only measure 8 mm square) can have as many as 1 million transistors!

### Digital ICs Started It

Digital circuits using bipolar transistors as their active devices were the application that propelled ICs into high-volume production, providing a wide variety of ICs at lower and lower cost. These ICs included only resistors, diodes and transistors, and had no need for capacitors, until memory ICs became popular. Increased circuit density and easier manufacturing soon caused manufacturers to substituted MOS transistors for bipolar transistors and their higher power requirements. Today, CMOS (Complementary-Metal-Oxide-Semiconductor) transistor circuitry is the widely-accepted standard for digital IC logic circuits. There are many early, bipolar IC designs that are still in use, and these will not be changed. But most new designs incorporate CMOS before bipolar; however, bipolar ICs are still used based on specific application needs and the special parameters offered by bipolar active devices.

### Analog ICs Used Are Based on Application

The use of particular active devices has not become as segregated in analog ICs as it has in digital ICs. The selection of the active devices used in an analog IC is more dependent on the application. For example, under present technology, an analog circuit application may:

A. Allow the IC to be designed using only CMOS devices for lower power consumption, highest density, and highest frequency.

B. Mix both bipolar and MOS devices on the same IC, either based on a MOS manufacturing process or on a special process that is tailored to making both bipolar and MOS devices.

C. Mix both digital and analog devices on the same IC.

D. Mix small-signal and power devices on the same IC.

E. Have the IC process dictated by the operating voltage of the circuit.

# How Does an IC Work?

## A Digital CMOS Circuit

Since CMOS technology is used most in IC design, the inverter circuit shown in *Figure 10-1a* will be used to answer the question, "How does an IC work?" Here is how the operation of this digital IC circuit is described in *Basic Digital Electronics*[1]:

"The P-channel MOSFET Q1 acts as a drain resistor for the N-channel MOSFET Q2, and the N-channel MOSFET Q2 acts as a drain resistor for the P-channel MOSFET Q1. A positive $V_{IN}$ voltage near $V_{DD}$ on the gates of Q1 and Q2 turns ON the N-channel transistor Q2, pulling $V_{OUT}$ to ground ($V_{SS}$). Q2 sinks current from the output to ground. The positive $V_{IN}$ voltage turns OFF the P-channel transistor Q1 so there is no current through it. When

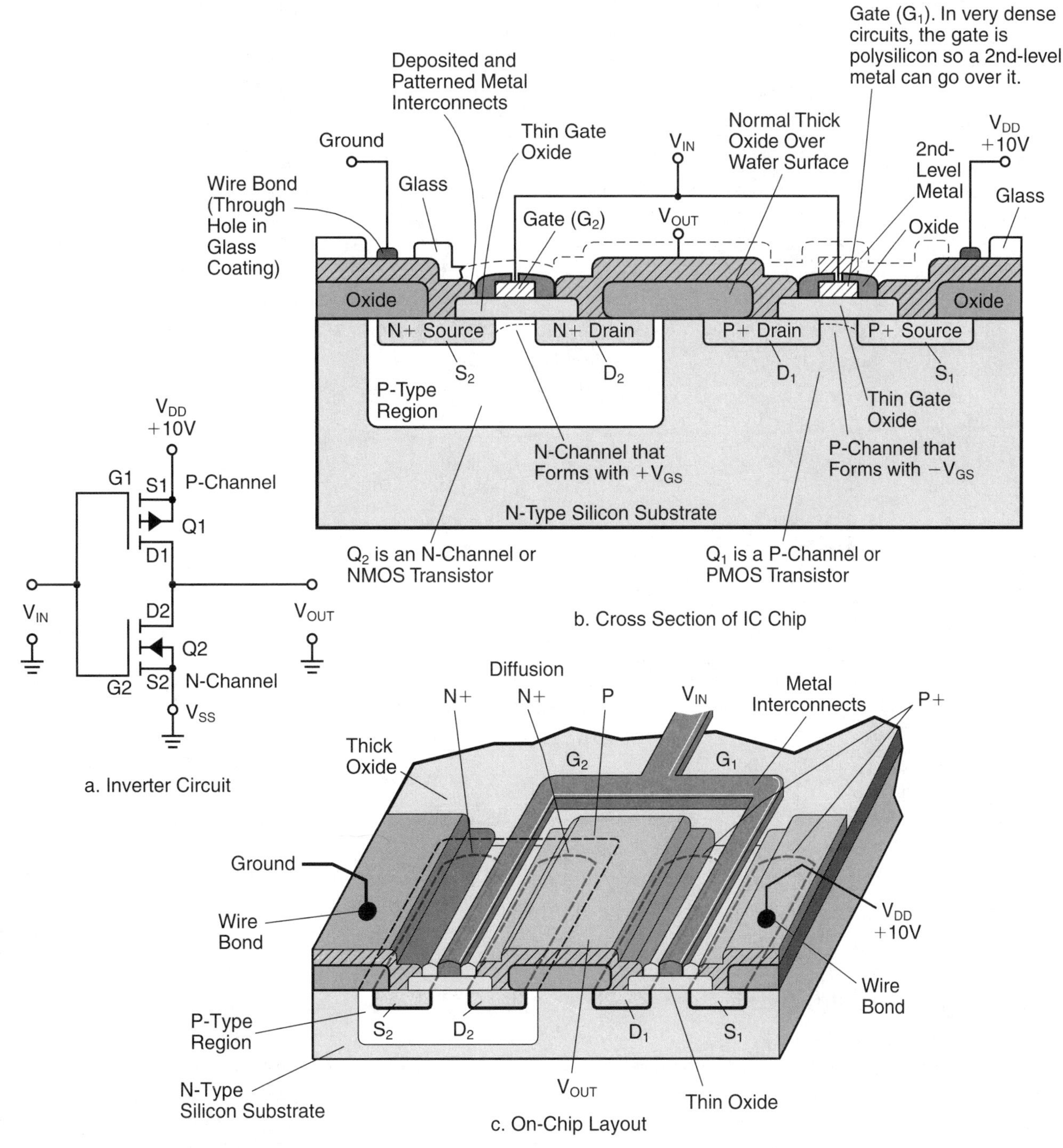


*Figure 10-1. A CMOS circuit.*
*Source: Basic Digital Electronics, A. Evans, ©1996, Master Publishing, Inc., Lincolnwood, IL.*

1 Basic Digital Electronics, A. E. Evans, ©1996 Master Publishing, Inc.

$V_{IN}$ is zero volts (near $V_{SS}$), Q1 is ON and the N-channel transistor Q2 is OFF. Now the current to the output is supplied through Q1 from $V_{DD}$, and $V_{OUT}$ is close to $V_{DD}$. A digital 1 is represented by a high-level $V_{OUT}$ close to $V_{DD}$, and a digital 0 is represented by a low-level $V_{OUT}$ close to ground ($V_{SS}$). Since Q2 is OFF when Q1 is ON, and Q1 is OFF when Q2 is ON, power is only used through the two transistors during the switching times. Power increases as switching speeds increase, but standby power is very low."

For analog IC circuits, the voltages on $V_{IN}$ vary continuously rather than being just two digital levels, and voltages in between $V_{DD}$ and $V_{SS}$ appear at $V_{OUT}$.

*Figure 10-1c* shows how the circuit is laid out as an integrated circuit on a silicon substrate. *Figure 10-1b* is a cross section showing how the various regions of the silicon chip are modified to form the drain and source of the N-channel and P-channel MOS devices, and shows how the device areas are connected together with a layer of aluminum deposited on the surface and then etched to make the interconnection pattern. The silicon chip is a substrate of N-type silicon. The P-channel MOS is formed from a P+ source and P+ drain diffused into selected areas of the N-type substrate. A gate electrode for G1 between the source and drain is electrically isolated from the N-type region below it with a layer of thin oxide. The P-channel is formed in the N-type region under the gate when a negative voltage is applied between gate and source.

In order to form the N-channel MOS, first a region of the N-type substrate is modified by diffusion to form a well of P-type material. Then an N+ diffusion is made into the well at selectively defined areas to make the N+ drain and N+ source. A gate electrode for G2 between the source and drain is isolated from the P-type region below it with thin oxide. A positive gate-to-source voltage on G2 forms an N-channel in the P-type material. The exploded view of *Figure 10-1c* shows in greater detail how the components are laid out on the chip, and how the aluminum is selectively removed to make connection between the various regions of the chip to form the circuit. Wires are bonded to the aluminum to provide the ground and $V_{DD}$ external connections for power to the circuit. At some point on the total circuit, $V_{OUT}$ from the end transistor of a circuit also will be connected to an external pin with a bonding wire.

## Bipolar Circuit

The simple circuit of *Figure 10-2a* will be used to describe how bipolar ICs are made. It consists of an NPN transistor with a resistor in the signal input to the base and a resistor from emitter to ground. The collector is tied to +5 V. The layout of the circuit components on the silicon chip is shown in *Figure 10-2c*, and a cross section of the chip, *Figure 10-2b*, shows how the regions of the original silicon substrate are modified to form the components.

The fabrication process begins using a P-type silicon substrate. As shown in *Figures 10-2b* and *c*, areas are selectively defined and N-type diffusions are made into the P-type substrate to form three N-type wells. In the N-type well area, P-type diffusions are made in selected areas to form the resistors $R_B$ and $R_E$ and the P-type base of the NPN transistor. A selected area of the P-type base of the transistor is diffused with a N+ diffusion to form the N-type emitter of the NPN transistor. At the same time, in a selected area of the N-type well forming the collector of the transistor, the N+ diffusion is made to provide a low-resistance connection to the collector. The resistor value formed by the P-type diffusion used for the NPN base is set by the length and width of the areas defined for the resistors.

After all diffusions are made, the chip is coated with silicon dioxide. Then, using a mask to define where the holes are positioned, holes are cut in the silicon dioxide. Next, aluminum is deposited on the surface and makes contact with the respective areas of the diffused components through the holes in the silicon dioxide. Another mask is placed over the aluminum and the aluminum is etched to define the interconnection pattern of the circuit, as shown in *Figure 10-2c*.

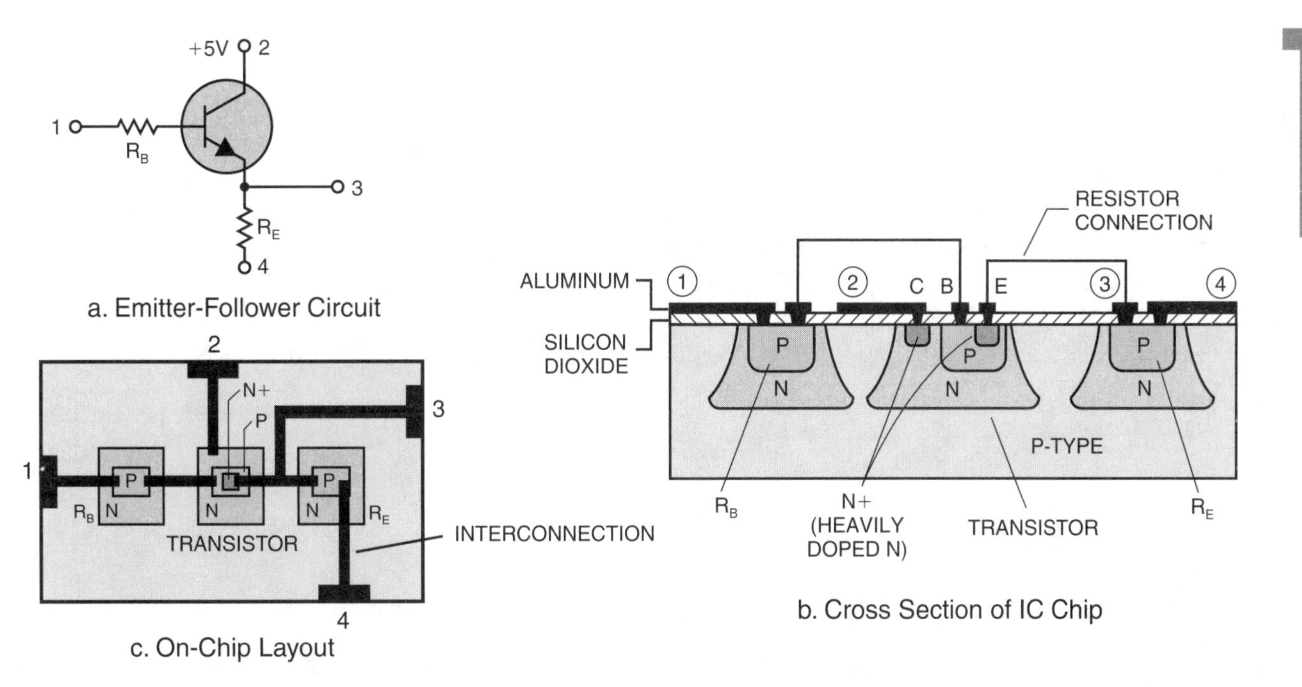


*Figure 10-2. A bipolar circuit.*

# How ICs are Made

## The Silicon Substrate (Wafers)

The fabrication process for making a silicon IC is shown in *Figure 10-5* on pages 188 and 189. The process begins with a slice of silicon that is sawed from a silicon ingot, as shown in *Figure 10-3*. This slice is about 8-10 mils thick and called a wafer. The wafer is ground and the top surface is polished to a mirror-like finish so that it is smooth and all imperfections are removed. Typically, the wafer is anywhere from 4″ to 12″ in diameter and about 6 mils (0.006″) thick. ( A mil is 0.001 of an inch.) It is then patterned with a matrix of IC layouts, and many hundreds of chips can be made at one time. At the end of the manufacturing process, the individual IC chips are cut from the wafer. All of the wafer processing steps occur inside ultra, ultra clean rooms. Any contamination — even just a few microscopic particles on the wafer surface or on the masks that are used to form the component patterns — can cause imperfections in the circuit layout and ruin the circuit.

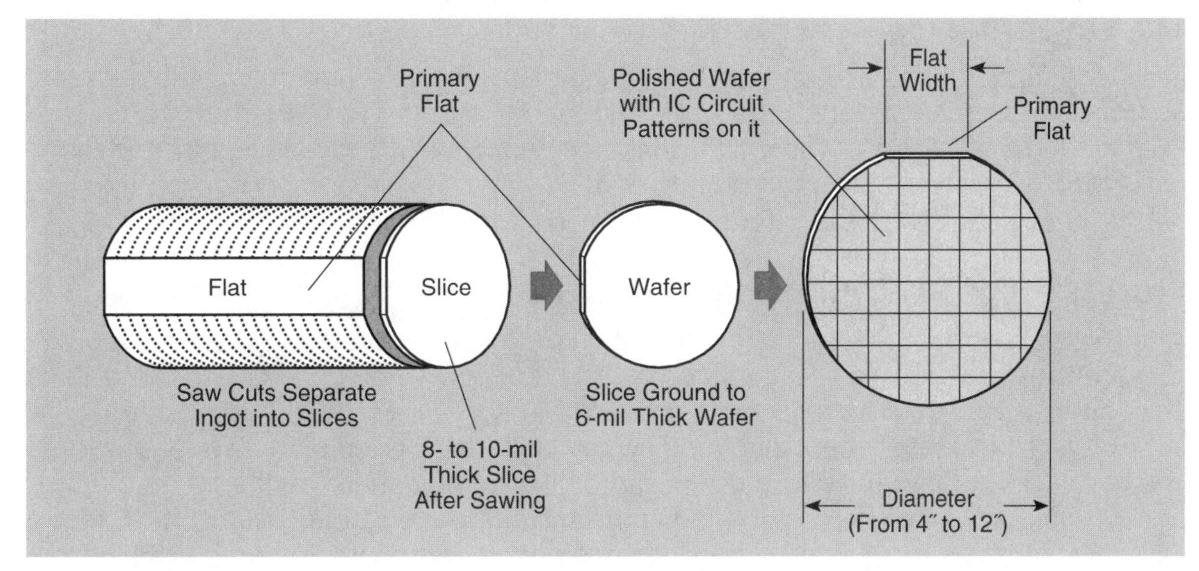


*Figure 10-3. Silicon slices.*

## The Circuit Design

The circuit layouts, such as those described in *Figures 10-1* and *10-2*, are designed using computer-aided systems that place the circuit components at their required size and position according to rules established for the manufacturing process. The computer design is produced at 400 to 500 times larger than the actual size of the circuit on the wafer so the circuit layout can be carefully examined and checked for any errors. The checking and comparison to the design layout and manufacturing rules also are done by computer to make the task efficient and reliable. Once the design layout is correct, the computer-aided design (CAD) system separates out the various masks that define a particular pattern for each layer, as shown in *Figure 10-4*.

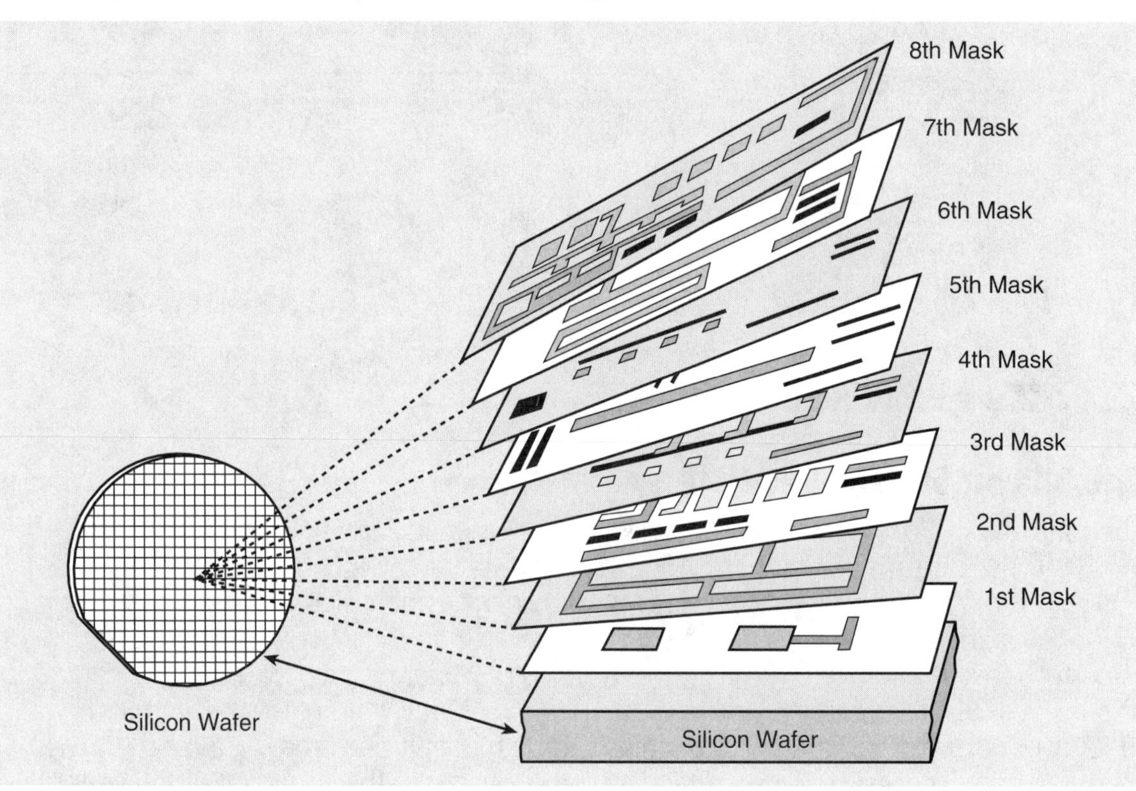


*Figure 10-4. Photomask set for IC.*
*(Source: Basic Digital Electronics, A. Evans, ©1996, Master Publishing, Inc., Lincolnwood, IL.)*

## The Layer Masks

Each mask pattern is etched into chrome-plated glass by an electron beam. The glass is accurately positioned on top of the wafer and used to transfer the pattern to the wafer. Each mask pattern defines where diffusions will modify the silicon wafer surface to form areas for diodes, transistors and resistors, or define areas for MOS gates or capacitors, or define metal patterns for capacitor plates or circuit interconnections.

## Process Steps to Modify Silicon Areas

After the polishing step in the wafer fabrication process of *Figure 10-5*, the next step is to place the wafer in a high-temperature furnace that has an oxygen-rich atmosphere. This causes a layer of silicon dioxide to grow on the wafer surface. This silicon dioxide layer protects the wafer from being modified by impurities placed on the wafer surface. Then holes are cut into the silicon dioxide according to the pattern on the mask so that dopants (impurities) will be able to modify the areas where the holes are cut. Next, the wafer, with the dopants either on its surface or in the atmosphere, is placed in a high-temperature furnace. The high temperature causes the impurities to diffuse into the wafer surface and forms the active and passive devices — bipolar transistor bases or emitters, MOS sources and drains, resistors, back plates of capacitors, etc. The type of device created is determined by the type of impurity and length of time of diffusion. At each step indicated in *Figure 10-5*, for each mask indicated in *Figure 10-4*, there is a subset of steps in the process as follows:

a) Grow silicon dioxide on wafer surface.

b) Cut holes in the dioxide according to a mask pattern to select wafer areas to be modified.

c) Place wafer in atmosphere that contains the selected dopant so it will be deposited, or deposit it directly, on the wafer.

d) Place wafer in furnace to diffuse the dopant into the wafer and modify selected areas.

e) Clean the wafer with very detailed cleaning techniques to prevent the diffusion dopant from being diffused at the next step.

f) Grow additional oxide to re-protect hole areas so they will not be affected by the next process step.

At each mask step, before the mask is placed on the silicon-dioxide covering the wafer surface, a material (a photoresist) that is sensitive to ultraviolet light is spread evenly over the wafer surface and baked at low temperature. The mask is then positioned very accurately on the surface and ultraviolet light exposes the surface of the wafer. A solvent dissolves the photoresist where it is not protected by the mask. The holes in the photoresist are the areas where holes in the oxide are cut with hot acid solutions. The same type of acid solutions are used at other steps in the process to clean the wafers. Any photoresist that is on the surface of the wafer protects the silicon dioxide from being etched by the acid. After the photoresist is removed, the silicon dioxide protects the surface of the wafer from any dopant that is being diffused, except where the holes have been etched. The diffusions can result in collectors, bases, or emitters for bipolar devices; or in drains or sources for MOS devices. Special processing produces a very thin oxide over the regions used for gates in MOS devices.

## Interconnecting the Circuit

When it is time to interconnect the circuit, a mask is used to define where holes will be cut in the silicon dioxide to allow aluminum deposited on the wafer surface to contact the selected silicon areas. The aluminum is insulated from the wafer surface by the silicon dioxide except where the holes have been cut. Finally, the surface aluminum is etched using another mask that defines the interconnection pattern that connects the components together as a circuit.

## Finding the Good ICs

Look again at *Figure 10-5*. After the interconnection of the circuits, the next step is the wafer probe. At wafer probe, a card containing thin wire probes is lowered into position on the wafer. The wire probes contact aluminum pads on the wafer. Electrical test voltages and currents are applied through the probe wires by computer-controlled equipment to check the IC against final test specifications. Any circuits that fail are marked with an ink dot, as shown in *Figure 10-5*, or produce a data base wafer map, which is stored in a computer and follows the wafer lot as it moves through the process to the assembly site. The wafer is then scribed with a diamond scribe and the individual chips broken apart. Any circuit without an ink dot or indicated as good in the data base is saved and passed on for assembly. The failed ICs are discarded.

a. Overview of Clean Room where ICs are Fabricated

b. ICs on a Wafer

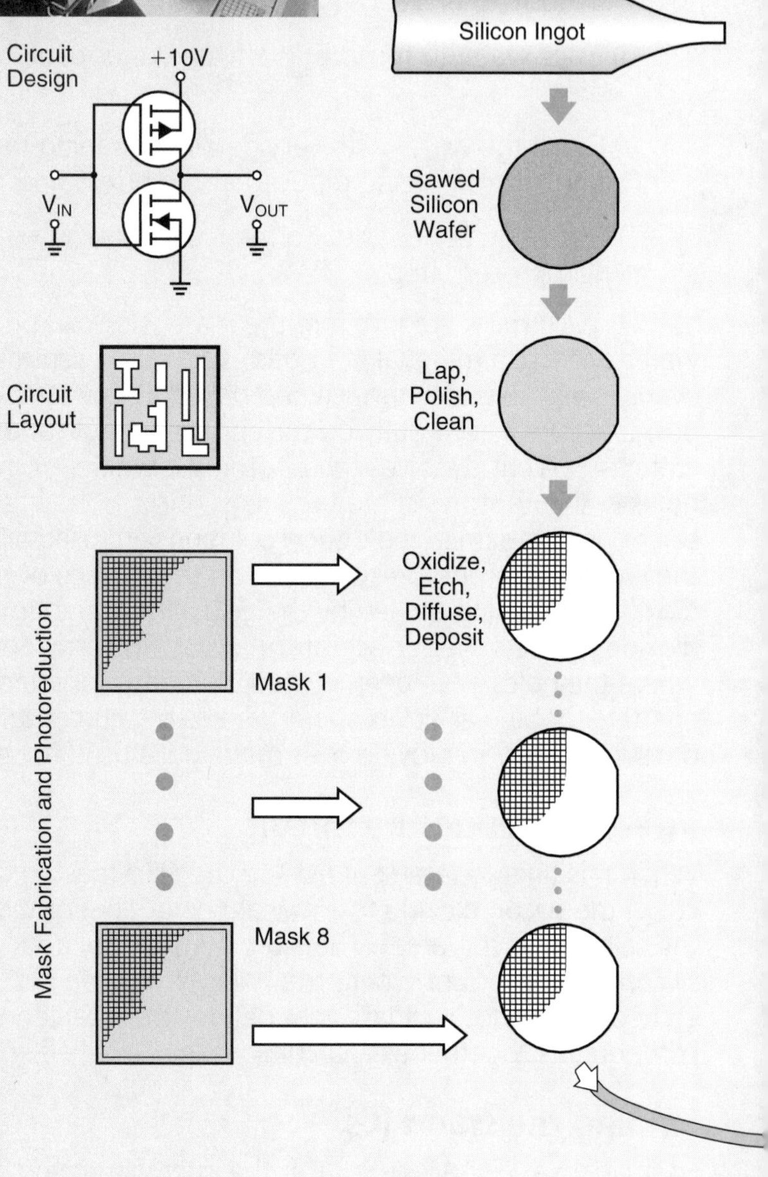


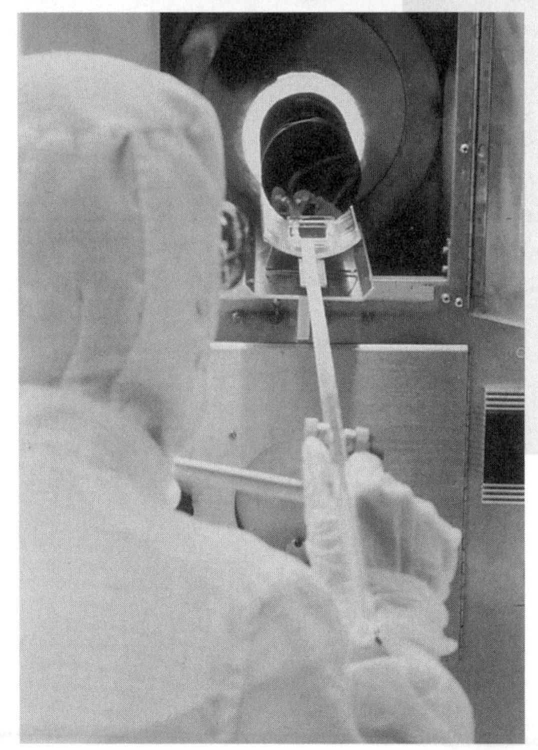

c. Inserting Wafers into a Diffusion Furnace

*Figure 10-5. IC fabrication process.*
*(Photos courtesy of Texas Instruments Corporation.)*

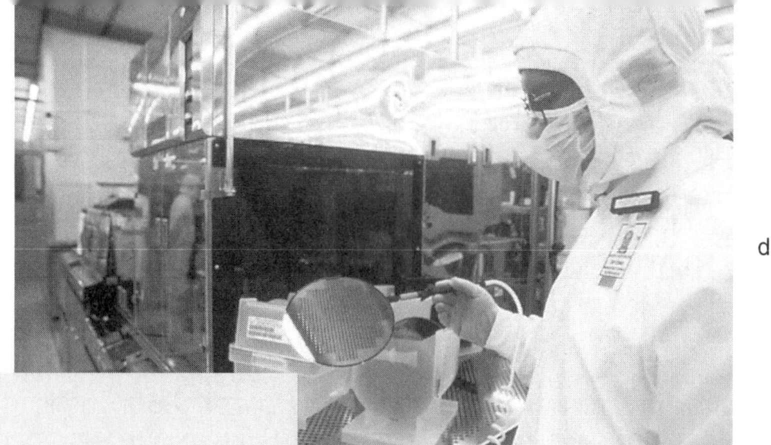

d. Clean Room Operator Positioning Wafer at a Mask Station

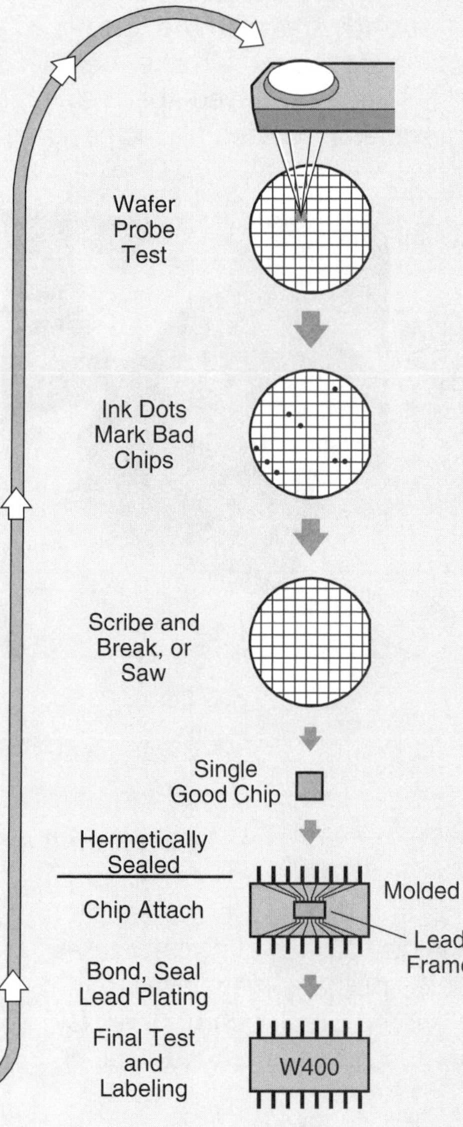


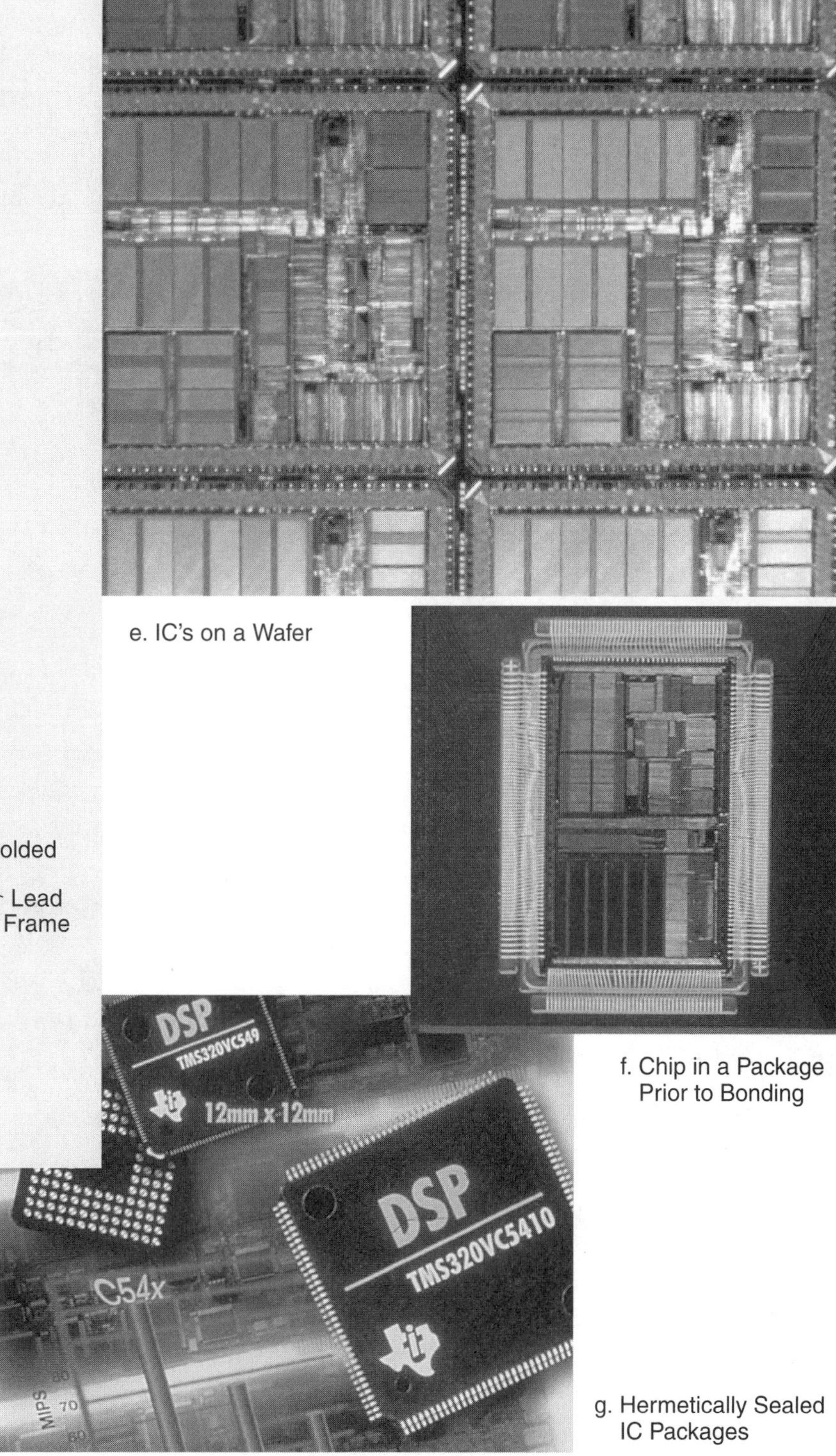


e. IC's on a Wafer

f. Chip in a Package Prior to Bonding

g. Hermetically Sealed IC Packages

## IC Assembly

All steps in the fabrication process prior to assembly use the complete wafer with all ICs in place. Assembly starts the steps in which each IC is handled individually and connections to each IC are made separately. Assembly of an IC chip into a molded plastic package is shown in *Figure 10-6*. The chip is mounted on a lead frame, shown in *Figure 10-6a*, which contains the number of package pins required. Thin gold wires are thermocompression-bonded to the aluminum pads on the IC and, at the other end of the wire, to the appropriate pin on the lead frame. The chip mounted to the lead frame with the bonding wires in place is shown in *Figure 10-6b*. The lead frames are manufactured so that a number of chips (from 6 to 8) are mounted, bonded and molded at the same time. Next, the lead frames with the bonded chips are placed in a mold. Plastic pellets are heated in the mold and the melted plastic flows around the IC to protect it environmentally. After curing in an oven, the pins are tinned to make them easy to connect into a printed circuit board. The tinned ICs are then sawed apart, cut from the frame, and the pins trimmed and formed. A completed molded plastic package is shown in *Figure 10-6b*.

a. Lead Frame Strip

b. Molded Package

c. Hermetically Sealed Package

*Figure 10-6. IC packages.*

If the package is a hermetically-sealed package, as shown in *Figure 10-6c*, then the chip is mounted to a plated pad in the center of the package, which acts as a mounting pad and a heat sink at the same time. The mounted chip is connected to the package pins typically with gold bonding wires, just as for the molded package. In many cases, aluminum bonding wires are used. The hermetically-sealed package may have straight pins out the bottom (See *Figure 10-6c*) to insert into a printed circuit board, or it may have leads out the side for "surface mounting" to a printed circuit board with solder reflow techniques.

## Burn-In and Final Test

Prior to final testing, ICs that will be used for special applications are placed in ovens with power applied and the oven temperature cycled between hot and cold limits. This "burn-in" occurs for varied times depending on the intended application for a particular IC. Commercial ICs typically bypass the burn-in step. Military units may be on burn-in for six weeks; special consumer units from a few days to several weeks. Burn-in stresses the IC beyond normal conditions and weeds out weak chips. ICs that survive burn-in are passed on to final test.

Final testing is by computer-controlled stations that select the type of tests required for the particular IC, including temperature tests, and run the tests and sort the passed ICs from those that failed. The ICs that pass are marked with the circuit type, the manufacturer's logo, and a code that tells when and where the IC was packaged and usually where it was tested. A visual inspection completes the assembly process and the IC is passed on for shipment. This completes the fabrication process.

## About Combination Bipolar and MOS ICs

We stated earlier that analog circuit applications dictate which active devices are used. As a result, many IC applications require a manufacturing process that uses a combination of bipolar and CMOS devices. The process also must accommodate other requirements of analog circuits being designed in integrated circuit form, such as:

A. The need to control matched resistor values to 1% tolerances. This tolerance is much tighter than bulk-diffused resistors used in digital ICs, whose tolerances are typically 10% to 20%

B. The need to have capacitors controlled to a 20% tolerance, rather than to have them greater than some minimum value.

C. The need to have the $h_{FE}$ of bipolar transistors controlled within a range, rather than to have it be above some minimum value.

D. The need to control the thermal characteristics of the components used.

E. The need to operate at average supply voltages of 10 V, at intermediate levels of 30 V, and even up to 60 V — voltages much greater than digital logic circuit voltages of 1, 3 and 5 V.

In addition to these component and parameter requirements, several special design techniques that are used regularly in analog circuit design must be accommodated. Here are several:

A. Feedback techniques to control gain, like those discussed for op-amps in Chapter 4.

B. Cross-coupling of devices for good matching of parameters, as shown in the MC1496 balanced modulator in Chapter 5.

C. The use of accurately-controlled current sources to set operating conditions, and the use of capacitors for controlled time constants.

The combination manufacturing process that we will now describe satisfies these needs. The process requires a larger number of steps than the simple bipolar and CMOS circuits that we described previously. The number of added steps usually depends on how accurately the bipolar parameters must be controlled but, more importantly, on the number of different components required and the number of different voltages needed for a design. Poly-silicon interconnection, which is a special combination of silicon and metal, is used rather than just aluminum. Poly-silicon has a higher ohmic value than aluminum, and the ohmic value can be controlled by exposing the poly-silicon to one or more of the diffusions normally used in the process.

The use of poly-silicon results in more accurately-controlled resistors and capacitors, which are required for analog applications. Matched resistors can be controlled to a 1% tolerance, and resistors with various temperature coefficients can be fabricated using both poly-silicon and bulk-diffused resistors. Capacitors with close tolerances can be made with two layers of poly-silicon separated by a thin nitride layer or, for lesser-controlled values, the capacitor can be made using poly-silicon for one plate and a heavily-doped bulk diffusion for the other plate separated by a thin layer of oxide.

Bipolar transistor parameters can be varied by using a diffused well for a base and the same diffused well to form digital CMOS devices, or by diffusing a separate base into a deeper diffused well that also is used for analog CMOS devices. And even further parameter variations are possible because transistor bases can be diffused into an epitaxial layer that is over the substrate used for the chip.

## Combination Process Details

Let's look at how a combination bipolar and CMOS process produces MOS devices for use as active devices in logic circuits, MOS devices for use in analog circuits, and bipolar devices with different parameters for use in analog circuits ranging from small-signal to power circuits — all on the same chip. To demonstrate the techniques used, we will use only cross sections of the IC chips and not refer to any specific circuit.

Let's look at *Figure 10-7*. Here the silicon chip substrate is a P-type but, in special defined areas where NPN transistors are to be formed, a heavily-doped N+ is diffused into the substrate. It is called a buried layer. This will form a low resistance path for collector current over the collector area. After these diffusions are made, a layer of single-crystal P-type silicon is deposited over the complete substrate. This is called an epitaxial layer and is a well-controlled layer designed to contribute specific characteristics to the active devices formed in the IC.

### *An NPN Using the Digital MOS Diffusions*

In the process used to make MOS devices for use in digital circuits, a shallow N-well (SN well) and a shallow P-well (SP well) are diffused into the epitaxial layer to receive diffusions for source and drain regions for MOS digital logic devices.If MOS digital devices alone are made, they would be minimum size and be dimensioned with the tightest layout rule tolerances that the process would allow. However, the combined devices use looser tolerances. They are completed by N+ diffusions into the P-well for N-channel devices, and P+ diffusions into the N-well for P-channel devices. Poly-silicon patterned to fit over the region between the source and drain, separated by a controlled thin-gate oxide, forms the gates of the MOS logic devices. As indicated in *Figure 10-7*, another oxide is placed over the wafer and aluminum patterned to connect the devices.

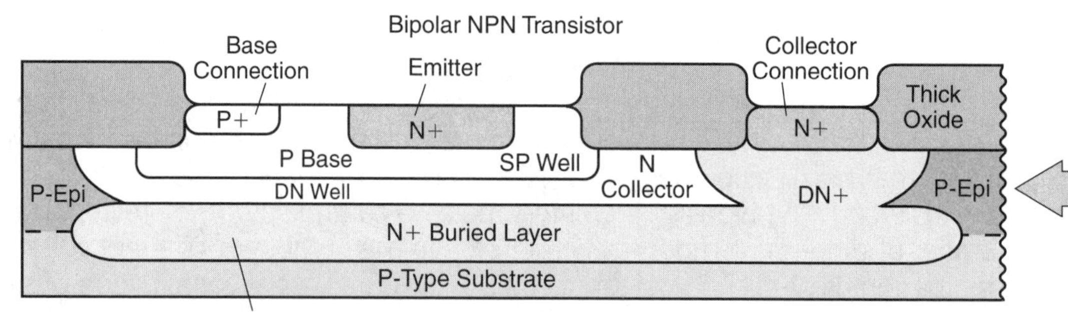


*Figure 10-7. Digital MOS and NPN transistor.*

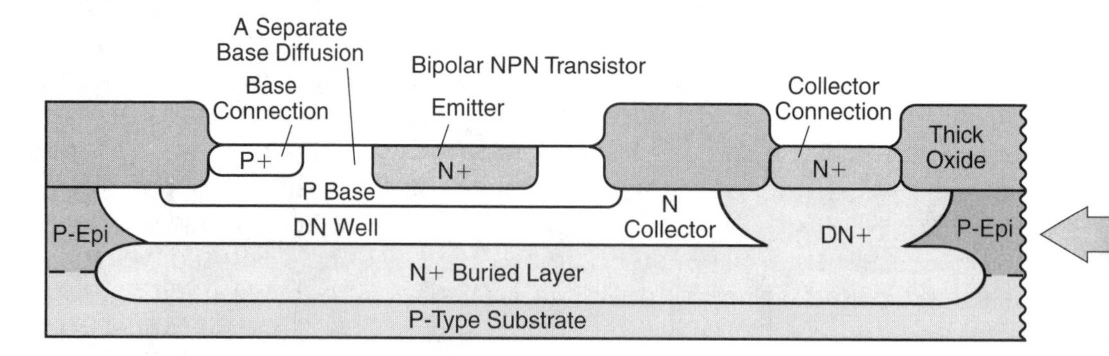


*Figure 10-8. Analog MOS and NPN transistor.*

An NPN transistor is made by using the basic P-well diffusion used for the logic devices as a base of the transistor. This is shown in *Figure 10-7*. However, prior to the P-well diffusion, a deeper N-well diffusion (DN well) is made as well as a deep heavily-doped N+ region (DN+) that reaches to the buried layer under the epitaxial layer. The N+ region makes a low-resistance connection from the surface to the buried layer under the N-type collector. A heavily-doped N+ diffusion is made into the P-well base of the NPN to form the N emitter. At the same time this N+ diffusion is added to the deep N+ region to make a better collector ohmic connection at the surface. A P+ diffusion, made when the source and drain diffusions are made, is added to the base to make a good ohmic connection at the surface. Thus, the result is an NPN transistor using the same diffusions as those used for the CMOS devices.

Note in *Figure 10-7* that a capacitor is shown made from one plate of poly-silicon and the other plate from a heavily-doped N+ diffusion separated by a thin layer of oxide. The capacitance value is controlled by the area of the plates and the thickness of the oxide. Also shown is a resistor made from poly-silicon. Its length and width determines the resistor value. Diffusions used for the CMOS or bipolar devices are used to set the resistivity of the poly-silicon and change the resistor value.

## An NPN Using the Analog MOS Diffusions

In *Figure 10-8*, the process steps are shown when the MOS devices are to be used for analog circuits. For the P-channel MOS, a deep N-well (DN well) is diffused in the selected area where the MOS device is to be formed. Then P+ regions for source and drain are diffused into the well. Patterned poly-silicon over a controlled thin oxide that is over the region between the source and drain forms the gate of the device. The N-channel analog MOS is made just with N+ diffusions into the P-type epitaxial layer. Again, a patterned poly-silicon over a controlled thin oxide over the region between the source and drain regions forms the gate.

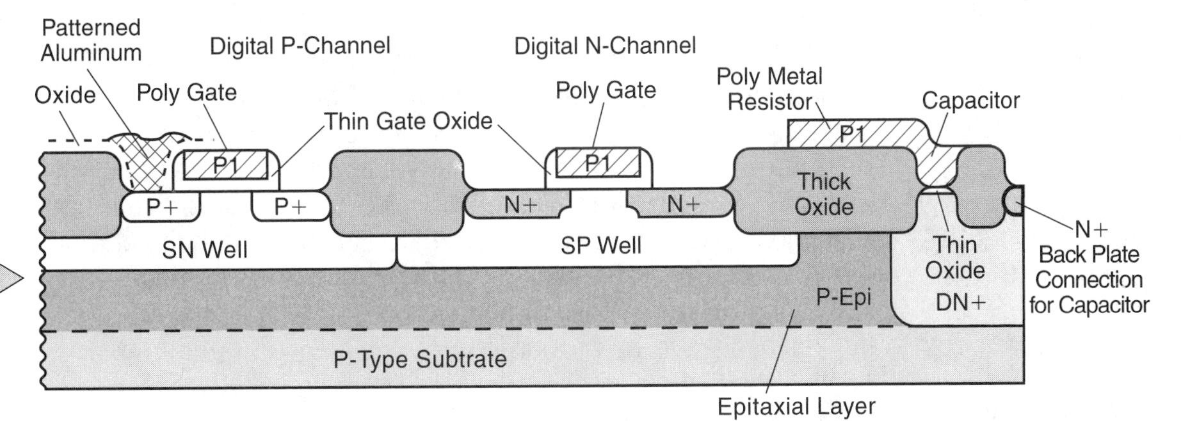


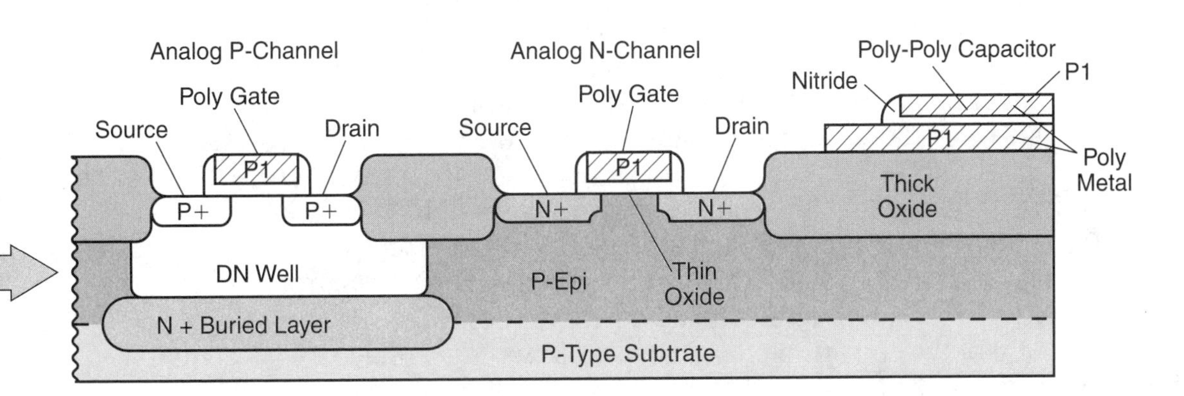

The NPN transistor uses the same steps as those in *Figure 10-7*, except there is a separate P-type diffusion to form the base region. This provides variable parameters from the NPN in *Figure 10-7*. Note in *Figure 10-8* there is a capacitor made from two plates of poly-silicon separated by a thin layer of nitride. This provides a capacitor with closer controlled tolerances. The capacitor value is determined by the area of the plates and the thickness of the nitride. In both of the cross sections of *Figures 10-7* and *10-8*, a layer of protective oxide through which holes are cut to allow the metal interconnect to make contact to the device regions is not shown, except for the small example section shown in *Figure 10-7*. This oxide would be present between each layer of the interconnection as one, two or three levels of interconnections are deposited and patterned to connect the circuit together. The interconnection can be aluminum or poly-silicon. Some power devices used in analog ICs can have very large metal areas to handle the current. Multiple bond wires have to be used in many applications to carry the current to external pins.

## Analog MOS Versus Digital MOS

Analog circuit MOS devices generally use larger devices for matching, and for higher voltage operation. The layout rules have wider tolerances than the highest packing density of the logic MOS devices. Using larger devices provides better control over device characteristics and gives more manageable parameters. MOS enhancement-mode devices are most common in analog IC designs, but depletion-mode devices are available and can be used if needed. As mentioned previously, operating voltages are typically 10 V, but can be up to 30 V and, with special processes, 60 V operation is possible. All varieties of IC packaging are available, many times dictated by the power that must be dissipated by the package. Where 8-, 16-, 24- and 32-pin packages are common for logic circuits, 100-pin packages are common for analog ICs that have a wide variety of analog and digital functions integrated together.

## A Bipolar and CMOS IC

*Figure 10-9* is an analog circuit that uses a process similar to the one shown in *Figure 10-8* to fabricate an IC that uses both bipolar and MOS active devices. The circuit is called a current bias cell. It has many applications in analog circuit design to set a particular constant current value. The output $V_{BIAS}$ couples a particular voltage to the gate of a neighboring circuit's P-channel MOS device. If the coupled device has the same dimensions as the P-channel Q6 and Q7, then the same constant current that is present in Q6 and Q7 will be present in the coupled device. This is particularly true if the devices are made close together on the same IC. By changing the relative dimensions of the coupled device, the constant current in the coupled device will be proportional to the ratio of the dimensions.

## Four Different Circuits

The analog circuit of *Figure 10-9* can be divided into four functional circuits, as shown. The NPN transistors Q9 and Q10 form what is called a "Delta $V_{be}$ Current Source." They set the amount of current in the circuit, and the combination of a poly-silicon resistor and a bulk-diffused resistor (R1 and R2) in the emitter of Q10 temperature-compensates the current to make its value proportional to the absolute temperature of the IC. The N-channel MOS Q8 in the circuit is called a "beta helper" and compensates for base current error that one would otherwise have between the currents into the bases of NPN Q9 and Q10.

The P-channel Q6 and Q7 are called a "current mirror." They assure that the current in the collectors of the NPN Q9 and Q10 are the same — they are mirrored — and the $V_{BIAS}$ output will mirror current to a coupled device, as described previously.

It turns out that the current reference circuit has two stable states — no current and the designed operating current. The starting circuit of the N-channel MOS Q1 and P-channel MOS $Q_2$ assures the current reference circuit comes up in the designed operating

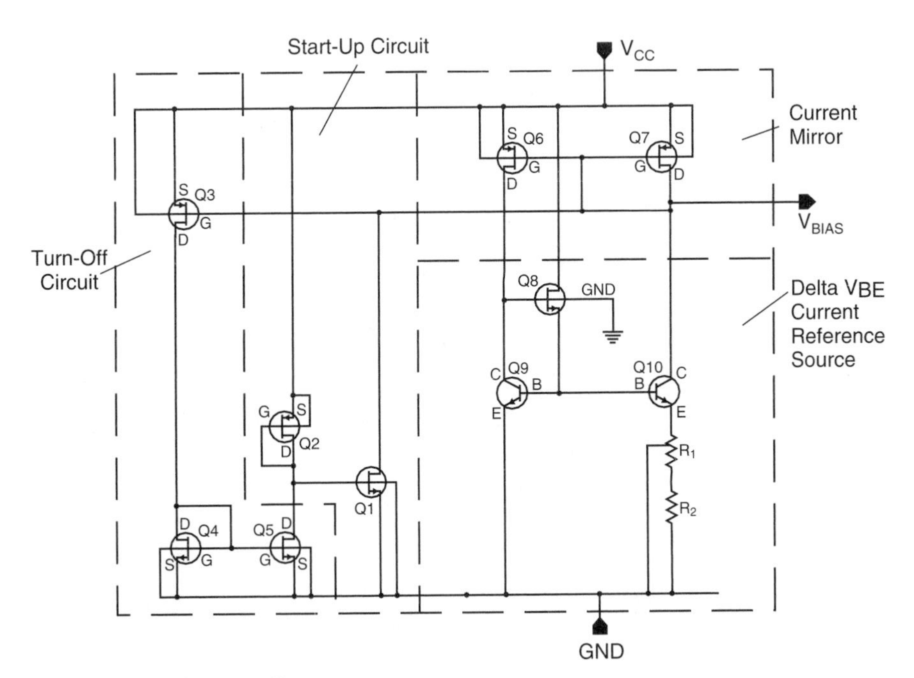


*Figure 10-9. Current bias cell.*

current condition. The P-channel Q2 acts just like a resistor from $V_{CC}$ to apply $V_{CC}$ to the gate of the N-channel Q1. As $V_{CC}$ increases from zero, Q1 will conduct current and pull the gates of the P-channel Q6 and Q7 so that they conduct current and stabilize Q9 and Q10 to the designed current operating condition. Once the operating point is reached, the P-channel MOS Q3 conducts current and pulls the gate of the N-channel Q5 towards $V_{CC}$. The N-channel Q5 conducts and pulls the gate of the N-channel Q1 to ground and turns Q1 off. All currents in the circuit are then stable, and the current value in the collectors of the NPN Q9 and Q10 remain constant over wide temperature changes. The IC is laid out such that particular values of R1 can be chosen to set the reference current to a design value.

## The IC Layout

*Figure 10-10* shows the layout of a complete IC in which a different bias cell circuit similar to that of *Figure 10-9* is used. The small portion of the total IC used for the current bias cell is identified with an overlay. A detailed layout of the circuit in *Figure 10-9* is shown in *Figure 10-11*. In the layout of *Figure 10-11* in several cases, multiple active devices are used in parallel. This is the case for the P-channel MOS Q6 and Q7, and the P-channel MOS Q3. It also is the case for the NPN transistors Q9 and Q10.

There are three levels of interconnect — a poly-silicon layer and two aluminum layers — used for component interconnections or for gates of MOS devices. The poly-silicon is used for the resistor $R_1$ while a diffused resistor is used for $R_2$.

The total IC of *Figure 10-10* is designed into a chip that is 10.5 mil by 5.5 mil. Comparing the photograph of the current bias cell IC shown in *Figure 10-11* with the total IC shown in *Figure 10-10* gives a good idea of the density of circuitry that has been designed into the total IC in the small area of 10.5 mil x 5.5 mil.

## Summary

In this chapter we have described what an IC is and how it is made. We showed the difference between MOS and bipolar ICs, and explained an IC manufacturing process that allows bipolar and MOS active devices to be used in the same analog IC. We found that analog ICs require many different devices to be mixed in order to satisfy the circuit requirements.

In the next Chapter, we will discuss digital signal processing.

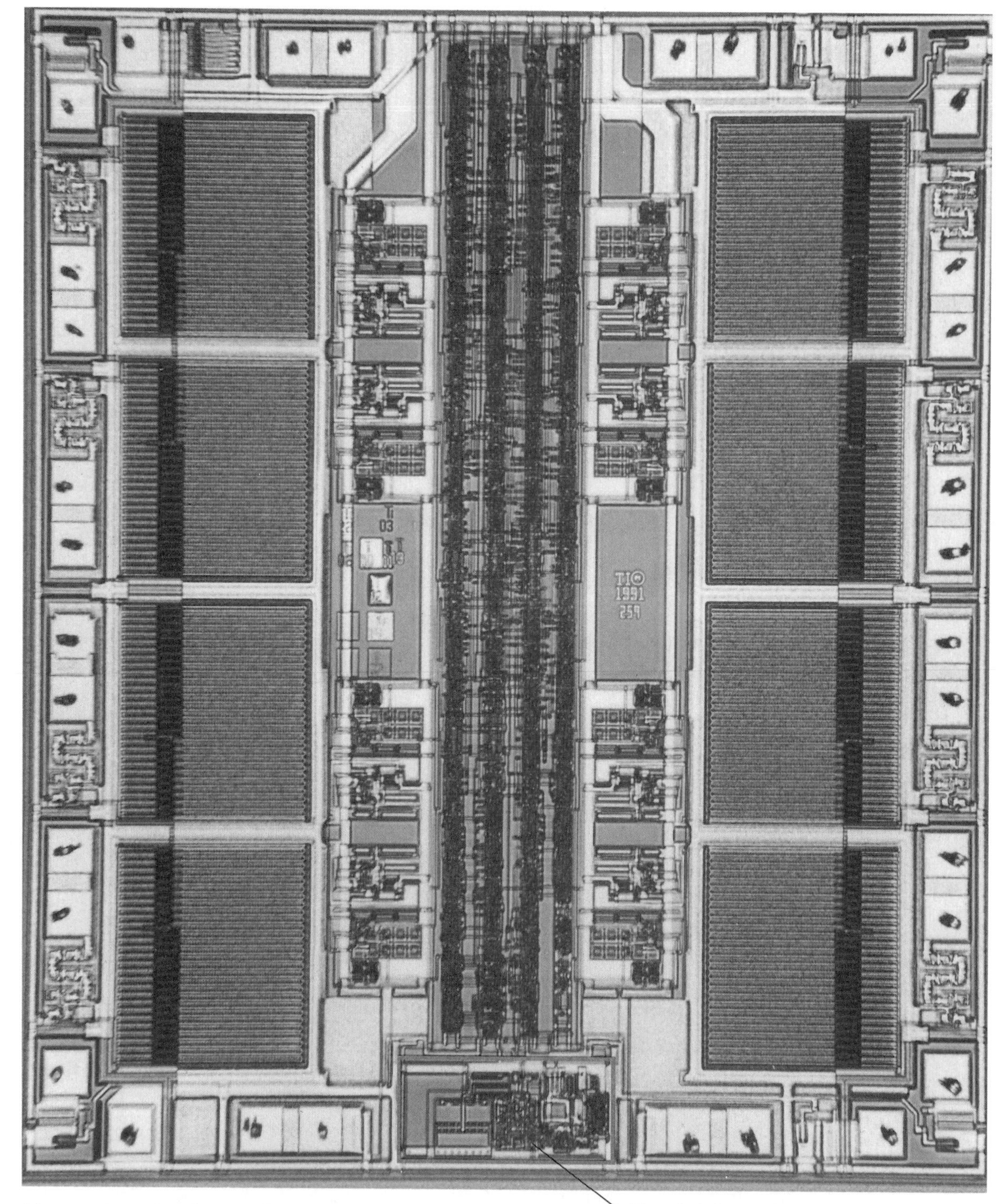


*Figure 10-10. Photo of analog IC.*
*(Courtesy Texas Instruments Incorporated)*

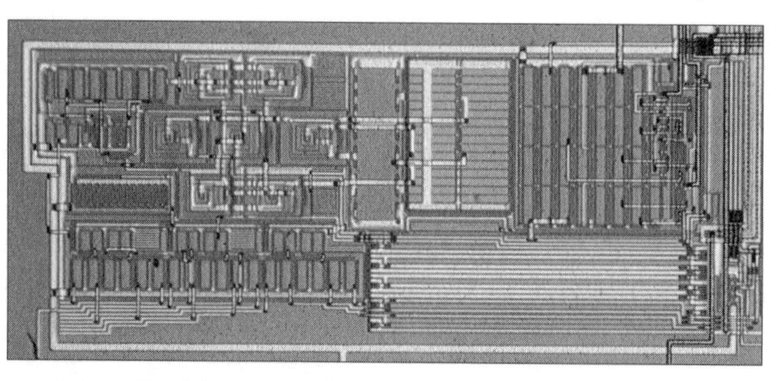

*Figure 10-11. Photo of detailed layout of current bias cell of Figure 10-9.*
*(Courtesy Texas Instruments Incorporated)*

# Quiz for Chapter 10

1. A bipolar IC uses what type of active devices?
   a) N channel MOS devices.
   b) P channel MOS devices.
   c) Bipolar junction transistors.
   d) PNP bipolar devices only.
2. The original high volume IC's were:
   a) Digital.
   b) Analog.
   c) Mixed analog and digital.
   d) None of the above.
3. MOSFETS are preferred for digital IC's because they…
   a) Are inexpensive to fabricate.
   b) Have low power dissipation.
   c) Have good performance characteristics.
   d) All of the above.
4. IC's are normally manufactured on a silicon wafer because…
   a) It is rare but gives the best performance.
   b) Only one high quality circuit can be made at a time.
   c) It is a very inexpensive material.
   d) Allows many high-quality IC's to be manufactured at the same time.
5. Metalizations are use in IC's to…
   a) Interconnect components.
   b) Bring connections to outside.
   c) Interconnect only active devices.
   d) None of the above.
6. A finished IC is attached and connected to ____________ and packaged.
   a) an IC carrier device.
   b) a leadframe.
   c) a plastic holder.
   d) a socket device.
7. Connecting wires are normally made from
   a) Gold and steel.
   b) Gold and aluminum.
   c) Steel and aluminum.
   d) Always gold.
8. Compared to analog ICs, digital ICs usually have…
   a) About the same number of components on the chip.
   b) Fewer components on the chip.
   c) More components on the chip.
   d) Are made from different material.
9. In general, when compared to digital ICs, analog ICs.
   a) Have tighter manufacturing tolerances.
   b) Have looser manufacturing tolerances.
   c) Require much thicker silicon substrates.
   d) Are smaller than digital ICs.
10. Analog ICs may be required to operate up to power supply voltages as large as
   a) +1V
   b) ±3V
   c) +5V
   d) 60V

**Answers:**
1 c, 2 a, 3 d, 4 d, 5 a, 6 b, 7 b, 8 c, 9 a, 10 d

## Questions & Problems for Chapter 10

1. An integrated circuit is ________________________.
2. Wafer probe is a step in the IC manufacturing process designed to do what?
3. What is the purpose of the burn-in step in IC manufacturing?
4. What is the widely-accepted IC circuitry for the highest density logic circuits?
5. ICs became popular because
   a) They provide a low-cost means of building complete circuits in a single manufacturing process.
   b) They have uniform circuit parameters and, in many cases, important matched-device characteristics.
   c) They provide small active devices for low-power and high-frequency operation.
6. When using both bipolar and MOS active devices for an analog application, which IC process is used?
7. What diameter are the silicon wafers used for IC manufacturing?
8. In the IC manufacturing process of *Figure 10-5,* at which step do the IC chips start to be handled individually?
9. What technique explained in Chapter 4 is used in analog IC design to control gain?
10. In the combination IC manufacturing process shown in in *Figures 10-7* and *10-8,* when is the N+ buried layer diffused into the IC wafer?

*(Answers on page 212.)*

# CHAPTER 11
# Digital Signal Processing

We identified the various analog functions in Chapter 2, and in subsequent chapters described the electronic circuits used to perform the functions and showed how these circuits are combined to make subsystems and complete systems. The application of the functions cause a modification of the baseband information signal — changing the amplitude, changing the frequency, moving the signal to a different position in the frequency spectrum, selecting a particular frequency or band of frequencies, mixing, modulating, and detecting signals. In each case, there is a modification of the signal from input to output.

Early in our book, we noted that digital circuits are taking over many system applications because they provide cost-effective solutions for system design. We also stated that there are still many applications where analog circuits are the least-expensive, best solution for particular designs. Yet, even when analog circuits provide the better solution, they still have some disadvantages, especially if the circuit is made up of discrete components. First and most important, analog circuits usually are not easily programmed; and second, they are component-sensitive — component values shift with voltage, temperature, and time to cause significant variations in the performance of the circuits.

## A New Technology Emerges

To overcome these disadvantages, a new technology has emerged that provides the programmability, stability, and cost advantages of digital technology while, at the same time, providing the ability to perform analog functions. This technology is called *digital signal processing* (DSP). The basic DSP system is shown as a block diagram in *Figure 11-1.*

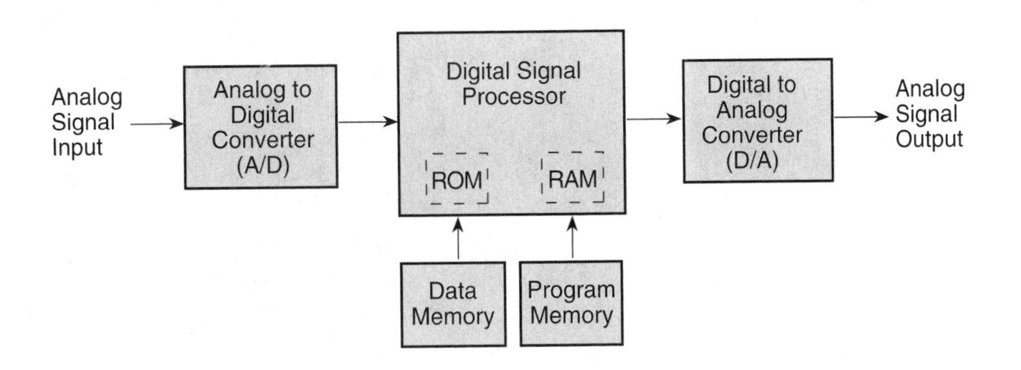


*Figure 11-1. Block diagram of a basic digital signal processing (DSP) system.*

The incoming analog signal, which might be a temperature measurement, a baseband signal, someone's voice, or video, is translated into a digital code of ones and zeros by a function called an *analog-to-digital converter* (A/D). These digital codes are easily manipulated by a software program running on a very fast microprocessor in order to modify the signal and accomplish the desired analog function. The output of the digital signal processor is fed to a function called a *digital-to-analog converter* (D/A), which restores the signal to analog form to drive a load, such as a temperature gauge, a motor, an audio speaker, or a controller.

One of the most important features of a DSP microprocessor is that it can do its computing operations so fast that the processing of the information from input to output occurs as things happen — that is to say, in "real time." This means that the processing is done in-line in continuous time, not taken off-line and manipulated on the side. Processing follows step-by-step as time progresses. In most cases, the need for real-time processing is the most important characteristic that dictates the use of DSP in a particular application.

DSP microprocessors are very-high-speed processors that execute most operations in their set of instructions in one clock cycle — other types of processors take several clock cycles. A DSP microprocessor can handle two billion operations per second. As an example of their speed, if it presently takes *10 minutes* to download a particular file from the Internet, the above DSP processor will reduce the download time to *five seconds!* All microprocessors perform such arithmetic operations as addition, subtraction, multiplication and division. One of these arithmetic operations of particular importance in many DSP applications is a routine of *multiply and accumulate* called a *MAC*. DSP microprocessors execute a complete MAC in one clock cycle. It takes other microprocessors many cycles to perform the same operation. It is this speed that makes real-time processing possible.

## Some Application Examples

One of the earliest, high-visibility, consumer applications of DSP was the Speak & Spell™ learning aid for children manufactured by Texas Instruments Incorporated. With the press of a button, human speech reinforced the decisions that children made in response to questions asked by the computer. The electronics had to generate audio signals modeled after human speech by microprocessor manipulations under the direction of a computer program. Such speech-recognition techniques have now been advanced to the point that we can dial a telephone by speaking the number, or speak to our computer to type a letter or memo without touching the keyboard.

DSP applications important to communication systems include filtering, frequency selection, and noise reduction. DSP systems are used extensively to form particular types of filters to filter input signals, to reject selected frequencies and select others and, coupled with amplification, to amplify signals to generate a particular type of signal response against time or frequency. DSP systems have been designed for use inside automobiles to cancel out exterior noise, thus providing a quieter interior environment for passengers. Many different support circuits that aid in the application of DSP are being developed along with specific DSP circuits. Again, those that are important to communication systems are RF products for receivers — especially digital cordless telephones, cellular telephones, and Personal Communication Systems (PCS). Other functions outside of RF are frequency synthesizers, modulators, and power amplifiers.

A very common application for DSP is a telephone line repeater. As a signal propagates down a telephone line, the quality of the signal degrades (it loses amplitude and changes amplitude with frequency). As shown in *Figure 11-2*, the DSP repeaters reconstruct the signal, restoring each repeater's output signal to that of the original signal, thus maintaining the same signal quality all along the telephone line.

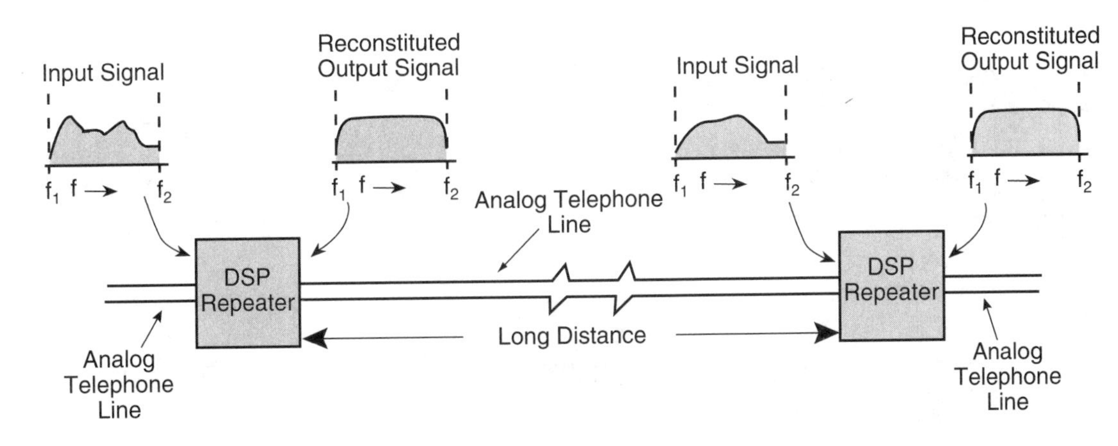


*Figure 11-2. A DSP repeater reconstructs the input signal, restoring it to that of original signal.*

## The Importance of Software

As mentioned previously, a big advantage of DSP is that the microprocessors are programmable; thus, the system can be changed readily without changing hardware. This means that DSP systems are like a personal computer — their operations are directed by a software program. Windows® is an operating system software program that runs more than 90% of all personal computers. In addition to the operating system, each personal computer application requires a software program to perform such functions as word processing, to interface with and search the Internet, draw artwork, develop and manipulate spreadsheets, and so on. Likewise, DSP systems require software programs to perform their tasks. To change the application and make the DSP system perform a different function, only the program needs to be changed. Instead of building new system hardware, applications are changed simply by changing software. Manufacturers of DSP microprocessors and third-party programming companies provide the software. Additional support services, such as development systems, software emulators, and testing aids, are provided to help system engineers with their designs.

## What Functions are Required?

### A/D and D/A

Looking again at *Figure 11-1*, we see that the input signal is changed by an analog-to-digital (A/D) converter. The function of the A/D is to accept the input information, sample it at regular intervals, and generate a digital code to represent the amplitude of the signal at the time it is sampled. *Figure 11-3* illustrates what happens. In order to make the explanation simpler, only a 4-bit code is used. Present systems use 8- or 16-bit codes.

*Figure 11-3a* also shows the digital-to-analog (D/A) function, which is at the end of the DSP system of *Figure 11-1*, after the DSP microprocessor. The D/A function reverses the A/D function. It takes the digital code at a particular sampling time and outputs a voltage value derived from the code. The voltage will be constant until the next sample. After filtering, the analog voltage represents the original analog input signal that has been modified by the computations of the microprocessor under the direction of the software program.

Let's examine *Figure 11-3a* closer. The input signal is sampled at regular intervals by a sampling pulse. At each point, a 4-bit digital code is generated that represents the amplitude of the input signal. The output of the D/A is a stream of digital codes

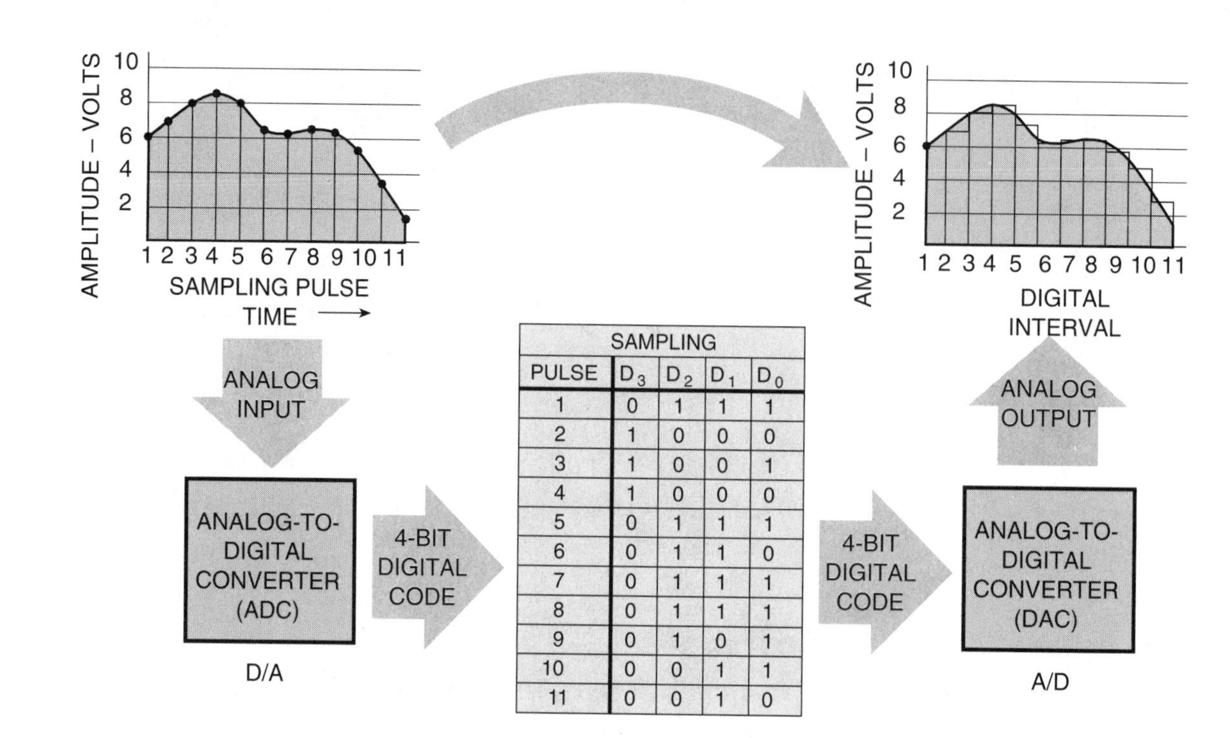


| SAMPLING | | | | |
|---|---|---|---|---|
| PULSE | $D_3$ | $D_2$ | $D_1$ | $D_0$ |
| 1 | 0 | 1 | 1 | 1 |
| 2 | 1 | 0 | 0 | 0 |
| 3 | 1 | 0 | 0 | 1 |
| 4 | 1 | 0 | 0 | 0 |
| 5 | 0 | 1 | 1 | 1 |
| 6 | 0 | 1 | 1 | 0 |
| 7 | 0 | 1 | 1 | 1 |
| 8 | 0 | 1 | 1 | 1 |
| 9 | 0 | 1 | 0 | 1 |
| 10 | 0 | 0 | 1 | 1 |
| 11 | 0 | 0 | 1 | 0 |

a. A/D to Code and Code to D/A

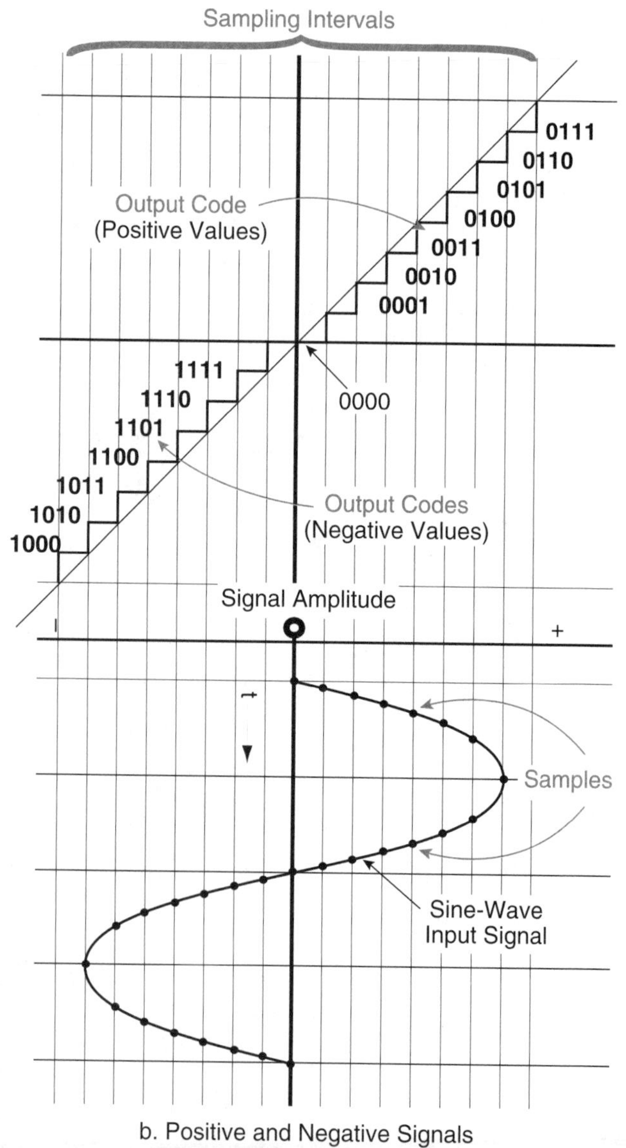


b. Positive and Negative Signals

*Figure 11-3. The A/D and D/A functions of digital signal processing.*

representing the analog signal at each of the sampling points. While it is beyond the scope of this book, it can be shown that no information is lost if a signal is sampled at least twice in a cycle. (The sampling rate limits the maximum high-frequency of the input signal.) This is an important result, since it guarantees that analog signals may be put in digital form without loss of information. The representative codes for the amplitude are shown in the table of *Figure 11-3a*. These coded values are sent to the DSP microprocessor and manipulated and modified by the processor following processing steps dictated by a software program.

The processor output is a digital code that represents the amplitude of the now-modified analog signal over each sampling interval. The code is the result of the processor computations to perform the desired function.

Once the digital signal is processed, it may be converted back to the analog form using the D/A converter. Let's suppose now that the codes shown in the sampling table of *Figure 11-3a* are the modified signal codes at the sampling points after modification by the processor computations dictated by the software program. The D/A converter decodes the processor output and produces an analog output waveform. Note that the decoded output waveform derived from the codes passing through the D/A function is a stair-stepped waveform whose value is constant between sampling intervals. Filtering smooths the signal into the final continuous analog signal. There is some error between the decoded stairstep value at the output and the actual, true analog output. These errors exist in DSP systems, but are very, very small — although in some applications such errors could limit the accuracy of the DSP system performance. *Figure 11-3b* shows how a sine-wave signal is coded and how the code changes when the amplitude changes from positive to negative.

Since the DSP is a real-time system, the signal processing must occur between the sampling pulses. This is the reason that DSP microprocessors are designed to have the highest computing speed of any microprocessors available. When we say that two billion operations are performed per second, that means that one operation is performed in one-half of a nanosecond — *one-half of a billionth of a second!* The faster the processor, the more processing that can occur between samples, and the higher the frequency that can be handled as an input signal.

## The DSP Microprocessor

The DSP microprocessor is really a self-contained digital computer. It has input/output coupling circuits, the arithmetic mathematical calculating unit, read-only memory (ROM), and random-access memory (RAM). Additional memory blocks attached to the microprocessor are shown in *Figure 11-1* — one to store the data for manipulation, and the other to store the program that tells the microprocessor what to do for the particular application. There are a wide variety of microprocessors to choose from to meet particular application goals. This variety allows a designer to balance performance and cost at the production volume required. *Figure 11-4* is a photograph of the detailed layout of the components of a DSP on a silicon chip. The chip is mounted in a package and connected to package pins to interface to the external circuitry. *Figure 11-5* shows a typical package for a DSP IC — in this case, a 144-pin package.

Figure 11-4. Photograph of the detailed layout of the DSP components on a silicon chip. (Photo courtesy of Texas Instruments Incorporated.)

Figure 11-5. A typical IC package for a DSP. (Photo courtesy of Texas Instruments Incorporated.)

## System Makeup

An expanded general system block diagram is shown in *Figure 11-6*. It is the same as the basic system of *Figure 11-1*, but with added components. Interface circuits are added at the input and output, and a controller is added to keep the complete system operating in time sequence. The data memory and program memory will vary in size depending on the system application. A power supply completes the system. The DSP is a single IC, as are the A/D and D/A. The interface circuits can be operational amplifiers used for amplification, or for summing, or for combining signals. The interface circuits also could be multiplexers to provide coupling to a number of signals. All are ICs. The controller usually is a special combination of ICs designed for the specific application, or it might be another DSP adapted to be a controller. As the system application expands, a variety of analog ICs and digital ICs usually are added for interstage coupling, timing, and signal modification, and power amplifiers are added to drive the required load.

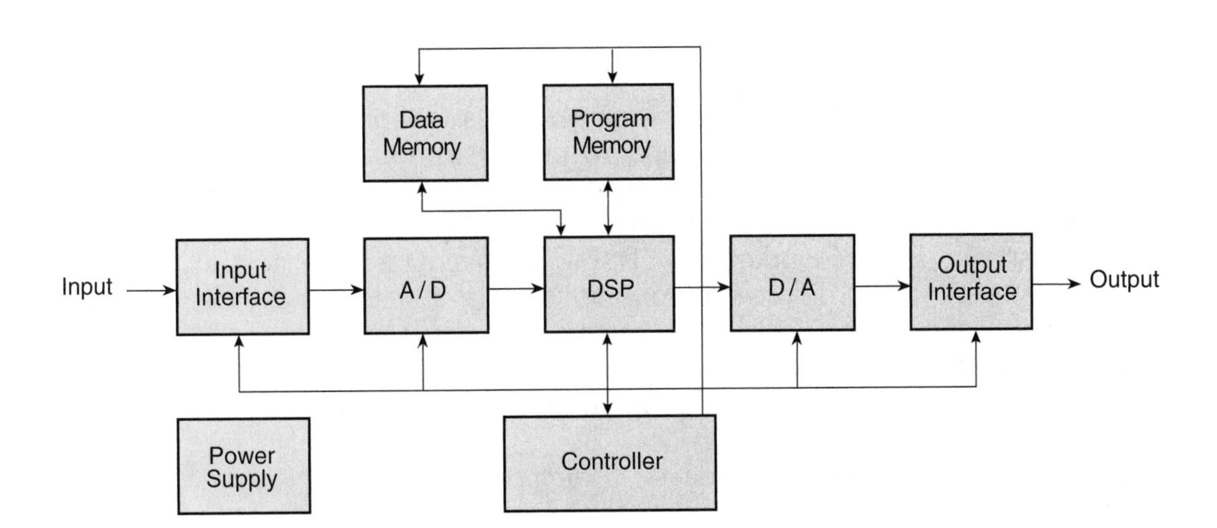


*Figure 11-6. Block diagram of a typical DSP system.*

## Typical Applications

Here is a list of some of the many applications areas of DSP technology:

***General Purpose***
- Digital and adaptive filtering
- Signal correlation
- Fast-Fourier transforms
- Waveform generation

***Voice/Speech***
- Voice mail
- Speech recognition, verification, enhancement, synthesis
- Text to speech
- Speech to text

***Telecommunications***
- Line repeaters
- Channel multiplexing
- Modems
- Dial-tone multiple frequency encoding/decoding
- Video conferencing
- Cellular telephones
- FAX
- Data encryption

***Instrumentation***
- Spectrum analysis
- Function generators
- Pattern matching
- Phase-locked loops

***Graphics/Imaging***
- 3-D image rotation
- Image compression/transmission
- Pattern recognition

***Computers***
- Hard disk control
- Laser printer control
- Scanner processing

***Industrial***
- Motor control
- Robot control
- Numeric machine control
- Security access
- Power line monitors

***Automotive***
- Engine control
- Anti-lock brakes
- Adaptive ride control
- Global positioning systems

***Medical***
- Hearing aids
- Ultrasound equipment
- Patient monitoring
- Diagnostic equipment

## Summary

In this chapter, we have discussed the emerging digital signal processing (DSP) technology that is having a significant impact on today's communication systems, and which will have an even greater impact in the future. Two new functions were introduced — A/D and D/A, which interface analog signals to digital, and digital signals to analog, respectively. We found out that DSP systems require the fastest microprocessors in order to do their computations in real time in the very-brief time between sampling intervals. DSP microprocessors are manufactured as ICs. We learned that DSP systems that perform analog functions overcome two disadvantages of analog circuits — component drift and lack of programmability — while providing cost-effective solutions for communications systems.

This chapter concludes our book, *Basic Communications Electronics*. We began with the objective of bringing you an understanding of the functions required in analog systems and the electronic circuits used to accomplish those functions, both in subsystems and in systems. We hope we have succeeded in our goal.

# Quiz for Chapter 11

1. Digital Signal Processing...
   a) Allows analog signals to be manipulated digitally.
   b) Converts analog signals into digital signals and back to analog.
   c) Provides a technology to overcome analog disadvantages.
   d) All of the above.
2. "MAC" stands for
   a) Multiple and accumulate, a digital arithmetic function.
   b) Macintosh computers, which are used for digital signal processing.
   c) Modulate and amplify, a DSP function.
   d) Manipulate and accomplish, a DSP function.
3. DSP microprocessors
   a) Are self-contained, miniature computers.
   b) Execute most operations in their set of instructions in just one clock cycle.
   c) Perform their operations in "real time."
   d) All of the above.
4. DSP software
   a) Allows designers to change the system by changing the software.
   b) Requires specific hardware for specific applications.
   c) Is not transferable to analog applications.
   d) Is fixed for every application.
5. An A/D converter
   a) Converts frequency to time signals.
   b) Converts amplitude variations to voltages.
   c) Converts analog signals into digital codes.
   d) Converts frequency variations to voltages.
6. DSP systems have been designed for use in automobiles to
   a) Control the steering.
   b) Reduce the interior noise.
   c) Control the lighting.
   d) Clean up the exhaust.
7. An D/A converter
   a) Converts frequency into noise.
   b) Converts digital codes to voltage values.
   c) Converts analog signals into digital codes.
   d) Converts frequency variations into voltages.
8. A/D codes for positive and negative values of input signals
   a) Don't change but remain the same.
   b) Change every four seconds to allow for easy decoding.
   c) Randomly change as a synch pulse is detected.
   d) Differ by the fact that a "1" or "0" appears in the most significant bit position.
9. A DSP microprocessor
   a) Is a very slow processor.
   b) Is not programmable.
   c) Performs its computations between sampling intervals.
   d) Must have a slow varying voltage as an input.
10. A/D and D/A converters presently use
    a) 8- to 16-bit codes.
    b) 320-bit codes.
    c) Only 4-bit codes.
    d) 128-bit codes.

**Answers:**
1 d, 2 a, 3 d, 4 a, 5 c, 6 b, 7 b, 8 d, 9 c, 10 a

## Questions & Problems for Chapter 11

1. The simplest DSP system consists of an ___________ at the input, a ___________, and an ___________ at the output.
2. Analog circuits, even though they provide the better solution, still have two disadvantages. What are they?
3. What are three DSP applications important to communication systems?
4. Why is software important to a DSP system?
5. What does "real time" processing mean in a DSP system?
6. Calculate the period of a timing square-wave that has a frequency of two billion cycles per second.
7. How many times must a 1-MHz analog signal be sampled in order to accurately reproduce its information?
8. What kind of an output voltage does a D/A converter output from the input digital codes?
9. A ___________ is required at the output of a D/A converter to output the desired continuous analog signal.
10. In addition to the power supply, ___________ and ___________ ___________ circuits and a ___________ are added to a simple DSP system to fill out the typical DSP block diagram of *Figure 11-6.*

*(Answers on page 213.)*

# Appendix

## Answers to Chapter Questions and Problems

### Chapter 1

1. Text, Data, Audio, and Video.
2. Transmitter, transmission link, and receiver.
3. Vectors.
4. Cycles per second or frequency.
5. Bandwidth.
6. Sine wave.
7. Audio 10-20 kHz; video 6 MHz.
8. Use the equation f = 1/T. Rearrange to T = 1/f.
   a) For 500 kHz, T = 1/500000 = .000002 seconds = 2 x $10^{-5}$ seconds = 2μS. μS means microseconds.
   b) For 20 kHz, T = 1/20000 = .00005 seconds = 0.5 x $10^{-4}$ seconds = 50μS.
   c) For 10 MHz, T = 1/10000000 = .0000001 seconds = 0.1 x $10^{-6}$ seconds = 0.1μS.
9. a) 6 kHz
   b) 25 MHz
   c) 1440 kHz
   d) 88 MHz
10. Use the equation f = 1/T
    a) For 0.001 second period, f = 1 / .001 = 1000 Hz
    b) For 0.05 second period, f = 1 / .05 = 20 Hz
    c) For .00001 second period, f = 1 / .00001 = 100000 Hz = 100 kHz
11. Use the equation λ (in meters) = 300 / f (in MHz).
    a) For 600 kHz = 0.6 MHz, λ = 300/0.6 = 500 meters
    b) For 100 MHz, λ = 300/100 = 3 meters
    c) 250 MHz, λ = 300/250 = 1.2 meters
12. a) 60°; lags
    b) 30°; lags
    c) 30°; leads
13. Amplitude Modulation (AM)
14. Frequency Modulation (FM)
15. Phase Modulation (PM)

### Chapter 2

1. Spectrum analyzer.
2. Noise.
3. Electrical fields and magnetic fields.
4. $f_c + f_m$ = 1000 kHz + 2 kHz = 1002 kHz for the upper sideband, and $f_c - f_m$ = 1000 kHz – 2 kHz = 998 kHz for the lower sideband.
5. $f_c + f_m$ = 100 kHz + 20 kHz = 120 kHz, and $f_c - f_m$ = 100 kHz – 20 kHz = 80 kHz.
6. 0.707 x 100 volts = 70.7 volts.
   0.707 x 500 volts = 353.5 volts
7. Power gain = $\frac{P_{out}}{P_{in}}$. $P_{out}$ = 40 x 1;
   $P_{in}$ = 5 x 0.1; Power gain = $\frac{40}{0.5}$ = 80
8. 1) on – off; 2) gain; 3) frequency; and 4) time relationships.
9. Band-pass.
10. Automatic gain control.

### Chapter 3

1. Use Ohm's Law, E = I x R.
   E = 50 ohms x 1.5 amperes = 75 volts.
2. Use Ohm's Law, R = E / I.
   R = 500 volts / 200 amps = 2.5 ohms.
3. Use Ohm's Law, I = E / R.
   I = 125 volts / 75 ohms = 1.67 amperes.
4. Solution: $40\text{dB} = 20 \log_{10} \frac{V_{out}}{V_{in}}$
   $$\frac{40}{20} = \log_{10} \frac{V_{out}}{V_{in}}$$
   $$2 = \log_{10} \frac{V_{out}}{V_{in}}$$
   $$\therefore 10^2 = \frac{V_{out}}{V_{in}}$$
   $V_{out}$ is 100 times $V_{in}$.
5. Use the power equation for this problem: dB = 10 $\log_{10} P_{OUT} / P_{IN}$:
   $$\text{dB} = 10 \log_{10} \frac{2000}{2} = 10 \log_{10} 1000.$$
   $10^3$ = 1000, therefore: $\log_{10} 1000 = 3$
   ∴dB = 10 x 3 = 30

6. Use $dB = 20 \log_{10} \frac{V_{out}}{V_{in}}$.

   $dB = 20 \log_{10} \frac{100}{1} = 20 \log_{10} 100$

   Since $10^2 = 100$, $\log_{10} 100 = 2$

   $\therefore dB = 20 \times 2 = 40$.

7. Remember $\pi$ is a constant equal to 3.14.

   $X_L = 2 \pi f_L = 2 \times 3.14 \times 2 \times 10^6 \times 5 \times 10^{-6} = 6.28 \times 10 = 62.8$ ohms.

   Since there is no resistance, $Z = X_L \angle 90°$; therefore $Z = 62.8 \angle 90°$, which means it is a vector with a magnitude of 62.8 ohms at +90° to the zero axis.

8. $X_c = \frac{1}{2\pi f_c} = \frac{1}{2 \times 3.14 \times 2 \times 10^6 \times 0.01 \times 10^{-6}}$

   $= \frac{1}{6.28 \times 0.02} = \frac{1}{6.28 \times 2 \times 10^{-2}} =$

   $\frac{1}{12.56 \times 10^{-2}} = \frac{1}{1.256 \times 10^{-1}} = 0.796 \times 10^1 =$

   7.96 ohms.

   Since there is no resistance $Z = X_c \angle -90°$; therefore $Z = 7.96 \angle -90°$, which means it is a vector with a magnitude of 7.96 ohms at -90° to the zero axis.

9. Since $Z_L$ for the inductor is a vector at +90° to the zero axis and $Z_c$ of the capacitor is a vector at -90° to the zero axis, to arrive at the total series impedance, Z, the two quantities are added vectorally. Since $X_L$ is at +90° and $Z_L$ is at -90°, they oppose each other directly and therefore, $Z = Z_L - Z_C$ or $Z = 62.8 - 7.96 = 54.84$ ohms $\angle 90°$.

10. Anode, cathode.

11. Light-emitting diodes (LEDs) and photo transistors.

12. $h_{fe} = \frac{I_c}{I_b} = \frac{1.8 \text{ ma}}{0.01 \text{ ma}} = \frac{1.8}{1 \times 10^{-2}} =$

    $1.8 \times 10^2 = 180$.

13. Negative, negative.

14. $gm = \frac{5 \text{ ma}}{1.25V} = 4 \times 10^{-3}$ μmhos or 4000 μmhos.

15. Depletion.

## Chapter 4

1. Solution:

   $V_B = \frac{R3}{R3 + R4} \times V_{CC} = \frac{4400}{4400 + 22000} \times 10 =$

   $\frac{4400}{26,400} \times 10 = 1.66V$

   $V_E = V_B - 0.7V = 1.66 - 0.7 = 0.96V$

   $I_E = \frac{0.96V}{1200\Omega} = 0.8$ mA

   $I_C \cong I_E \therefore V_C = 10V - (0.8 \times 10^{-3} \times 4.7 \times 10^3) = 10V - 3.76V = 6.24V$

2. $h_{FE} = \frac{\Delta I_C}{\Delta I_B}$ large-signal current gain.

   $\beta = \frac{\Delta i_c}{\Delta i_b}$ small-signal current gain

3. Down 3 dB or 33 dB.

4. Emitter follower.

5. By cascading the stages.

6. Disadvantages are: a) low gain with resistive loads, and b) they are difficult to bias.

7. $A = -gm \times R_L = (8 \times 10^{+3} \times 10^{-6}) \times (10 \times 10^{+3}) = .008 \times 10 = 80$.

8. $R_P = QX_L \qquad X_L = 2\pi fL$

   $= 6.28 \times (20 \times 10^{+6}) \times (20 \times 10^{-6})$

   $= 6.28 \times 400 = 2512$ ohms

   $R_P = 20 \times 2512 = 50,240$ ohms

9. $f_r = \frac{1}{2\pi \sqrt{LC}}$ or $f_r^2 = \frac{1}{(2\pi)^2 LC}$

   $\therefore C = \frac{1}{(2\pi)^2 L f_r^2} =$

   $\frac{1}{(6.28)^2 (20 \times 10^{-6}) \times (20 \times 10^6)^2} =$

   $\frac{1}{(6.28)^2 \times 8 \times 10^7} = \frac{1}{3.94 \times 8 \times 10^8} =$

   $\frac{1}{3.152 \times 10^9} = 0.315 \times 10^{-9} =$

   $315 \times 10^{-12} = 315$ pF

10. $BW = \frac{f_r}{Q} = \frac{200 \text{ MHz}}{50} = 4$ MHz

11. $-\frac{Rf}{Rin} = -\frac{200 \text{ k}\Omega}{2 \text{ k}\Omega} = -100$. The minus sign means a 180° phase shift.

12. Class B.
13. Class C.
14. An additional 180° is required.
15. The error voltage input.

## Chapter 5

1. Information.
2. The sidebands frequencies are $f_c + f_m$ and $f_c - f_m$, or 1 MHz + 4 kHz = 1.004 MHz, and 1 MHz – 4 kHz = 0.996 MHz (or 996 kHz).
3. Since $P_{OUT} = P_C (1 + \frac{m^2}{2})$ and m = 0.5, the total Power, $P_{OUT} = P_C (1 + \frac{(0.5)^2}{2})$ or $P_{OUT} = P_C (1 + 0.125)$. Of the total Power Out, 1000 watts are in the carrier and 125 watts are in the sidebands. Each sideband has 62.5 watts.
4. The total bandwidth is the range of frequencies from $f_{low}$ to $f_{high}$. The modulation results in a frequency range from 1 MHz – 1 kHz = 0.999 MHz to 1 MHz + 1 kHz = 1.001 MHz. The difference between $f_{high}$ and $f_{low}$ is 1.001 MHz – 0.999 MHz = 2 kHz.
5. Carrier.
6. Balanced modulator.
7. Phase modulation (PM); frequency modulation (FM); and Angle modulation.
8. 1000 watts. The total amount of power in an FM signal is constant, regardless of modulation.
9. Use $M_{FM} = f_{c\,max} / f_{m\,max}$ to calculate the deviation ratios, as follows:

| $f_{c\,max}$ | $\div f_{m\,max}$ | $= M_{FM}$ |
|---|---|---|
| 120 kHz | ÷ 8 kHz | = 15 |
| 200 kHz | ÷ 20 kHz | = 10 |
| 70 kHz | ÷ 10 kHz | = 7 |
| 60 kHz | ÷ 12 kHz | = 5 |

10. Capture.

## Chapter 6

1. $F_{OUT}$ = 17Mhz + 5Mhz = 22Mhz and 17Mhz – 5Mhz = 12Mhz.
2. $F_{osc}$ = 1.8MHz + 0.455MHz = 2.255 MHz, or 1.8 MHz – 0.455 MHz = 1.245 MHz
3. $F_{img}$ = 11MHz + 2 x (9MHz) = 11 MHz + 18 MHz = 29MH
4. No. Diodes have no gain.
5. FETs have both a non-linear characteristic and gain required for mixing.
6. A <u>very stable</u>, <u>tunable</u> <u>local</u> oscillator.
7. 100 GHz.
8. –1.7V gate-to-source voltage allows 2mA of drain-to-source current; therefore, the bias resistor is:

   $R_Z = \frac{1.7V}{2mA} = \frac{1.7}{2 \times 10^{-3}} = 0.85 \times 10^3 =$ 850 ohms
9. A spectrum analyzer will show the signals at:

   100 MHz – 4 MHz = 96 MHz
   100 MHz – 2 MHz = 98 MHz
   100 MHz + 2 MHz = 102 MHz
   100 MHz + 4 MHz = 104 MHz
10. A <u>full-wave</u> <u>balanced</u> <u>multiplication</u> produces the mixing.

## Chapter 7

1. Frequency stability, signal quality, and reliability.
2. Class-C tuned amplifier.
3. To improve stability of the output signal.
4. AM modulation is produced at high power levels; SSB at low power levels.
5. <u>The</u> <u>frequency</u> <u>of</u> <u>the</u> <u>variable</u> <u>frequency</u> <u>oscillator</u> <u>(VFO)</u>.
6. To increase the signal frequency to the proper transmitting frequency.
7. In the power amplifier stage.
8. Since the output freqency = 108 MHz

   the divider ratio = $\frac{81.9 \times 10^6}{3.2 \times 10^5} =$ $25.6 \times 10^1 = 256$.
9. The driver and power output amplifiers must be linear amplifiers for SSB.
10. With three triplers, the total frequency multiplication is 9. Therefore, the VCO output frequency needs to be:

    $\frac{108\text{ MHz}}{9} = 12\text{ MHz}$

## Chapter 8

1. Noise Figure (NF).
2. The AGC maintains the receiver output to the detector at a near-constant level even though the input RF signal varies widely.
3. A beat-frequency oscillator (BFO).
4. Two mixers and two IF strips.
5. It limits the amplitude of the input signal to the detector, which eliminates any amplitude variations that might cause noise.
6. Recall from Chapter 6 that the result of mixing is the sum and difference frequencies. Thus, the local oscillator frequencies can be: 6.25 MHz + 2.5 MHz = 8.75 MHz, and 6.25 MHz – 2.5 MHz = 3.75 MHz.
7. Recall from Chapter 6 that image frequencies occur at the desired frequency plus 2 times the Intermediate Frequency (IF). Thus: 12.0 MHz + (455kHz x 2) = 12.91 MHz.
8. The audio output signals are the recovered baseband frequencies only; in this case 1kHz and 3kHz.
9. 14.7 MHz and 6.4 MHz
10. Error voltage.

## Chapter 9

1. By induction.
2. The electric field determines polarization.
3. A vertical antenna also is known as a Marconi antenna; a dipole also is sometimes called a Hertz antenna.
4. 1/2 wavelength and 1/4 wavelength.
5. The 11-year solar flux (or sun spot) cycle. High solar flux activity generally improves sky-wave propagation.
6. MUF is Maximum Usable Frequency.
7. 22,300 miles.
8. The impedance of the transmission line and antenna must be matched in order to assure efficient transfer of energy from transmitter to antenna.
9. A VSWR of 1.5 or less indicates efficient transfer of energy by a transmission line.
10. Since $Z_A = Z_L = 0.5\ Z_o$

    $$\rho = \frac{0.5 - 1}{0.5 + 1} = \frac{0.5}{1.5} = 0.333, \text{ and}$$

    $$0.333 = \frac{\text{sigma} - 1}{\text{sigma} + 1}$$

    $$\therefore \text{sigma} = \frac{1.333}{0.667} = 2$$

    VSWR = 2
11. Coupled power = Incident Power – Reflected Power

    850 = 1000 – 150

    Reflected Power = 150 watts
12. Losses per 100 ft. @ 400 MHz:

    | RG-8X | RG-8U | Difference |
    |---|---|---|
    | 8dB | 4.1dB | 3.9dB |

    For 1,000 ft. RG-8X will have 39 dB more losses than RG-8U at 400 MHz.

## Chapter 10

1. A miniature electronic circuit produced on and within a single piece of semiconductor material, usually silicon.
2. A test used to identify bad ICs so that only good ICs are passed along to the remaining manufacturing steps.
3. Burn-in is a test period in which the individually-packaged IC is placed under an electrical load and stressed to maximum and minimum limits at various temperatures. It is intended to weed-out weak ICs.
4. CMOS (Complementary-Metal-Oxide-Semiconductor) circuitry.
5. All of the reasons cited.
6. Either one based on an MOS process, or on a special process tailored to making both bipolar and MOS devices.
7. 4 to 12 inches.
8. Assembly is the step where the IC chips are handled individually for the first time.
9. Negative feedback from output to input.
10. The N+ buried layer is diffused into the silicon wafer substrate before the epitaxial layer is applied.

## Chapter 11

1. A/D converter, DSP Microprocessor, D/A converter.
2. Analog circuits are not programmable and are component-sensitive.
3. Filtering, frequency selection, and noise reduction.
4. Software allows the same hardware to be used to solve many different applications by changing the program that directs the microprocessor.
5. The modifying of the input signals by the DSP microprocessor under the direction of a program and outputting an analog signal occurs as the input signal changes with continuous time.
6. The period of a wave form is $T = \frac{1}{f}$. Therefore, for a frequency of two billion cycles per second, $T = \frac{1}{2 \times 10^9} =$ $0.5 \times 10^{-9}$. $10^{-9}$ is nanoseconds. Therefore, T = 0.5nS.
7. Since a signal must be sampled at least twice per cycle to reproduce it accurately, 2 MHz.
8. A stair-step output voltage increasing and decreasing in value as the codes change.
9. Smoothing filter.
10. Input and output interface circuits and controller are added.

# Schematic Symbols

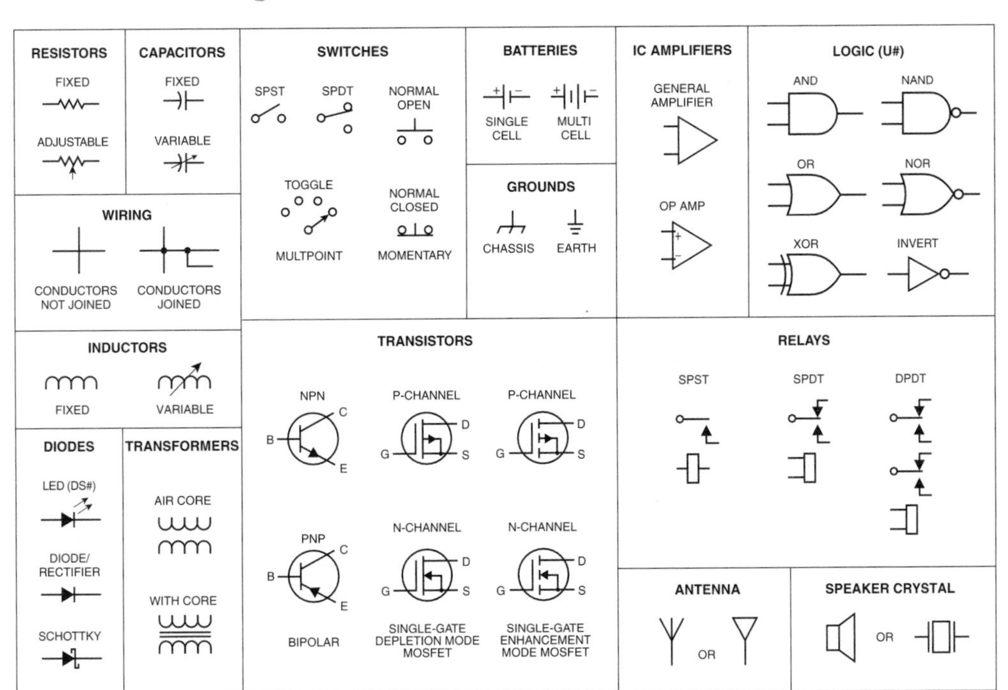


# Scientific Notation

| Prefix | Symbol | | | Multiplication Factor |
|---|---|---|---|---|
| exa | E | $10^{18}$ | = | 1,000,000,000,000,000,000 |
| peta | P | $10^{15}$ | = | 1,000,000,000,000,000 |
| tera | T | $10^{12}$ | = | 1,000,000,000,000 |
| giga | G | $10^{9}$ | = | 1,000,000,000 |
| mega | M | $10^{6}$ | = | 1,000,000 |
| kilo | k | $10^{3}$ | = | 1,000 |
| hecto | h | $10^{2}$ | = | 100 |
| deca | da | $10^{1}$ | = | 10 |
| (unit) | | $10^{0}$ | = | 1 |

| Prefix | Symbol | | | Multiplication Factor |
|---|---|---|---|---|
| deci | d | $10^{-1}$ | = | 0.1 |
| centi | c | $10^{-2}$ | = | 0.01 |
| milli | m | $10^{-3}$ | = | 0.001 |
| micro | μ | $10^{-6}$ | = | 0.000001 |
| nano | n | $10^{-9}$ | = | 0.000000001 |
| pico | p | $10^{-12}$ | = | 0.000000000001 |
| femto | f | $10^{-15}$ | = | 0.000000000000001 |
| atto | a | $10^{-18}$ | = | 0.000000000000000001 |

# Metric Conversions

## International System of Units (SI) – Metric Units

### Power

| | Btu/h | ft – lb/s | hp | kW | Watt |
|---|---|---|---|---|---|
| 1 British thermal unit per hour = | 1 | 0.2161 | 3.929 $\times 10^{-4}$ | 2.930 $\times 10^{-4}$ | 0.2930 |
| 1 foot-pound per second = | 4.628 | 1 | 1.818 $\times 10^{-3}$ | 1.356 $\times 10^{-3}$ | 1.356 |
| 1 horsepower = | 2545 | 550 | 1 | 0.7457 | 745.7 |
| 1 kilowatt = | 3413 | 737.6 | 1.341 | 1 | 1000 |
| 1 Watt = | 3.413 | 0.7376 | 1.341 $\times 10^{-3}$ | 0.001 | 1 |

### Mass

| | gm | kg | oz | lb |
|---|---|---|---|---|
| 1 gram = | 1 | 0.001 | 3.527 $\times 10^{-2}$ | 2.205 $\times 10^{-3}$ |
| 1 Kilogram = | 1000 | 1 | 35.27 | 2.205 |
| 1 ounce = | 28.35 | 2.835 $\times 10^{-2}$ | 1 | 6.250 $\times 10^{-2}$ |
| 1 pound = | 453.6 | 0.4536 | 16 | 1 |

### Length

| | cm | Meter | km | in. | ft | mi |
|---|---|---|---|---|---|---|
| 1 centimeter = | 1 | $10^{-2}$ | $10^{-5}$ | 0.3937 | 3.281 x $10^{-2}$ | 6.214 $\times 10^{-6}$ |
| 1 Meter = | 100 | 1 | $10^{-3}$ | 39.3 | 3.281 | 6.214 $\times 10^{-4}$ |
| 1 kilometer = | $10^5$ | 1000 | 1 | 3.937 $\times 10^4$ | 3281 | 0.6214 |
| 1 inch = | 2.540 | 2.540 $\times 10^{-2}$ | 2.540 $\times 10^{-5}$ | 1 | 8.333 x $10^{-2}$ | 1.578 $\times 10^{-5}$ |
| 1 foot = | 30.48 | 0.3048 | 3.048 $\times 10^{-4}$ | 12 | 1 | 1.894 $\times 10^{-4}$ |
| 1 mile = | 1.609 $\times 10^5$ | 1609 | 1.609 | 6.336 $\times 10^4$ | 5280 | 1 |

### Force

| | dyne | Newton | lb |
|---|---|---|---|
| 1 dyne = | 1 | $10^{-5}$ | 2.248 $\times 10^{-6}$ |
| 1 Newton = | $10^5$ | 1 | 0.2248 |
| 1 pound = | 4.448 $\times 10^5$ | 4.448 | 1 |

# 0°–90° Trigonometric Functions

| ANGLE | SIN | COS | TAN |
|---|---|---|---|
| 0° | 0.0000 | 1.000 | 0.0000 |
| 1 | .0175 | .9998 | .0175 |
| 2 | .349 | .9994 | .0349 |
| 3 | .0523 | .9986 | .0524 |
| 4 | .0698 | .9976 | .0699 |
| 5 | .0872 | .9962 | .0875 |
| 6 | .1045 | .9945 | .1051 |
| 7 | .1219 | .9925 | .1228 |
| 8 | .1392 | .9903 | .1405 |
| 9 | .1564 | .9877 | .1584 |
| 10 | .1736 | .9848 | .1763 |
| 11 | .1908 | .9816 | .1944 |
| 12 | .2079 | .9781 | .2126 |
| 13 | .2250 | .9744 | .2309 |
| 14 | .2419 | .9703 | .2493 |
| 15 | .2588 | .9659 | .2679 |
| 16 | .2756 | .9613 | .2867 |
| 17 | .2924 | .9563 | .3057 |
| 18 | .3090 | .9511 | .3249 |
| 19 | .3256 | .9455 | .3443 |
| 20 | .3420 | .9397 | .3640 |
| 21 | .3584 | .9336 | .3839 |
| 22 | .3746 | .9272 | .4040 |
| 23 | .3907 | .9205 | .4245 |
| 24 | .4067 | .9135 | .4452 |
| 25 | .4226 | .9063 | .4663 |
| 26 | .4384 | .8988 | .4877 |
| 27 | .4540 | .8910 | .5095 |
| 28 | .4695 | .8829 | .5317 |
| 29 | .4848 | .8746 | .5543 |
| 30 | .5000 | .8660 | .5774 |

| ANGLE | SIN | COS | TAN |
|---|---|---|---|
| 31° | .5150 | .8572 | .6009 |
| 32 | .5299 | .8480 | .6249 |
| 33 | .5446 | .8387 | .6494 |
| 34 | .5592 | .8290 | .6745 |
| 35 | .5736 | .8192 | .7002 |
| 36 | .5878 | .8090 | .7265 |
| 37 | .6018 | .7986 | .7536 |
| 38 | .6157 | .7880 | .7813 |
| 39 | .6293 | .7771 | .8098 |
| 40 | .6428 | .7660 | .8391 |
| 41 | .6561 | .7547 | .8693 |
| 42 | .6691 | .7431 | .9004 |
| 43 | .6820 | .7314 | .9325 |
| 44 | .6947 | .7193 | .9657 |
| 45 | 0.7071 | 0.7071 | 1.0000 |
| 46 | .7193 | .6947 | 1.0355 |
| 47 | .7314 | .6820 | 1.0724 |
| 48 | .7431 | .6691 | 1.1106 |
| 49 | .7547 | .6561 | 1.1504 |
| 50 | .7660 | .6428 | 1.1918 |
| 51 | .7771 | .6293 | 1.2349 |
| 52 | .7880 | .6157 | 1.2799 |
| 53 | .7986 | .6018 | 1.3270 |
| 54 | .8090 | .5878 | 1.3764 |
| 55 | .8192 | .5736 | 1.4281 |
| 56 | .8290 | .5592 | 1.4826 |
| 57 | .8387 | .5446 | 1.5399 |
| 58 | .8480 | .5299 | 1.6003 |
| 59 | .8572 | .5150 | 1.6643 |
| 60 | .8660 | .5000 | 1.7321 |

| ANGLE | SIN | COS | TAN |
|---|---|---|---|
| 61° | .8746 | .4848 | 1.8040 |
| 62 | .8829 | .4695 | 1.8807 |
| 63 | .8910 | .4540 | 1.9626 |
| 64 | .8988 | .4384 | 2.0503 |
| 65 | .9063 | .4226 | 2.1445 |
| 66 | .9135 | .4067 | 2.2460 |
| 67 | .9205 | .3907 | 2.3559 |
| 68 | .9272 | .3746 | 2.4751 |
| 69 | .9336 | .3584 | 2.6051 |
| 70 | .9397 | .3420 | 2.7475 |
| 71 | .9455 | .3256 | 2.9042 |
| 72 | .9511 | .3090 | 3.0777 |
| 73 | .9563 | .2924 | 3.2709 |
| 74 | .9613 | .2756 | 3.4874 |
| 75 | .9659 | .2588 | 3.7321 |
| 76 | .9703 | .2419 | 4.0108 |
| 77 | .9744 | .2250 | 4.3315 |
| 78 | .9781 | .2079 | 4.7046 |
| 79 | .9816 | .1908 | 5.1446 |
| 80 | .9848 | .1736 | 5.6713 |
| 81 | .9877 | .1564 | 6.3138 |
| 82 | .9903 | .1392 | 7.1154 |
| 83 | .9925 | .1219 | 8.1443 |
| 84 | .9945 | .1045 | 9.5144 |
| 85 | .9962 | .0872 | 11.43 |
| 86 | .9976 | .0698 | 14.30 |
| 87 | .9986 | .0523 | 19.08 |
| 88 | .9994 | .0349 | 28.64 |
| 89 | .9998 | .0175 | 57.29 |
| 90 | 1.0000 | .0000 | ∞ |

# Glossary



**Active device:** A device capable of producing power gain or control; for example: bipolar transistors, field-effect transistors, vacuum tubes, and saturable reactors.
**Alternating current (ac):** An electrical current that periodically changes in magnitude and direction.
**Alternation:** Either half of a cycle of alternating current. It is the time period during which the current increases from zero to its maximum value (in either direction) and decreases to zero.
**Ampere (A):** The unit of measurement for electrical current in coulombs ($6.25 \times 10^{18}$ electrons) per second. One ampere results in a circuit that has one ohm resistance when one volt is applied to the circuit. One ampere is 1 coulomb of change per second.
**Amplification:** See Gain.
**Amplifier:** An electrical circuit designed to increase the current, voltage, or power of an applied signal.
**Amplitude:** The magnitude of an electrical signal or waveform above or below some reference, often ground or zero. Usually measured in volts or current.
**Analog:** Being continuous or having a continuous range of values, as opposed to having discrete values.
**Analog-to-Digital Conversion or Converter (ADC or A/D):** The process of converting an analog signal to a digital code that represents the amplitude of the original signal.
**Anode:** The positive electrode of an electrical device.
**ASCII:** American Standard Code for Information Interchange. The most widely used alphanumeric code. A commonly used 7-bit code which represents numbers, letters, and control characters. It is used for transferring information between digital systems, especially those which otherwise would be incompatible.
**Audio and audio frequency (AF):** The range of frequencies normally heard by the human ear. Typically, about 20 to 20,000 Hz.

**Balanced Detector:** A type of demodulator used in FM systems that has tuned circuits above and below the carrier frequency.
**Balanced Modulator:** A circuit used in SSB in which only the upper and lower sidebands are output and the carrier is suppressed.
**Band-pass filter:** A device or circuit that allows passage of a band of signals between two frequencies and attenuates signals above and below those frequencies.
**Bandwidth:** A specified range of frequencies from $f_{low}$ to $f_{high}$ required to accurately represent the information in a signal.
**Base:** The region between the emitter and collector which received minority carriers injected from the emitter.
**Baseband signal:** A signal with all of the frequencies that contain the information.
**Beta (β):** The ac or small-signal current gain of a transistor when connected in a common-emitter circuit configuration.
**Bias:** In an electronic circuit, a voltage or current applied to an active device (bipolar transistor, FET, etc.) to set the steady-state operating point of the circuit.
**Binary system:** A system of mathematical computation based on powers of two.
**Binary system logic:** The use of two binary states, usually "one" and "zero" or ON and OFF, to form electronic logic circuits.
**Bipolar:** A semiconductor device having both majority and minority carriers within the transistor structure.
**BJT:** Bipolar junction transistor. A semiconductor device used for switching or amplification. A BJT has two junctions, the base-emitter junction and the base-collector junction.
**Block diagram:** A system diagram which shows the relationship between the main functional units of the system represented by blocks.
**Breakdown:** The condition for a reverse-biased semiconductor junction when its high resistance, under the reverse bias, suddenly decreases, causing excessive current. Not necessarily destructive.
**Bridge rectifier:** A full-wave rectifier in which the rectifier diodes are connected in a bridge circuit to allow current to the load during both the positive and negative alternation of the supply voltage.

**Capacitance (C):** The capability to store a charge in an electrostatic field. It can be expressed as equal to the charge (Q) in coulombs that is stored divided by the voltage (E) in volts that supplied the charge. Capacitance tends to oppose any change in voltage. The unit is farad.
**Capacitive reactance ($X_c$):** The opposition that a capacitor offers to a time changing signal or supplied voltage. Its value is $X_c = 1/2\pi fC$
**Capacitor (C):** A device made up of two metallic plates separated by a dielectric or insulating material. Used to store electrical energy in the electrostatic field between the plates.
**Carrier:** An electronic signal of constant amplitude, frequency, and phase before modulation. The amplitude, frequency, or phase of the wave is modified to superimpose information upon it (modulated) for transmission. Thus the signal becomes the carrier of the information.
**Cathode (K):** The negative electrode of an electrical device.
**Circuit:** A complete path that allows electrical current from one terminal of a voltage source to the other terminal.
**Clock or Clock generator:** An electronic circuit that generates accurate and precisely controlled, regularly occurring synchronizing or timing signals called clock signals.
**Clock rate:** The frequency of oscillation of the master clock, or oscillator, in a system.
**CMOS (Complementary-Metal-Oxide Semiconductor):** A type of transistor circuit which uses both p-type and n-type field-effect transistors.
**Coil:** The component that is formed when several turns of wire are wound on a cylindrical form. The core may be air or ferromagnetic material.
**Collector (C):** The element in a transistor that collects the moving electrons or holes, and from which the output usually is obtained. Analagous to the plate of a triode vacuum tube.
**Collector cutoff:** The operating condition of a transistor when collector current is reduced to the leakage current of the collector-base junction.
**Collector junction:** The junction between the base and collector regions of a transistor, normally biased in the reverse direction.
[1]**Common-base amplifier:** An amplifier with the transistor connected into the external circuit so that the base electrode is common to the input and output circuits.
[1]**Common-collector amplifier:** An amplifier with the transistor connected into the external circuit so that the collector electrode is common to the input and output circuits.
[1]**Common-emitter amplifier:** An amplifier with the transistor connected into the external circuit so that the emitter electrode is common to the input and output circuits.
**Comparator:** An electronic circuit or IC that compares the magnitudes of two input signals and produces an output indicating the relationship of the signal amplitudes
**Component:** The individual parts that make up a circuit, a function, a subsystem or a total piece of equipment.
**Composite Signal:** All of the elements of a signal that make up a complete signal. In video, the composite signal includes audio, video, and timing signals.
**Conductivity:** The ability of a material to conduct current.
**Conductor:** A substance through which electrons flow with relative ease.
**Continuity:** A continuous electrical path.
**Coulomb (C):** The unit of electrical charge, made up of a quantity of $6.25 \times 10^{18}$ electrons.
**Current (I):** The flow of electrons, measured in amperes. One ampere results when one volt is impressed on a circuit that has a resistance of one ohm.

**DC operating point:** The dc values of element voltages and currents of a transistor with no signal applied.
**Decibel (dB):** One tenth of a bel. The standard unit for expressing the ratio between powers $P_1$ and $P_2$. $dB = 10\log_{10}P_1/P_2$.
**Depletion Mode:** A type of MOSFET in which the already existing current from drain to source is turned off (depleted) when a voltage is applied to the gate.
**Detector:** A stage or circuit in a radio or TV receiver that extracts (demodulates) the information from the composite signal.
**Dielectric:** The non-conducting material used to separate the plates

[1]Common means "connected to" or "grounded"

of a capacitor or for insulating electric contacts.
**Digital signal:** A signal whose level has only discrete values, like on or off, 1 or 0, +5V or +0.2V.
**Digital to Analog Conversion (or Converter) DAC or D/A):** A circuit that converts a digital input signal to an analog output signal.
**Diode:** A device which has two terminals and has a high resistance to current in one direction and a low resistance to current in the other direction. It acts as a one-way valve for current.
**DIP:** Dual In-Line Package with pins in line on each side of the package, either plastic or ceramic, used to package ICs.
**Direct Current (dc):** Current in a circuit in one direction only. Usually a non-changing or average value.
**Discriminator:** A circuit that demodulates an FM signal; thus, an FM detector.
**Doping:** In semiconductor fabrication, the addition of controlled minute amounts of impurities to create a particular type of semiconductor.
**Drain:** The element in field-effect transistors which is roughly analagous to the collector of a bipolar transistor.

**Effective value:** The value of ac current that will produce the same heating effect in a load resistor as an equivalent dc current.
**Electricity:** A form of energy produced by the flow of electrons through materials and devices under the influence of an electromotive force produced electrostatically, mechanically, chemically or thermally.
**Electrolytic capacitor:** A capacitor whose electrodes are immersed in a wet electrolyte or dry paste.
**Electromagnetic radiation:** Radiation or a wave made up of both electric and magnetic fields that travel in free space at the speed of light. Radio waves, light, X-rays, and infrared waves are examples of electromagnetic radiation.
**Electromotive force (E):** The force which causes an electrical current in a circuit when there is a difference in potential. Synonym for voltage.
**Electron:** The basic atomic particle having a negative charge that rotates around a positively charged nucleus of an atom.
**Electrostatic field:** The electrical field or force surrounding objects that have an electrical charge.
**Emitter (E):** The semiconductor material in a transistor that emits carriers into the base region when the emitter-base junction is forward biased.
**Emitter junction:** The junction between the emitter and base regions of a transistor, normally biased in the forward direction to induct carriers in the base region.
**Enhancement Mode:** A type of MOSFET in which current is produced and increased as an increasing voltage is applied from gate to source.

**Farad (F):** The basic unit for capacitance. A capacitor has a value of one farad when it has stored one coulomb of charge with one volt across it.
**Field coil:** An electromagnet formed from a coil of insulated wire wound around a soft iron core. Commonly used in motors and generators.
**Feedback – positive and negative:** Returning a portion of a signal from the output to the input. Positive feedback reinforces the signal. Negative feedback cancels the signal.
**Field-Effect Transistor (FET):** A 3-terminal semiconductor device where current is from source to drain due to a conducting channel formed by a voltage field between the gate and the source.
**Filament:** The heated element in an incandescent lamp or vacuum tube.
**Filter:** A circuit element or group of components which passes signals of certain frequencies while blocking signals of other frequencies.
**Forward current:** The net current that flows across the forward-biased PN junction.
**Forward voltage (or bias):** A voltage applied across a semiconductor junction in order to permit forward current through the junction and the device.
**Frequency (F or f):** The number of completed cycles of a periodic waveform during one second. The unit of frequency is the Hertz.
**Frequency Response:** A plot of how a circuit or device responds to different frequencies. Usually refers to the amplitude response of the output as a constant amplitude, varying frequency signal is applied to the input.

**Gain (G):** 1. Any increase in the current, voltage or power level of a signal. 2. The ratio of output to input signal level of an amplifier.
**Gate electrode:** The control electrode of a FET, MOS transistor, or other semiconductor devices like controlled rectifier, or triac.
**Ground (or Grounded):** 1. The common return path for electric current in electronic equipment. Called electrical ground. 2. A reference point connected to, or assumed to be at zero potential with respect to the earth.

**Heat sink:** A material (usually a metal) placed in contact with a hot body to increase the flow of heat away from the body.
**Henry (H or h):** The unit of inductance. The inductance of a coil of wire in henries is a function of the coil's size, the number of turns of wire and the type of core material.
**Hertz (Hz):** One cycle per second.
**Heterodyne:** A method of signal mixing in which a signal from a variable-frequency local oscillator (VFO) is mixed with the incoming signal to create a signal at an intermediate frequency, which is then processed, amplified, and demodulated.
**$h_{FE}$:** The dc or large-signal current gain of a transistor connected in a common-emitter circuit.
**High-pass filter:** A device or circuit that allows passage of signals above a certain frequency and attenuates signals below that frequency.

**IF Amplifier:** The amplifier stage in a receiver that amplifies intermediate frequency signals.
**Impedance (Z):** In a circuit, the opposition that circuit elements present to alternating current. The impedance includes both resistance and reactance.
**Inductance (L):** The capability of a coil to store energy in a magnetic field surrounding it which results in a property that tends to oppose any change in the existing current in the coil.
**Inductive reactance ($X_L$):** The opposition that an inductance offers when there is an ac or pulsating dc in a circuit. $X_L = 2\pi fL$.
**Input:** The signal or line going into a circuit. A signal that controls the operation of a circuit.
**Input impedance:** The impedance seen by a source when a device or circuit is connected across the source.
**Integrated circuit (IC):** A complex semiconductor structure that contains all the circuit components for a high-functional-density analog or digital circuit interconnected together on a single chip of silicon.

**Junction:** The region separating two layers in a semiconductor material; e.g., a p-n junction.
**Junction diode:** A two-terminal, small-area PN junction used for electronic-circuit functions (as opposed to a junction rectifier).
**Junction transistor:** A PNP or NPN transistor formed from three alternate regions of p and n type material. The alternate materials are formed by diffusion or ion implantation.

**Leakage (or Leakage current):** The undesired flow of electricity around or through a device or circuit. In the case of semiconductors, it is the current across a reverse-biased semiconductor junction.
**Light-emitting diode (LED):** A PN junction that emits light when biased in the forward direction.
**Limiter:** An over-driven amplifier which clips the top and bottom of input sine-wave signals. It is used to remove amplitude variations from an FM signal.
**Linear amplifier:** A class A or B amplifier whose output signal is directly proportional to the input signal. The output is an exact reproduction of the input except for the increased gain.
**Load:** Any component, circuit, subsystem or system that consumes power delivered to it by a source of power.
**Load line:** A line drawn on the collector characteristic curves of a transistor on which the operating point of the transistor moves as collector current changes. It is called a load line because the slope of the line depends on the value of the collector load resistance.
**Loop:** A closed path around which there is a current or signal.
**Low-pass filter:** A device or circuit that allows passage of signals below a certain frequency and attenuates signals above that frequency.
**Low-pass filter:** An L-C network that allows only the frequencies below a specified cutoff point to pass.

**Magnetic Field:** The force field surrounding a magnet.
**Magnetic lines of force:** The imaginary lines called flux lines used to indicate the directions of the magnetic forces in a magnetic field.
**Magnitude:** The amplitude or value of a quantity.

**Megohm (MΩ):** A million ohms. Sometimes abbreviated meg.
**Metal-oxide semiconductor (MOS) transistor:** An active semiconductor device in which conduction is controlled in a region between two electrodes by a voltage applied to an insulated electrode over the region.
**Microampere (mA):** One millionth of an ampere.
**Microelectronics:** A broad term covering the entire field of silicon integrated circuits, thick- and thin-film circuits, and any special component of an extremely small nature.
**Microfarad (mF):** One millionth of a farad.
**Microprocessor:** An IC that can be programmed to perform arithmetic, logic, and other operations and to process data in a specified manner. It is the nerve center of a digital system.
**Milliampere (mA):** One thousandth of an ampere.
**Millihenry (mH):** One thousandth of a henry.
**Milliwatt (mW):** One thousandth of a watt.
**Mixer:** The circuit in a receiver that mixes a local oscillator signal with an incoming signal to produce an intermediate frequency (IF) signal.
**Mixing:** The function of multiplication of signals of different frequencies to produce signals at the sum and difference frequencies of the original signals.
**Modulation:** The process of impressing information on a carrier signal by varying the carrier's amplitude, frequency, or phase.
**Monolithic:** Formed within a single body of material.
**Multiplexer (MUX):** An electronic circuit capable of arranging signals from several sources in time or frequency on a single line or transmission path.

**N-channel FET:** A field-effect transistor in which a channel in the P-type region between an N-type source and an N-type drain is formed by applying voltage between gate and source so that there is current between drain and source.
**Noise:** Any unwanted signal or electromagnetic radiation, particularly that which disrupts normal operation, that is added to signals containing original information.
**NPN Transistor:** A bipolar transistor with a p-type base sandwiched between an n-type emitter, and an n-type collector.
**N-type semiconductor material (N):** A semiconductor material in which the majority carriers are electrons, and there is an excess of electrons over holes.

**Ohm (Ω):** The unit of electrical resistance. A circuit component has a resistance of one ohm when one volt applied to the component produces a current of one ampere.
**Ohms-per-volt:** The sensitivity rating for a voltmeter. Also expresses the impedance (resistance) presented to a circuit by the meter when a voltage measurement is made.
**Open circuit:** An incomplete path for current.
**Operating point:** The steady-state or no-signal operating point of a circuit or active device. Also called the quiesence point.
**Operational amplifier (OP AMP):** A high-gain analog amplifier with an inverting and non-inverting input and one output.
**Oscillation:** A sustained condition of continuous operation where the circuit outputs a constant signal at a frequency determined by circuit constants and as a result of positive or regenerative feedback.

**Parallel circuit:** A circuit connected so that there is current in two or more parallel branches. The branches are said to be in parallel when they have common connections at each end and the same voltage across the branches.
**Passive device:** A device not capable of producing gain; for example, resistors and capacitors.
**Period:** The time required for a periodic waveform to complete a cycle and repeat itself.
**P-channel FET:** A field-effect transistor in which a channel in the N-type region between P-type source and P-type drain is formed by applying voltage between gate and source so that there is current between drain and source.
**Phase:** The time relationship of signals to each other; graphically represented with a vector.
**PLL (Phase Lock Loop):** A closed-loop electronic circuit that automatically adjusts and locks the frequency of an oscillator to the correct frequency.
**Phase reversal:** A 180-degree change in phase.
**Photodiode:** A junction diode that utilizes the photosensitivity of a PN junction.
**Phototransistor:** A photodetector that incorporates transistor action to give an amplified output.
**Pi (π):** The mathematical constant which is equal to the ratio of the circumference of a circle to its diameter. Approximately 3.14.
**Picofarad (pF):** A unit of capacitance that is $1 \times 10^{-12}$ farads or one millionth of a millionth of a farad.
**PN junction:** The region of transition between P-type and N-type material in a semiconductor crystal.
**PNP Transistor:** A bipolar transistor with an n-type base sandwiched between a p-type emitter and a p-type collector.
**Polarity:** The description of whether a voltage is positive or negative with respect to some reference point.
**Potential difference:** The voltage difference between two points, calculated algebraically.
**Power (P):** The time rate of doing work.
**Power (reactive):** The product of the voltage and current in a reactive circuit measured in volt-amperes (apparent power).
**Power (real):** The power dissipated in the purely resistive components of a circuit measured in watts.
**Power dissipation:** The dispersion or giving up of power in the form of heat.
**Power supply:** A defined unit that is the source of electrical power for a device, circuit, subsystem or system.
**Program:** A list of computer instructions arranged to achieve a specific result. A generic name is software.
**P-type semiconductor material (P):** A semiconductor material in which holes are the majority carriers and there is a deficiency of electrons.
**Push-pull circuit:** A two transistor output stage in which the outputs of the transistors are 180 degrees out of phase but the outputs are in parallel.

**Q, Quality Factor:** A measure of the sharpness of frequency selectivity of timed circuit's frequency response. In a parallel resonant circuit it is equal to $R_P/X$, and in a series resonant circuit it is equal to $X/R_S$.

**Reactance (X):** The opposition that a pure inductance or a pure capacitance provides to current in an ac circuit.
**Rectification:** The process of converting alternating current into pulsating direct current.
**Relay:** A device in which a set of contacts is opened or closed by a mechanical force supplied by turning on current in an electromagnet. The contacts are isolated from the electromagnet.
**Resistance (R):** A characteristic of a material that opposes the flow of electrons. It results in loss of energy in a circuit dissipated as heat.
**Resistivity:** The resistance of a material expressed as ohms per square.
**Resistor (R):** A circuit component that provides resistance to current in the circuit.
**Reverse bias:** An external voltage applied in the nonconducting, reverse-biased direction of a PN junction. The connections are opposite to those for forward bias.
**Reverse-breakdown voltage:** The reverse-biased voltage across a PN juction that exceeds the value that produces a sharp increase in reverse current without a significant increase in voltage.
**Reverse current:** The current when a semiconductor junction is reverse biased.
**Root-Mean-Square (RMS):** See effective value. The RMS value of an ac sinusoidal waveform is 0.707 of the peak amplitude of the sine wave.

**Saturation:** The operating condition of a transistor when an increase in base current produces no further increase in collector current.
**Selectivity:** The ability of a communications circuit or system to select the desired signal from those not wanted.
**Sensitivity:** The ability of a communications circuit or system to detect "weak" input signals with an acceptable S/N ratio.
**Semiconductor:** One of the materials falling between metals as good conductors and insulators as poor conductors in the periodic chart of the elements.
**Series Circuit:** In electrical and electronic circuits, arranging the circuit components so that there is the same current through all components in the circuit.
**Sideband – upper and lower:** The frequency bands formed on both sides of a modulated carrier by the sum and difference of the information frequency and the carrier frequency.
**Signal:** In electronics, the information contained in electrical quantities of voltage or current that forms the input, timing, or output of a device, circuit, or system.
**Silicon Controlled Rectifier (SCR):** A semiconductor diode in which current through a third element, called the gate, controls turn-on, and the anode-to-cathode voltage controls turn-off.

**Sine (sinusoidal) wave:** A waveform whose amplitude at any time through a rotation of an angle from 0° to 360° is a function of the sine of an angle.
**SMT:** Surface mount technology. An IC packaging technique in which the packages are smaller than DIPs. The leads from the package are designed to be mounted directly to the printed surface of the circuit board by solder reflow.
**Source:** The second region of a field-effect transistor separated by a gate region from the drain region.
**Spectrum:** The complete range of electromagnetic radiations, from the longest wavelength to the shortest.
**Step-down transformer:** A transformer in which the secondary winding has fewer turns than the primary.
**Step-up transformer:** A transformer in which the secondary winding has more turns than the primary.

**Thermistor:** A resistor whose resistance changes with temperature.
**Threshold voltage:** The minimum voltage applied to the gate electrode of a MOS transistor that initiates conduction between the source and drain.
**Transformer:** A set of coils of wire wound together on an iron core in which a magnetic field couples energy between coils.
**Transistor:** A three-terminal semiconductor device used in circuits to amplify electrical signals or to perform as a switch to provide digital functions.
**Triac:** A device, similar to the controlled rectifier, in which both the forward and reverse characteristics can be triggered from blocking to conducting.
**Truth table:** A chart which tabulates and summarizes all the combinations of possible states of the inputs and outputs of a circuit. It tabulates what will happen at the output for a given input combination.
**Tuned-circuit Amplifier:** An LC (inductor-capacitor) circuit with one or more variable devices that can be adjusted to resonate at a desired frequency.
**Turns ratio:** The ratio of secondary winding turns to primary winding turns of a transformer.

**Vector:** A line representing the magnitude and time phase of some quantity, plotted on rectangular or polar coordinates.
**Voltage (or Volt):** The unit of electromotive force that causes current when included in a closed circuit. One volt causes a current of one ampere through a resistance of one ohm.
**Voltage drop:** The difference in potential between two points caused by a current through an impedance or resistance.

**Watt (W):** The unit of electrical power in joules per second, equal to the voltage drop (in volts) times the current (in amperes) in a resistive circuit.

**Zener diode:** A junction diode with a sharp reverse breakdown at low voltages, between 3 and 15V. Commonly used as a voltage reference.

# Index


A/D (analog-to-digital) converter: 200, 201
AC power: 31, 54
Active devices: 34, 50-53
- Bipolar transistors: 50
  - NPN: 50
  - PNP: 51
- FET (Field Effect Transistors): 52
  - Depletion mode: 52
  - Enhancement mode: 52
- MOSFET (Metal-Oxide-Semiconductor FET): 52
  - Depletion mode: 53
  - Enhancement mode: 53

Amplifiers:
- BJT tuned amplifier: 72
- Cascaded: 65
  - with FETs: 65
- Class A: 76
- Class B: 76
- Class C: 78
- Common-base: 64
- Common-collector: 64
- Common-emitter: 60
  - Frequency response: 64
  - Rules for biasing: 61
- FET tuned: 70
- High-frequency: 67
- Linear RF: 80
- Op Amp: 35, 72
- Power amplifier: 33, 76
  - Complementary-symmetry: 76
  - Linear RF: 80
  - Push-pull: 78
- Quality Factor (Q): 69
- Resonance: 67
- RF tuned: 71
- Small signal: 34, 59
  - FETs in: 65
- Stage: 60
- $r_e'$ (re prime): 62
- Tuned circuit: 67

Amplitude Modulation (AM): 18, 87-92
Analog defined: 1
Analog vs. digital ICs: 194
Angle Modulation: 97
Antennas: 157-166
- 4-square phased array: 165
- Beam: 161
- Characteristics of: 160
- Dipole: 161
- Directivity: 164
- Driven element: 159
- Electromagnetic fields: 157
- Element size: 161
- Field strength: 166
- Gain: 163
- Impedance: 166
- Isotropic radiator: 163
- Polarization: 160
- Propagation: 167
- Radiation patterns: 162
- Receiving: 158
- Transmitting: 157
- Vertical 1/4 wave: 161, 166
- Wavelength: 161
- Yagi-Uda: 161, 164

Automatic gain control (AGC): 39, 141

Band pass: 39
Bandwidth: 11, 103
Barkhausen Criterion: 80
Baseband amplifier: 124, 141
Baseband signal defined: 23
Beat-frequency oscillator (BFO): 94
Biasing: 34, 53
Bipolar transistors: 50
BJT tuned amplifier: 72
Buffer: 122, 131
Burn-in: 190

Capacative Reactance: 48
Capacitors: 48
Capture: 104
Carrier: 17, 26
Carson's Rule: 103
Cascade: 65
Characteristic impedance: 173-175
- Equations: 173

Coaxial cable: 171
Colpitts oscillator: 82
Common-base: 64
Common-collector: 64
Common-emitter: 60
Communication system defined: 1
Converter: 109
Coupling:
- AC: 54
- DC: 54
- Optical: 55
- Transformer: 54

Current: 32
CW (Continuous Wave): 87

D/A (digital-to-analog) converter: 200, 201
DC no-signal point: 34
DC power: 31, 54
Decibel: 46
Depletion mode: 52
Detection: 30, 141, 146
Detector: 141
Diffusion process: 186
Digital CMOS: 183
Digital defined: 1
Digital Signal Processing (DSP): 199
- A/D (analog-to-digital) converter: 200, 201
- applications: 200, 205
- block diagram: 199
- D/A (digital-to-analog) converter: 200, 201
- MAC (multiply and accumulate): 200
- microprocessors: 200, 203
- sampling illustrated: 202
- sampling rate: 203
- software: 201

Diode: 50, 112
Diode-balanced modulator: 97
Dipole: 161
Direct waves: 167
Directivity: 164
Discriminator: 146
Driven element: 159
Driver stage: 123
Dual-conversion receiver: 141
Dynamic range: 138

Electromagnetic fields: 157
Element size: 161
Enhancement mode: 52

Feedback: 36
FET (Field Effect Transistors): 52
Fiber optic: 172
Filters:
- Band pass: 39
- High Pass: 39
- Low Pass: 39

Frequency: 7
- Harmonics: 9
- Image: 116
- Intermediate (IF): 109
- Multiplication: 131
- Response: 39, 62
- Selection: 39
- Spectrum: 15, 25
- Stability: 121
- Wavelength: 15

Frequency Modulation (FM): 18, 99
- Bandwidth: 103
- Capture: 104
- Carson's Rule: 103
- Commercial FM: 103
- Deviation: 99
- FM versus AM: 102
- Narrow-band FM: 103
- Power output: 103
- Processing gain: 104
- Quieting: 104
- S/N ratio: 104

Frequency synthesizer: 83, 127
Front end: 140

Gain: 39, 71, 163
- Current: 32
- Power: 32
- Voltage: 32

gm (transconductance): 53, 71
Ground waves: 168

Hartley oscillator: 82
Heterodyning: 109, 139

IF amp: 140
Image frequency: 116
Image rejection: 139
Impedance, characteristic: 173
Impedance matching: 176
Incident (forward) Power: 176
Inductive Reactance: 48
Inductors: 48
Information, types of: 5
- Text: 5
- Data: 6
- Audio: 6
- Video: 7

Integrated Circuits (ICs):
- Analog: 182
- Analog vs. digital: 194
- Assembly: 190
- Bipolar: 184
- Burn-in: 190
- Circuit design: 186
- Circuit interconnect: 187
- Combination analog/digital: 192
- Combination bipolar & MOS: 191
- Combination process: 192
- Defined: 181
- Diffusion process: 186
- Digital: 182
- Digital CMOS: 183

  - Fabrication: 185-191
  - Integrated Circuit:81
  - Layer masks: 186
  - Lead frame: 190
  - Packages: 35
  - Probe test: 187
  - Process illustrated: 188
  - Silicon substrate: 185
  - Wafer: 185
- Intermediate frequency (IF): 109
- Isotropic radiator: 163

- JFET: 113

- Large-deviation FM: 133
- Limiter: 150
- Linear RF Amplifier: 80
- Low Pass: 39

- MAC (multiply and accumulate): 200
- Mixers, types of:
  - Diode: 112
  - Dual Gate MOSFET: 115
  - IC: 117
  - JFET: 113
- Mixing: 28, 108
- Modulation: 17
  - Amplitude (AM): 18, 87-92
    - High-level: 91
    - Low-level: 91
  - Angle: 97
  - Continuous Wave (CW): 17, 87
  - Frequency (FM): 18, 99
    - Deviation: 99
  - Phase modulation (PM): 101
  - Single Side Band (SSB): 93
    - Diode-balanced modulator: 97
    - Balanced modulator: 94
  - Types of: 28
- Morse Code: 18, 38
- MOSFET (Metal-Oxide-Semiconductor FET): 52, 115

- Narrow-band FM: 103
- Noise: 25, 137
- Noise Figure: 138
- NPN transistors: 50

- Ohm's Law: 45
- Op Amps (Operational Amplifiers): 35, 72-75
- Operating Point: 34, 53
- Optical coupling: 55
- Oscillators: 35, 80-83
  - Barkhausen Criterion: 80
  - BFO: 94
  - Colpitts: 82
  - Crystal: 82
  - Discrete component: 82
  - Frequency synthesizer: 83
  - Hartley: 82
  - IC (Integrated Circuit):81
  - Phase-shift: 80
  - PLL (Phase-lock-loop): 83
  - VCO: 99

- Passive devices: 48
  - Capacitors: 48
  - Inductors: 48
  - Resistors: 48
- Period: 15
- Phase: 13
- Phase modulation (PM): 101
- Phase-shift: 80
- PLL (Phase-lock-loop): 83
  - discriminator: 148
- PNP transistors: 51
- Polarization: 160
- Power: 32, 33
- Power amplifier: 76
- Power output: 103
- Power supply: 41
- Processing gain: 104
- Propagation (see RF propagation)
- Push-pull amplifier: 78

- Quality Factor (Q): 69
- Quieting: 104

- Radiation patterns: 162
- Receiver: 37
  - Dynamic range: 138
  - Heterodyning: 139
  - Image rejection: 139
  - Noise: 137
  - Noise Figure: 138
  - Selectivity: 138
  - Sensitivity: 137
  - Stability: 138
- Receivers:
  - AM receivers:
    - Automatic gain control (AGC): 141
    - Baseband amp: 141
    - Detail: 149
    - Detection: 142
    - Detector: 141
    - Dual-conversion: 141
    - Front end: 140
    - General: 139-145
    - IF amp: 140
    - Schematic: 149
  - FM receivers:
    - Detail: 150
    - Detection: 146
    - Discriminator: 146
    - Limiter: 150
    - PLL discriminator: 148
    - Schematic: 152
    - Slope detector: 147
  - SSB receivers:
    - Detail: 150
    - General: 146
    - Schematic: 150
- Reflected Power: 176
- Resistors: 48
- Resonance: 67
- RF Propagation:
  - Direct waves: 167
  - Ground waves: 168
  - Patterns: 167
  - Sky waves: 168
- RF tuned amplifier: 71
- RMS (Root Mean Square) values: 31
- Rules for biasing: 61

- S/N (signal-to-noise ratio): 104
- Satellite transmission links: 168
- Sidebands: 87
  - Power in: 90
- Signal control: 38
- Signal phase: 13
- Silicon substrate: 185
- Sine-wave: 8
  - Frequency: 15
  - Period: 15
- Sky waves: 168
- Slope detector: 147
- Spectrum: 15, 25
- Spectrum analyzer: 23
- SSB (Single Side Band): 93, 128
- Stage: 60

- Time division multiplexing (TDM): 40
- Transconductance (gm): 53
- Transducer: 37
- Transformer: 54
- Transmission link: 25
- Transmission lines: 169-173
  - Characteristic impedance: 173
  - Coaxial cable: 171
  - Fiber optic: 172
  - Impedance matching: 176
  - Open-wire: 171
  - Single-wire: 171
  - Standing Wave Ratio (SWR): 175
  - Twin-lead line: 171
  - Twisted pair: 171
  - Two-wire shielded cable: 171
  - Wave guide: 172
- Transmitters: 36, 121
  - AM: 122
    - Baseband amplifier: 124
    - Buffer: 122, 131
    - Driver stage: 123
    - Final amplifier: 123
    - Oscillator: 122
    - Frequency synthesizer: 126
    - Modulator: 123
    - Schematic: 124
  - FM: 130
    - Buffer amplifier: 131
    - Driver: 132
    - Frequency multiplier: 131
    - Large-deviation FM: 133
    - Modulator: 130
    - Schematic: 132
  - SSB: 128
    - Schematic: 130
- Tuned circuit: 67

- Variable capacitor: 126
- Variable inductor: 126
- Vector: 8
- Vertical 1/4 wave: 161, 166
- Voltage: 32
- Voltage-controlled oscillator (VCO): 99, 127
- VSWR (Voltage Standing Wave Ratio): 175

- Wafer: 185
- Wave guide: 172
- Wavelength: 15, 161

- Yagi-Uda: 161, 164


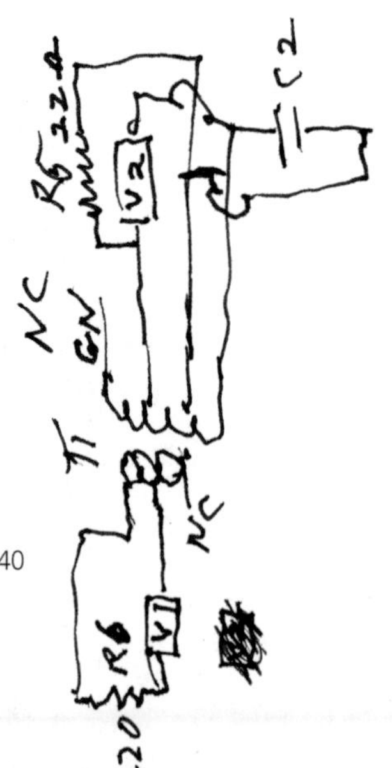